**WITHDRAWN**

# Structural Analysis of Composite Beam Systems

**HOW TO ORDER THIS BOOK**

BY PHONE: 800-233-9936 or 717-291-5609, 8AM–5PM Eastern Time

BY FAX: 717-295-4538

BY MAIL: Order Department
Technomic Publishing Company, Inc.
851 New Holland Avenue, Box 3535
Lancaster, PA 17604, U.S.A.

BY CREDIT CARD: American Express, VISA, MasterCard

# Structural Analysis of Composite Beam Systems

A. M. Skudra
F. Ya. Bulavs
M. R. Gurvich
A. A. Kruklinsh

**TECHNOMIC**
PUBLISHING CO., INC.
LANCASTER · BASEL

## Structural Analysis of Composite Beam Systems
a TECHNOMIC® publication

*Published in the Western Hemisphere by*
Technomic Publishing Company, Inc.
851 New Holland Avenue
Box 3535
Lancaster, Pennsylvania 17604 U.S.A.

*Distributed in the Rest of the World by*
Technomic Publishing AG

Copyright © 1991 by Technomic Publishing Company, Inc.
All rights reserved

No part of this publication may be reproduced, stored in a retrieval system, or transmitted, in any form or by any means, electronic, mechanical, photocopying, recording, or otherwise, without the prior written permission of the publisher.

Printed in the United States of America
10 9 8 7 6 5 4 3 2 1

Main entry under title:
 Structural Analysis of Composite Beam Systems

A Technomic Publishing Company book
Bibliography: p. 293

Library of Congress Card No. 91-58004
ISBN No. 87762-837-8

# CONTENTS

*Preface* ix

PART I. STRUCTURAL MECHANICS OF A MATERIAL

**Chapter 1. Structural Mechanics of a Fabric-Reinforced Plastic** ........................... 3
    1.1 Calculation Model for a Fabric-Reinforced Plastic    3
    1.2 Elastic Characteristics of a Fabric-Reinforced Plastic    7
    1.3 Algorithm for Determining Technical Elastic Properties of a Fabric-Reinforced Plastic    13
    1.4 Stress State of Structural Parts of a Fabric-Reinforced Plastic    15
    1.5 Strength of a Fabric-Reinforced Plastic Under Uniaxial Tension    20
    1.6 Algorithm for Determining the Failure Stages in a Fabric-Reinforced Plastic Under Uniaxial Tension    26
    1.7 Strength of a Fabric-Reinforced Plastic Under Uniaxial Compression    26
    1.8 Deformation Diagram of a Fabric-Reinforced Plastic. Bimodularity    27

**Chapter 2. Structural Mechanics of Hybrid Composites** .... 31
    2.1 Elastic Properties of a Hybrid Composite    31
    2.2 Failure Peculiarities of a Hybrid Composite    33

**Chapter 3. Thermal Strains and Stresses** ............... 43
    3.1 Function of Thermal Expansion    43
    3.2 Thermal Stresses in Plies of a Balanced Reinforced Plastic    55

**Chapter 4. Relative Damping** .........................65
    4.1 Introductory Remarks   65
    4.2 Damping in a Unidirectionally Reinforced Ply   67
    4.3 Damping in a Reinforced Plastic Laminate   69
    4.4 Determination of an Algorithm for Relative Damping   76

**Chapter 5. Structural Theory of Creep** ..................77
    5.1 Viscoelastic Properties of Components   77
    5.2 Creep in a Unidirectionally Reinforced Ply   87
    5.3 Viscoelastic Properties of Laminated Reinforced Plastics Under Sustained Plane Stress State   100

**Chapter 6. Stress State of Reinforced Plastics Under Long-Term Loading** ...............................115
    6.1 Introductory Remarks   115
    6.2 Stress State of Unidirectionally Reinforced Plastics Components in Longitudinal Shear   117
    6.3 Stress State of Unidirectionally Reinforced Plastics Components Under Transversal Loading   123
    6.4 Stress State of Reinforced Plastic Laminates Under Long-Term Loading   136

**Chapter 7. Structural Theory of Long-Term Strength** .....143
    7.1 Criteria of Long-Term Strength of Reinforced Plastics Components   143
    7.2 Long-Term Strength of a Unidirectionally Reinforced Ply with the Polymer Matrix Failure   153
    7.3 Dependence of Long-Term Strength of a Unidirectionally Reinforced Ply upon the Fiber Strength Properties and the Inter-Component Bond   162
    7.4 Continuity Loss of Laminated Reinforced Plastics Under Long-Term Plane Stress State   170
    7.5 Ultimate State of Reinforced Plastic Laminates Under Long-Term Plane Stress State   179

PART II. MECHANICS OF ELASTIC COMPOSITE BEAM

**Chapter 8. Transverse Bending of Beam** ................187
    8.1 Normal Stresses in the Individual Plies   187
    8.2 Shear Stresses in the Individual Plies   196
    8.3 Strength Criteria of a Ply   198

## Chapter 9. Torsion of a Laminated Beam ............207
9.1 The Complex Nature of Inter-Ply Shear in Reinforced Plastic Laminates   207
9.2 Torsion of a Flat Laminated Beam   218
9.3 Torsion of a Thin-Wall Tabular Laminated Beam   232

## Chapter 10. Elastic Properties of Laminated Thin-Wall Beams with Open Profile ............................237
10.1 Formulation of the Problem and the Main Assumptions   237
10.2 Free Torsion   242
10.3 Axial Loading and Pure Bending   243
10.4 Beam Deformation Law Subjected to Shear Forces   246

PART III. MECHANICS OF BEAM SYSTEMS

## Chapter 11. Elastic Displacement of Laminated Beam Systems ........................................255
11.1 Potential Energy of a Laminated Beam   255
11.2 Castigliano's Method   257
11.3 Mohr's Method   258

## Chapter 12. Viscoelastic Displacements of Laminated Beam Systems............................263
12.1 Structural Method for the Determination of Rheological Response Characteristics   263
12.2 The Effect of Loading Duration on the Stress-Strain Response of Laminated Beam Systems   268

## Chapter 13. Numerical Analysis of Statically Indeterminate Beam Systems Under Long-Term Loading ............273
13.1 Analysis by the Method of Forces of Beam Systems Consisting of Reinforced Plastics Laminates   273
13.2 Specific Application of the Method of Displacements   277
13.3 The Effect of Loading Duration on Internal Stresses   282

*Appendix: List of Basic Symbols*   291

*References*   293

*Biographies*   299

# PREFACE

The basic advantages of composite materials are their relatively high strength-weight properties and stiffnesses compared to conventional structural materials. These properties make composites an efficient material, especially in the mass-sensitive structures of today.

The most advanced fields of machine-building have been developed by the application of various beam systems made of reinforced plastics. It should be noted that sometimes engineering and economy requirements in beam system designs can only be met by highly efficient types of reinforced plastics. Examples of such structures are some types of designs adapted for structures in outer space. Nowadays reinforced plastics are widely used and promising composite materials. The main feature of reinforced plastics design (volume content of fibers, packaging and orientation, type of fibers and matrices) is that it may be adapted to every specific case. This feature demands development of a methodology based on research and employed by design engineers and technology engineers working with various new reinforced plastics materials and structures.

The present paper includes research data on structural mechanics for systems made of composite materials discussed in the works of well-known scientists, such as N. A. Alfutov, V. V. Bolotin, G. A. Vanin, V. V. Vasilyev, I. G. Zhigun, S. T. Mileiko, I. F. Obraztsov, B. E. Pobedrya, G. G. Portnov, V. D. Protasov, A. Puk, Yu. N. Rabotnov, V. S. Strelyayev, Yu. V. Suvorova, V. P. Tamuzh, Yu. M. Tarnopolskii, G. A. Teters, M. Uemura, T. Hayashi, K. T. Herakovich, and S. W. Tsai. A broad outline of achievements in the field of composite mechanics made by Soviet scientists is found in the work by Yu. M. Tarnopolskii [68].

Structural mechanics of beam systems made of reinforced plastics is composed of two broad aspects—structural mechanics of the material itself and mechanics of beams. The first seven chapters in this book are devoted to problems of structural mechanics of reinforced plastics that have not been investigated widely. The first chapter discusses the foundation of micromechanics of fabric-reinforced plastics. The second chapter analyzes deformation and failure peculiarities in hybrid plastics depending upon the loading mode—this analysis is very important for correct interpretation of test data. The third chapter gives a new aspect in analyzing thermal deformations and stresses in reinforced plastics. The novelty here is the introduction of the thermal expansion function instead of the constant thermal expansion coefficient. The fourth chapter develops the structural approach to the prediction of relative damping in reinforced plastic laminates. The fifth chapter is devoted to the development of a basic structural theory of creep for reinforced plastics that makes it possible to predict their rheological properties on the basis of the rheological characteristics of their components. The sixth chapter offers a methodology for determining the stress state in the plastic laminate components under sustained load. The seventh chapter is devoted to discussion of the poorly investigated problem of the structural theory of long-term strength of reinforced plastics under simple types of loading.

The next three chapters contain investigations of the stress-strain state of an elastic composite beam. The eighth chapter is devoted to the development of engineering methods for the determination of the stress state in the plies of a laminated beam caused by the bending moment, shear and axial forces; structural criteria of the ply strength are also proposed. The ninth chapter involves the investigation of the stress state of a laminated beam in torsion. The tenth chapter proposes a methodology for determining the stress-deformation state of an open-profile thin-wall laminated beam.

The remaining three chapters deal with the mechanics of beam systems. The eleventh and the twelfth chapters contain the development of engineering methods for the determination of elastic and viscoelastic displacements in laminated beam systems. The design methods for statically indeterminate laminated beam systems under long-term loading are explained and the specific applications of the method of forces and the method of displacements to the beam systems of reinforced plastics are shown in the thirteenth chapter.

The present monograph does not claim to be an exhaustive discussion of the problems concerning structural mechanics of composite beam systems. The book deals primarily with the research data obtained at the Structural Mechanics Laboratory of the Department of Structural Analysis at the Riga Technical University. The research was performed by I. G. Radinsh, A. A. Skudra, K. K. Kalvish, O. V. Sbitnevs, Yu. R. Olengovich and A. E. Paeglitis. Chapters 2, 3, 8 and 11 were written by professor A. M. Skudra; Chapters 5 and 6, by professor F. Ya. Bulavs; Chapters 1, 4, 9 and 10 are by docent A. A. Kruklinsh; and Chapters 7, 12, and 13, by senior research associate M. R. Gurvich.

Alberts Skudra

# Part I
# STRUCTURAL MECHANICS OF A MATERIAL

CHAPTER 1

# Structural Mechanics of a Fabric-Reinforced Plastic

## 1.1 CALCULATION MODEL FOR A FABRIC-REINFORCED PLASTIC

Fabric-reinforced plastics belong to a very complicated class of composites. This complication is explained by the fact that the stiffness and stress state of fabric-reinforced plastics vary within the limits of repeated structural elements as the cross-section varies due to the interweaving of filaments. Within the limits of any cross-section the stress distribution is quite complicated and nonuniform. The present work is an attempt to determine, approximately, the stresses in the structural components of a fabric-reinforced plastic considering the weave structure and the stepwise character of the material failure. The stress-strain state of a fabric-reinforced plastic is studied by means of a structural calculation model presented in Fig. 1.1 where the directions of the warp, the weft, and the normal to the plane of the ply are denoted by symbols $o$, $y$ and $z$. The following assumptions lie in the basis of the proposed calculation model for a fabric-reinforced plastic:

(1) The material is regularly structured and all its components deform in a linear mode.
(2) Filament undulations in the adjacent plies are in phase.
(3) Filament undulations change only negligibly during the loading of the material.
(4) The curved fiber axis becomes broken and is characterised by parameters ($T_o$, $T_y$, $c_o$, $c_y$, $\beta_o$ and $\beta_y$) shown in Fig. 1.1. All the given parameters are determined from the microphotographs of the material structure.

**4** STRUCTURAL MECHANICS OF A MATERIAL

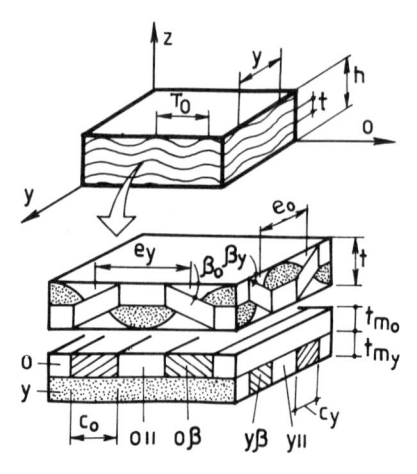

FIGURE 1.1. Structure model of a fabric reinforced plastic.

(5) An individual ply is represented as consisting of two conditional monolayers of the warp and the weft denoted by $o$ and $y$. Filament curvature is regarded as alternating diagonnally and longitudinally reinforced bands of the conditional monolayers. In Fig. 1.1 the warp monolayer is marked by $o\beta$ and $o\|$ but the weft monolayer is marked by $y\beta$ and $y\|$. The relative widths of diagonally reinforced bands are found from the expressions

$$n_o = 2c_o/T_o; \quad n_y = 2c_y/T_y$$

(6) The individual fibers are uniformly distributed across the bulk of the conditional monolayers that, as a result, contain relatively the same amount of fibers at all points, which is equal to the average relative volume content of fibers throughout the material $\psi$. These fibers have relative thicknesses $m_o$ and $m_y$ to be calculated according to the formulas

$$m_o = \frac{t_o}{t} = \frac{1}{1 + k_y f_y p_y / k_o f_o p_o}$$

$$m_y = \frac{t_y}{t} = \frac{1}{1 + k_o f_o p_o / k_y f_y p_y} \quad (1.1)$$

$$m_o + m_y = 1$$

where $k_o$ and $k_y$ are the number of individual fibers in warp and weft filaments (varying with the thread size); $f_o$ and $f_y$ denote the average cross-sectional area of individual fibers; $p_o$ and $p_y$ are the number of filaments per unit length. If the warp and weft filaments are the same type, the expressions in Equation System (1.1) are written as

$$m_o = \frac{1}{1 + \frac{p_y}{p_o}} \; ; \quad m_y = \frac{1}{1 + \frac{p_o}{p_y}}$$

(7) When the material is loaded, equal mean strains that are equivalent throughout the whole layer, appear in the conditional monolayers

$$\langle \epsilon_o^o \rangle = \langle \epsilon_o^y \rangle = \langle\langle \epsilon_e \rangle\rangle$$

$$\langle \epsilon_y^o \rangle = \langle \epsilon_y^y \rangle = \langle\langle \epsilon_y \rangle\rangle \quad (1.2)$$

$$\langle \gamma_{oy}^o \rangle = \langle \gamma_{oy}^y \rangle = \langle\langle \gamma_{oy} \rangle\rangle$$

Here and henceforward the subscripts denote strain or stress directions whereas superscripts refer to the structural element.

(8) Mean stresses in every layer are a combination of mean stresses in the conditional monolayers determined by the mixture law

$$\langle\langle \sigma_o \rangle\rangle = m_o \langle \sigma_o^o \rangle + m_y \langle \sigma_o^y \rangle$$

$$\langle\langle \sigma_y \rangle\rangle = m_o \langle \sigma_y^o \rangle + m_y \langle \sigma_y^y \rangle \quad (1.3)$$

$$\langle\langle \tau_{oy} \rangle\rangle = m_o \langle \tau_{oy}^o \rangle + m_y \langle \tau_{oy}^y \rangle .$$

(9) There is an interrelation between mean strains in the conditional monolayers and in the diagonally and in the longitudinally reinforced bands

$$\langle \epsilon_o^o \rangle = n_o \langle \epsilon_o^{o\beta} \rangle + (1 - n_o) \langle \epsilon_o^{o\|} \rangle$$

$$\langle \epsilon_y^y \rangle = n_y \langle \epsilon_y^{y\beta} \rangle + (1 + n_y) \langle \epsilon_y^{y\|} \rangle$$

$$\langle \epsilon_y^o \rangle = \langle \epsilon_y^{o\beta} \rangle = \langle \epsilon_y^{o\|} \rangle$$

$$\langle \epsilon_o^y \rangle = \langle \epsilon_o^{y\beta} \rangle = \langle \epsilon_o^{y\|} \rangle \quad (1.4)$$

$$\langle \gamma_{oy}^o \rangle = n_o \langle \gamma_{oy}^{o\beta} \rangle + (1 - n_o) \langle \gamma_{oy}^{o\|} \rangle$$

$$\langle \gamma_{oy}^y \rangle = n_y \langle \gamma_{oy}^{y\beta} \rangle + (1 - n_y) \langle \gamma_{oy}^{y\|} \rangle$$

(10) Mean stresses in the conditional monolayers and in the diagonally

and in the longitudinally reinforced bands are interrelated by the following expressions

$$\langle \sigma_o^o \rangle = \langle \sigma_o^{o\beta} \rangle = \langle \sigma_o^{o\|} \rangle$$

$$\langle \sigma_y^y \rangle = \langle \sigma_y^{y\beta} \rangle = \langle \sigma_y^{y\|} \rangle$$

$$\langle \sigma_y^o \rangle = n_o \langle \sigma_y^{o\beta} \rangle + (1 - n_o)\langle \sigma_y^{o\|} \rangle$$

$$\langle \sigma_o^y \rangle = n_y \langle \sigma_o^{y\beta} \rangle + (1 - n_o)\langle \sigma_o^{y\|} \rangle \qquad (1.5)$$

$$\langle \tau_{oy}^o \rangle = \langle \tau_{oy}^{o\beta} \rangle = \langle \tau_{oy}^{o\|} \rangle$$

$$\langle \tau_{oy}^y \rangle = \langle \tau_{oy}^{y\beta} \rangle = \langle \tau_{oy}^{y\|} \rangle$$

(11) Proceeding from the conclusion of Kruklinsh [37], which states that the warping effect of the conditional package consisting of unbalanced plies fades rapidly as the number of plies increases, and also considering that an individual ply in a fabric-reinforced plastic behaves as part of the package, it is assumed that the stress-strain state of conditional monolayers of the warp and the weft are uniform across their thickness.

(12) Stress-strain state of the conditional diagonally reinforced bands depends upon the interaction between the interlaced filaments in the places where they cross each other (Fig. 1.2). This interaction is found taking into consideration the following conditions for the diagonally reinforced bands:
condition of strain continuity

$$\langle \gamma_{oz}^{o\beta} \rangle c_o = -\langle \gamma_{yz}^{y\beta} \rangle c_y \qquad (1.6)$$

condition of equilibrium

$$\langle \tau_{oz}^{o\beta} \rangle e_o m_o = \langle \tau_{yz}^{y\beta} \rangle e_y m_y \qquad (1.7)$$

where $e_o$ and $e_y$ are band widths of the conditional monolayers of the warp and the weft, respectively, reinforced by one type of fibers ($e_o = 1/p_o$, $e_y = 1/p_y$).

(13) Normal stresses in the $z$-axis direction may be ignored.

(14) All the similar structural elements of a fabric-reinforced plastic fail simultaneously. All the assumptions made are applicable to linen (see Figs. 1.1 and 1.2) twill and satin weaves of fabrics.

Structural Mechanics of a Fabric-Reinforced Plastic 7

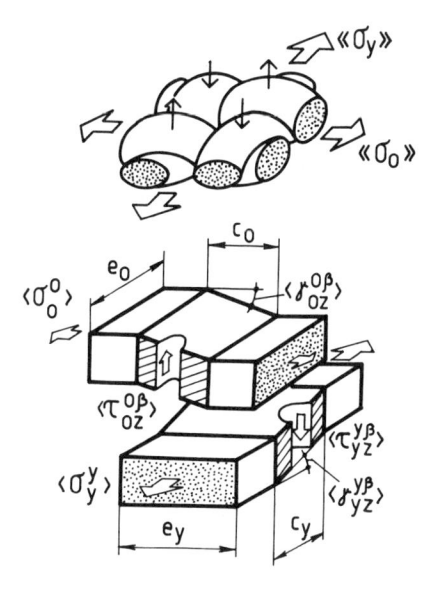

FIGURE 1.2. Interaction schematic of interlaced filaments.

## 1.2 ELASTIC CHARACTERISTICS OF A FABRIC-REINFORCED PLASTIC

The fabric-reinforced laminate represented in Fig. 1.1 is an orthotropic material for which the deformation law under load applied along elastic symmetry axes is written as

$$\left\{ \begin{array}{c} \langle\langle \sigma_o \rangle\rangle \\ \langle\langle \sigma_y \rangle\rangle \\ \langle\langle \tau_{oy} \rangle\rangle \end{array} \right\} = \begin{bmatrix} Q_{11}^F & Q_{12}^F & \\ Q_{21}^F & Q_{22}^F & \\ & & Q_{66}^F \end{bmatrix} \left\{ \begin{array}{c} \langle\langle \epsilon_o \rangle\rangle \\ \langle\langle \epsilon_y \rangle\rangle \\ \langle\langle \gamma_{oy} \rangle\rangle \end{array} \right\} \quad (1.8)$$

where superscript $F$ stands for a fabric-reinforced plastic. In accordance with the adopted calculation model, the deformation law in Equation (1.8) is also true for an individual ply incorporated in the package.

From Equations (1.2) and (1.3) it follows that the components of the given elasticity matrix $Q_{ij}^F$ in the deformation law are determined according to the law of mixture:

$$Q_{ij}^F = m_o Q_{ij}^o + m_y Q_{ij}^y \quad (1.9)$$

where the components of the given elasticity matrices of the conditional monolayers $Q_{ij}^o$ and $Q_{ij}^y$ should be found with respect to the diagonally reinforced bands and the interaction between the interlaced filaments as stated by assumptions 7 and 9 through 14 given in Section 1.1. As illustrated in Fig. 1.2 the interaction between the interlaced filaments finds its expression in the following way: when conditional normal stresses act, for instance, only upon the warp monolayer, strains are transferred also to the weft monolayer through its inclined reinforced bands due to their monoclinic anisotropy and in accordance with statements given in Equations (1.6) and (1.7). In the first approximation it is possible to ignore the interaction between the interlaced filaments under shear acting in the plane of the plies. Then, defining $Q_{ij}^o$ and $Q_{ij}^y$ in Equation (1.9) by means of the technical elastic characteristics of monolayers unidirectionally reinforced with straight fibers, we obtain [35, 37]

$$Q_{11}^F = \frac{m_o}{\omega a_o} + m_y E_\perp^y$$

$$Q_{22}^F = m_o E_\perp^o + \frac{m_y}{\omega a_y}$$

$$Q_{66}^F = m_o G_{\|\perp}^o + m_y G_{\|\perp}^y \qquad (1.10)$$

$$Q_{21}^F = -\frac{1}{\omega}\left(m_o E_\perp^o \frac{b_o}{a_o} + m_y E_\perp^y \frac{b_y}{a_y} + m_y \frac{d_o}{a_o a_y}\right)$$

$$Q_{12}^F = -\frac{1}{\omega}\left(m_o E_\perp^o \frac{b_o}{a_o} + m_y E_\perp^y \frac{b_y}{a_y} + m_o \frac{d_y}{a_o a_y}\right)$$

where

$$\omega = 1 - \frac{d_o d_y}{a_o a_y}$$

$$a_i = n_i\left[S_{11}^{i\beta} - \frac{(S_{15}^{i\beta})^2 c_i}{Gm_i e_i}\right] + \frac{1 - n_i}{E_\|^i}$$

$$b_i = n_i\left[S_{12}^{i\beta} - S_{15}^{i\beta} S_{25}^{i\beta} \frac{c_i}{Gm_i e_i}\right] - (1 - n_i)\frac{\nu_{\|\perp}^i}{E_\|^i}$$

$$d_i = -\Omega \frac{T_i}{m_j e_j}$$

$$\Omega = n_o n_y \frac{S_{15}^{o\beta} S_{15}^{y\beta}}{2G}$$

$$G = \frac{S_{55}^{o\beta} a_o}{m_o e_o} + S_{55}^{y\beta} \frac{c_y}{m_y e_y}$$

$$S_{11}^{i\beta} = \frac{\cos^4 \beta_i}{E_\parallel^i} + \frac{\sin^4 \beta_i}{E_\perp^i} + \left(\frac{1}{G_{\parallel\perp}^i} - 2\frac{\nu_{\parallel\perp}^i}{E_\parallel^i}\right) \cos^2 \beta_i \sin^2 \beta_i$$

$$S_{55}^{i\beta} = \frac{1}{G_{\parallel\perp}^i} + 4\left(\frac{1 + 2\nu_{\parallel\perp}^i}{E_\parallel^i} + \frac{1}{E_\perp^i} - \frac{1}{G_{\parallel\perp}^i}\right) \cos^2 \beta_i \sin^2 \beta_i$$

$$S_{12}^{i\beta} = -\frac{\nu_{\parallel\perp}^i}{E_\parallel^i} \cos^4 \beta_i - \frac{\nu_{\perp\parallel}^i}{E_\perp^i} \sin^4 \beta_i$$

$$S_{15}^{i\beta} = \left[\left(\frac{1}{G_{\parallel\perp}^i} - 2\frac{\nu_{\parallel\perp}^i}{E_\parallel^i}\right)(\cos^2 \beta_i - \sin^2 \beta_i)\right.$$

$$\left. - 2\left(\frac{\cos^2 \beta_i}{E_\parallel^i} - \frac{\sin^2 \beta_i}{E_\perp^i}\right)\right] \cos \beta_i \sin \beta_i$$

$$S_{25}^{i\beta} = 2\left(\frac{\nu_{\parallel\perp}^i}{E_\parallel^i} - \frac{\nu_{\perp\parallel}^i}{E_\perp^i}\right) \cos \beta_i \sin \beta_i$$

$$i, j = o, y; \quad i \neq j; \quad \beta_i > 0$$

Technical elastic characteristics $E_\parallel^i$, $E_\perp^i = E_\perp^i$, $\nu_{\parallel\perp}^i = \nu_{\parallel\perp}^i$, $\nu_{\perp\parallel}^i$, $G_{\parallel\perp}^i$ ($i = o, y$) are obtained from relationships given in, for example, papers by Skudra et al. [55, 57]. Indices $i$ and $j$ in Equation System (1.10) refer to the case when different fibers are used for the warp and the weft. In addition, formulas in Equation (1.10) have been slightly simplified without significantly affecting the final result.

Technical elastic characteristics of a fabric-reinforced plastic before the material loses its continuity are found according to the expressions:

$$E_o = Q_{11}^F - \frac{Q_{12}^F Q_{21}^F}{Q_{22}^F} \approx \frac{m_o}{a_o} + m_y E_\perp^y$$

(1.11)

$$E_y = Q_{22}^F - \frac{Q_{12}^F Q_{21}^F}{Q_{11}^F} \approx m_o E_\perp^o + \frac{m_y}{a_y}$$

$$\nu_{oy} = \frac{Q_{21}^F}{Q_{22}^F}$$

(1.12)

$$\nu_{yo} = \frac{Q_{12}^F}{Q_{11}^F}$$

$$G_{oy} = Q_{66}^F = m_o G_{\|\perp}^o + m_y G_{\|\perp}^y \qquad (1.13)$$

In relationships (1.10) $m_y d_o = m_o d_y$, and this enables us to state that fabric-reinforced plastics are characterized by symmetry of elastic properties before the loss of continuity, i.e., $Q_{12}^F = Q_{21}^F$.

Fabric-reinforced plastics lose continuity quite readily. Even low tensile stresses, if they act transversely to the reinforcement, may cause cracking in the conditional monolayers and thus alter the elastic qualities of the whole material. It is assumed that in the presence of cracks the injured conditional monolayers do not withstand shearing and tensile stresses acting respectively parallel and perpendicular to the crack plane. Therefore, in order to determine elastic characteristics of a fabric-reinforced plastic by means of the conditional cracked monolayer, one should assume $E_\perp^+ = \nu_{\|\perp}^+ = \nu_{\perp\|}^+ = \nu_{\perp\perp}^+ = G_{\perp\|} = 0$, where the plus sign denotes tension. After unloading and reloading by compression the cracks do not practically influence the capability of the monolayer to resist compressive stresses. Thus the fabric-reinforced plastic possesses bimodular properties after loss of continuity. Under tension or shear in the elastic symmetry direction the plastic may have the following technical elastic characteristics depending upon the type of continuity loss:

- when cracks are formed in the weft monolayer

$$E_o^+ = \frac{m_o}{a_o}$$

$$E_y^+ = E_y$$

$$G_{oy}^+ = m_o G_{\|\perp}^o \qquad (1.14)$$

$$\nu_{oy}^+ = \frac{m_o E_\perp^o b_o + m_y \dfrac{d_o}{a_y}}{\omega Q_{22}^F a_o}$$

$$\nu_{yo}^+ = E_\perp^o b_o + \frac{d_y}{a_y}$$

- when cracks develop in the warp monolayer

$$E_o^+ = E_o$$

$$E_y^{(+)} = \frac{m_y}{a_y}$$

$$G_{oy}^{(+)} = m_y G_{\|\perp}^y \qquad (1.15)$$

$$\nu_{oy}^{(+)} = E_\perp^y b_y + \frac{d_o}{a_o}$$

$$\nu_{yo}^{(+)} = \frac{m_y E_\perp^y b_y + m_o \dfrac{d_y}{d_o}}{\omega Q_{11}^F a_y}$$

- when cracks appear in the warp and the weft monolayers

$$E_o^{+(+)} = E_o^{(+)}$$

$$E_y^{+(+)} = E_y^{(+)}$$

$$G_{oy}^{+(+)} = 0 \qquad (1.16)$$

$$\nu_{oy}^{+(+)} = \frac{d_o}{a_o}$$

$$\nu_{yo}^{+(+)} = \frac{d_y}{a_y}$$

## 12  STRUCTURAL MECHANICS OF A MATERIAL

*Table 1.1. Elasticity Moduli and Poisson's Ratios for a Cracked Fabric Reinforced Plastic.*

| Sign of Stress | | The Cracked Conditional Monolayers | | |
|---|---|---|---|---|
| $\langle\langle\sigma'_o\rangle\rangle$ | $\langle\langle\sigma_y\rangle\rangle$ | $o$ | $y$ | $o$ and $y$ |
| − | − | $E_o E_y \nu_{oy} \nu_{yo}$ | $E_o E_y \nu_{oy} \nu_{yo}$ | $E_o E_y \nu_{oy} \nu_{yo}$ |
| + | + | $E_o E_y^{(+)} \nu_{oy}^{(+)} \nu_{yo}^{(+)}$ | $E_o^+ E_y^+ \nu_{oy}^+ \nu_{yo}^+$ | $E_o^+ E_y^+ \nu_{oy}^{+(+)} \nu_{yo}^{+(+)}$ |
| + | − | $E_o E_y \nu_{oy} \nu_{yo}$ | $E_o^+ E_y^+ \nu_{oy}^+ \nu_{yo}^+$ | $E_o^+ E_y^{(+)} \nu_{oy}^{(+)} \nu_{yo}^{(+)}$ |
| − | + | $E_o E_y^{(+)} \nu_{oy}^{(+)} \nu_{yo}^{(+)}$ | $E_o E_y \nu_{oy} \nu_{yo}$ | $E_o E_y^{(+)} \nu_{oy}^{(+)} \nu_{yo}^{(+)}$ |

If compression is applied in the directions of the elastic symmetry, the primary moduli of elasticity and Poisson's ratio of a fabric-reinforced plastic are not dependent upon the type of continuity loss. In the generalized case of plane stress state the above two characteristics do depend upon the type of continuity loss and also upon the relation between the applied stresses $\langle\langle\sigma_o\rangle\rangle$ and $\langle\langle\sigma_y\rangle\rangle$ or strains $\langle\langle\epsilon_o\rangle\rangle$ and $\langle\langle\epsilon_y\rangle\rangle$, i.e., whether the cracks open or they are compressed. In the first approximation, when selecting elasticity moduli and Poisson's ratios depending on the continuity loss of the fabric-reinforced plastic, it is sufficient to consider the signs for stresses $\langle\langle\sigma_o\rangle\rangle$ and $\langle\langle\sigma_y\rangle\rangle$ and to refer to Table 1.1.

It should be noted that cracking and interaction between the interwoven filaments are responsible for the asymmetry of the elastic properties (i.e., inequality $\nu_{oy} E_y \neq \nu_{yo} E_o$) of fabric-reinforced plastics [37]. It is assumed that the shear modulus of fabric-reinforced plastics in the elastic symmetry axes depends only on the type of continuity loss of the material, and for any kind of stress-strain state the modulus is determined according to Equations Systems (1.14)–(1.16).

By using the technical elastic characteristics of a fabric-reinforced plastic, which were obtained from Equation Systems (1.14)–(1.16), it is possible to find all the components of its compliance and the elasticity matrices with the type of continuity loss considered:

$$[S_{ij}^F] = \begin{bmatrix} \dfrac{1}{E_o} & -\dfrac{\nu_{yo}}{E_y} & \\ -\dfrac{\nu_{oy}}{E_o} & \dfrac{1}{E_y} & \\ & & \dfrac{1}{G_{oy}} \end{bmatrix}$$

$$[Q_{ij}^F] = \begin{bmatrix} \dfrac{E_o}{1 - \nu_{oy}\nu_{yo}} & \dfrac{\nu_{yo}E_o}{1 - \nu_{oy}\nu_{yo}} & \\ \dfrac{\nu_{oy}E_y}{1 - \nu_{oy}\nu_{yo}} & \dfrac{E_y}{1 - \nu_{oy}\nu_{yo}} & \\ & & G_{oy} \end{bmatrix}$$

To transform the compliance and the elasticity matrix components at a rotation of the coordinate axes, we apply the transformation formula

$$[\bar{S}_{ij}^F] = [T]^T[S_{ij}^F][T]$$

$$[\bar{Q}_{ij}^F] = [T]^{-1}[Q_{ij}^F][T]^{-1T} \qquad (1.17)$$

where

$$[T] = \begin{bmatrix} \cos^2\beta & \sin^2\beta & 2\cos\beta\sin\beta \\ \sin^2\beta & \cos^2\beta & -2\cos\beta\sin\beta \\ -\cos\beta\sin\beta & \cos\beta\sin\beta & \cos^2\beta - \sin^2\beta \end{bmatrix}$$

Fig. 1.3 illustrates the effect of cracking upon the elasticity matrix components of a fabric-reinforced plastic. It should be remarked that with the cracking of both monolayers of the fabric-reinforced plastic (when $G_{oy} \approx 0$) all the components of its compliance matrix at an angle to the elastic symmetry directions of the material tend to infinity ($\bar{S}_{ij}^F \to \infty$).

## 1.3 ALGORITHM FOR DETERMINING TECHNICAL ELASTIC PROPERTIES OF A FABRIC-REINFORCED PLASTIC

The calculation of the technical elastic characteristics of a fabric-reinforced plastic is made as follows.

(1) Technical elastic characteristics of the matrix and the fibers and their relative volume content are found.
(2) By means of known formulas given in, for example, a paper by Skudra [57], technical elastic characteristics are obtained for a unidirectionally reinforced medium.
(3) The following structural parameters of a fabric reinforced plastic described in Section 1.1 are found: $\beta_o$, $\beta_y$, $n_o$, $n_y$, $e_o$, $e_y$, $m_o$ and $m_y$.

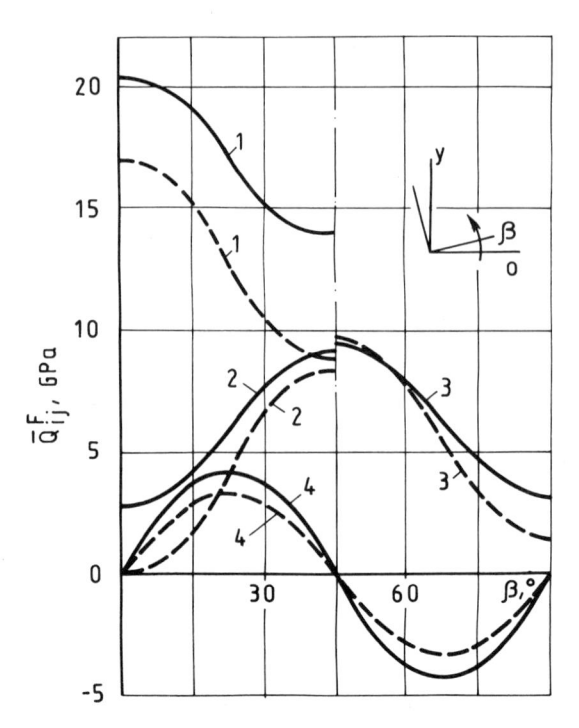

**FIGURE 1.3.** The dependence of stiffness matrix components of a fabric reinforced plastic before (———) and after (– – –) cracking of impregnated weft and warp filaments upon the rotation angle of the coordinate axes: 1-$\bar{Q}_{11}^F$; 2-$\bar{Q}_{66}^F$; 3-$\bar{Q}_{12}^F$; 4-$\bar{Q}_{16}^F$. The curves have been constructed according to formula (1.17) with the following initial data: $\psi = 0.45$; $B_o = B_y = 10°$; $m_o = m_y = 0.5$; $n_o = n_y = 0.2$; $E_o = 75$ GPa; $\gamma_B = 0.22$; $E_A = 3$ GPa; $\gamma_A = 0.35$.

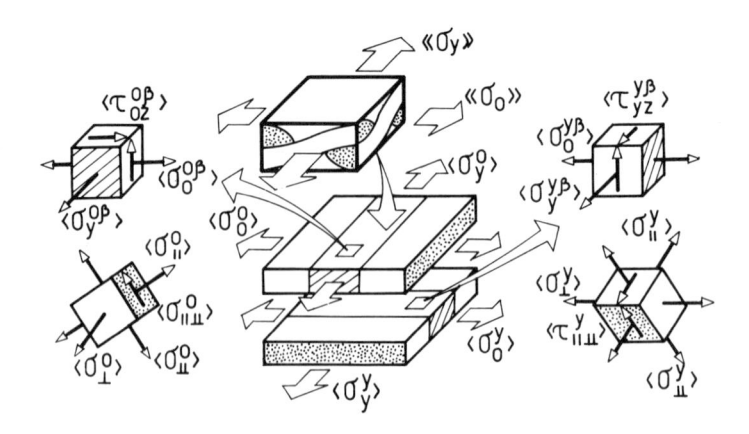

**FIGURE 1.4.** Stress state of structural components of a fabric reinforced plastic under biaxial tension.

(4) From Equation Systems (1.10)–(1.13) of Section 1.2 technical elastic characteristics of a fabric-reinforced plastic are determined considering continuity loss.
(5) Equation Systems (1.14)–(1.17) enable us to determine the technical elastic characteristics of a fabric-reinforced plastic considering the type of continuity loss.
(6) The calculation of technical elastic characteristics of a fabric-reinforced plastic after continuity loss is made according to Table 1.1, depending upon the cracking mode and the sign of the stresses $\langle\langle\sigma_o\rangle\rangle$ and $\langle\langle\sigma_y\rangle\rangle$.

## 1.4 STRESS STATE OF STRUCTURAL PARTS OF A FABRIC-REINFORCED PLASTIC

The strength of fabric-reinforced plastics is governed by the following factors:

- the strength of matrix-impregnated filaments, which is the function of the matrix, fibers, and bond strength properties
- the interaction mechanism between the impregnated warp and weft filaments, which determines their stress state to a large extent

In order to predict fabric-reinforced plastic strength under simple loads (like tension) it is necessary to determine the stress state of impregnated filaments in regions of fiber orientation which are parallel to the fabric plane and at an angle, taking into consideration the peculiar behaviour of filaments within a composite.

For modelling the behaviour of the impregnated filaments within the fabric-reinforced plastic and for determining their stress state, the material structure model as shown in Fig. 1.1 and the assumptions of Section 1.1 are used. According to this model the impregnated filaments are supposed to be monolayers of the warp and the weft in which the mean stresses (Fig. 1.4) under axial loading in the direction of elastic symmetry axis of the fabric-reinforced plastic are obtained from the relationships

$$\begin{Bmatrix} \langle\sigma_o^o\rangle \\ \langle\sigma_y^o\rangle \end{Bmatrix} = \begin{bmatrix} Q_{11}^o & Q_{12}^o \\ Q_{21}^o & Q_{22}^o \end{bmatrix} \begin{Bmatrix} \langle\langle\epsilon_o\rangle\rangle \\ \langle\langle\epsilon_y\rangle\rangle \end{Bmatrix}$$

$$\begin{Bmatrix} \langle\sigma_o^y\rangle \\ \langle\sigma_y^y\rangle \end{Bmatrix} = \begin{bmatrix} Q_{11}^y & Q_{12}^y \\ Q_{21}^y & Q_{22}^y \end{bmatrix} \begin{Bmatrix} \langle\langle\epsilon_o\rangle\rangle \\ \langle\langle\epsilon_y\rangle\rangle \end{Bmatrix}$$

(1.18)

where

$$Q^o_{11} = \frac{1}{\omega a_o}$$

$$Q^o_{22} \approx E^o_\perp$$

$$Q^o_{12} = -\frac{1}{\omega a_o}\left(b_o E^o_\perp + \frac{d_y}{a_y}\right)$$

$$Q^o_{21} = -\frac{b_o E^o_\perp}{\omega a_o}$$

$$Q^y_{11} \approx E^y_\perp$$

$$Q^y_{22} = \frac{1}{\omega a_y}$$

$$Q^y_{12} = -\frac{b_y E^y_\perp}{\omega a_y}$$

$$Q^y_{21} = -\frac{1}{\omega a_y}\left(b_y E^y_\perp + \frac{d_o}{a_o}\right)$$

Diagonally reinforced bands $o\beta$ and $y\beta$ are in the most disadvantageous stress state. Normal stresses in these bands are taken as equal to the mean stresses acting upon the whole monolayer, i.e.,

$$\langle \sigma^{o\beta}_o \rangle = \langle \sigma^o_o \rangle$$
$$\langle \sigma^{y\beta}_y \rangle = \langle \sigma^y_y \rangle \quad \text{(a)}$$

$$\langle \sigma^{o\beta}_y \rangle \approx \langle \sigma^o_y \rangle$$
$$\langle \sigma^{y\beta}_o \rangle \approx \langle \sigma^y_o \rangle \quad \text{(b)}$$

where equalities (b) have been assumed considering that

$$E^{y\beta}_o \approx E^y_o \approx E^y_\perp$$

and

$$E_y^{o\beta} \approx E_y^o \approx E_\perp^o.$$

Tangential stresses $\langle \tau_{oz}^{o\beta} \rangle$ and $\langle \tau_{yz}^{y\beta} \rangle$ (see Fig. 1.4) appear due to the interaction between the interwoven filaments (see Fig. 1.2) and are determined with the assumptions in Equations (1.6) and (1.7) by means of external stresses $\langle\langle \sigma_o \rangle\rangle$ and $\langle\langle \sigma_y \rangle\rangle$. Expressing the additional strains $\langle\langle \sigma_o \rangle\rangle$ and $\langle\langle \sigma_y \rangle\rangle$ in Equation (1.8) by means of $\langle\langle \epsilon_o \rangle\rangle$ and $\langle\langle \epsilon_y \rangle\rangle$ and taking into account (a) and (b), we obtain expressions for the determination of the stresses in the most hazardous places, namely, in diagonally reinforced bands of the warp and the weft monolayers:

$$\begin{Bmatrix} \langle \sigma_o^{o\beta} \rangle \\ \langle \sigma_y^{o\beta} \rangle \\ \langle \tau_{oz}^{o\beta} \rangle \end{Bmatrix} = \begin{Bmatrix} \langle \sigma_o^{o} \rangle \\ \langle \sigma_y^{o} \rangle \\ \langle \tau_{oz}^{o\beta} \rangle \end{Bmatrix} = \begin{bmatrix} g_{11}^o & g_{12}^o \\ g_{21}^o & g_{22}^o \\ g_{15}^o & g_{25}^o \end{bmatrix} \begin{Bmatrix} \langle\langle \sigma_o \rangle\rangle \\ \langle\langle \sigma_y \rangle\rangle \end{Bmatrix}$$

$$\begin{Bmatrix} \langle \sigma_y^{y\beta} \rangle \\ \langle \sigma_o^{y\beta} \rangle \\ \langle \tau_{yz}^{y\beta} \rangle \end{Bmatrix} = \begin{Bmatrix} \langle \sigma_y^{y} \rangle \\ \langle \sigma_o^{y} \rangle \\ \langle \tau_{yz}^{y\beta} \rangle \end{Bmatrix} = \begin{bmatrix} g_{11}^y & g_{12}^y \\ g_{21}^y & g_{22}^y \\ g_{15}^y & g_{25}^y \end{bmatrix} \begin{Bmatrix} \langle\langle \sigma_o \rangle\rangle \\ \langle\langle \sigma_y \rangle\rangle \end{Bmatrix}$$

(1.19)

where

$$g_{11}^o = \frac{1}{\omega a_o E_o}\left[1 + \nu_{oy}\left(b_o E_\perp^o + \frac{d_y}{a_y}\right)\right]$$

$$g_{22}^o = \frac{1}{E_y}\left(1 + \frac{\nu_{yo}b_o}{\omega a_o}\right)E_\perp^o$$

$$g_{12}^o = -\frac{1}{\omega a_o E_y}\left(\nu_{yo} + b_o E_\perp^o + \frac{d_y}{a_y}\right)$$

$$g_{21}^o = -\frac{1}{E_o}\left(\nu_{oy} + \frac{b_o}{\omega a_o}\right)E_\perp^o$$

$$g_{15}^o = \frac{1}{Ge_o m_o}(c_o g_{11}^o S_{15}^{o\beta} + c_y g_{11}^y S_{15}^{y\beta})$$

$$g_{25}^o = \frac{1}{Ge_o m_o}(c_o g_{12}^o S_{15}^{o\beta} + c_y g_{12}^y S_{15}^{y\beta})$$

## 18 STRUCTURAL MECHANICS OF A MATERIAL

$$g^y_{11} = -\frac{1}{\omega a_y E_o}\left(\nu_{oy} + b_y E^y_\perp + \frac{d_o}{a_o}\right)$$

$$g^y_{22} = -\frac{1}{E_y}\left(\nu_{yo} + \frac{b_y}{\omega a_y}\right)E^y_\perp$$

$$g^y_{12} = \frac{1}{\omega a_y E_y}\left[1 + \nu_{yo}\left(b_y E^y_\perp + \frac{d_o}{a_o}\right)\right]$$

$$g^y_{21} = \frac{1}{E_o}\left(1 + \nu_{oy}\frac{b_y}{\omega a_y}\right)E^y_\perp$$

$$g^y_{15} = \frac{1}{e_y m_y}g^o_{15}e_o m_o$$

$$g^y_{25} = \frac{1}{e_y m_y}g^o_{25}e_o m_o$$

After the material has lost its continuity, the coefficients $g^o_{ij}$ and $g^y_{ij}$ as well as the values of the technical elastic characteristics (see Section 1.2) have to be determined taking into account the cracking mode and the ratio between the deformations $\langle\langle\epsilon_o\rangle\rangle$ and $\langle\langle\epsilon_y\rangle\rangle$ or the stresses $\langle\langle\sigma_o\rangle\rangle$ and $\langle\langle\sigma_y\rangle\rangle$. In the first approximation the selection of coefficients $g^o_{ij}$ and $g^y_{ij}$, which include the cracking of the material, may be made considering only the signs of stresses $\langle\langle\sigma_o\rangle\rangle$ and $\langle\langle\sigma_y\rangle\rangle$ and using Table 1.2 where the coefficients $(g^o_{ij})^+$, $(g^y_{ij})^+$, $(g^o_{ij})^{(+)}$, $(g^y_{ij})^{(+)}$, $(g^o_{ij})^{+(+)}$ and $(g^y_{ij})^{+(+)}$ refer to the cracked material and are found from formulas:

Table 1.2. Coefficients $g^o_{ij}$ and $g^y_{ij}$ for a Cracked Fabric Reinforced Plastic.

| Sign of Stress || Cracked Conditional Monolayers |||
|---|---|---|---|---|
| $\langle\langle\sigma'_o\rangle\rangle$ | $\langle\langle\sigma_y\rangle\rangle$ | o | y | o and y |
| − | − | $g^o_{ij}g^y_{ij}$ | $g^o_{ij}g^y_{ij}$ | $g^o_{ij}g^y_{ij}$ |
| + | + | $(g^o_{ij})^{(+)}(g^y_{ij})^{(+)}$ | $(g^o_{ij})^+(g^y_{ij})^+$ | $(g^o_{ij})^{+(+)}(g^y_{ij})^{+(+)}$ |
| + | − | $g^o_{ij}g^y_{ij}$ | $(g^o_{ij})^+(g^y_{ij})^+$ | $(g^o_{ij})^+(g^y_{ij})^+$ |
| − | + | $(g^o_{ij})^{(+)}(g^y_{ij})^{(+)}$ | $g^o_{ij}g^y_{ij}$ | $(g^o_{ij})^{(+)}(g^y_{ij})^{(+)}$ |

$$\left.\begin{aligned}
(g_{11}^o)^+ &= \frac{1}{\omega m_o}\left[1 + \nu_{oy}^+\left(b_o E_\perp^o + \frac{d_y}{a_y}\right)\right] \\
(g_{22}^o)^+ &= \frac{1}{E_y}\left(1 + \nu_{yo}^+ \frac{b_o}{\omega a_o}\right) E_\perp^o \\
(g_{12}^o)^+ &= -\frac{1}{\omega a_o E_y}\left(\nu_{yo}^+ + b_o E_\perp^o + \frac{d_y}{a_y}\right) \\
(g_{21}^o)^+ &= -\frac{1}{m_o}\left(\nu_{oy}^+ a_o + \frac{b_o}{\omega}\right) E_\perp^o \\
(g_{11}^y)^+ &= -\frac{1}{\omega a_y m_o}(\nu_{oy} a_o + d_o) \\
(g_{22}^y)^+ &= (g_{21}^y)^+ = 0 \\
(g_{12}^y)^+ &= \frac{1}{\omega a_y E_y}\left(1 + \nu_{yo}^+ \frac{d_o}{a_o}\right)
\end{aligned}\right\} \quad (1.20)$$

$$\left.\begin{aligned}
(g_{11}^o)^{(+)} &= \frac{1}{\omega a_o E_o}\left(1 + \nu_{oy}^{(+)} \frac{d_y}{a_y}\right) \\
(g_{22}^o)^{(+)} &= (g_{21}^o)^{(+)} = 0 \\
(g_{12}^o)^{(+)} &= -\frac{1}{\omega a_o m_y}(\nu_{yo}^{(+)} a_y + d_y) \\
(g_{11}^y)^{(+)} &= -\frac{1}{\omega a_y E_o}\left(\nu_{oy}^{(+)} + b_y E_\perp^y + \frac{d_o}{a_o}\right) \\
(g_{22}^y)^{(+)} &= -\frac{1}{m_y}\left(\nu_{yo}^{(+)} a_y + \frac{b_y}{\omega}\right) E_\perp^y \\
(g_{12}^y)^{(+)} &= \frac{1}{\omega m_y}\left[1 + \nu_{yo}^{(+)}\left(b_y E_\perp^y + \frac{d_o}{a_o}\right)\right] \\
(g_{21}^y)^{(+)} &= \frac{1}{E_o}\left(1 + \nu_{oy}^{(+)} \frac{b_y}{\omega a_y}\right) E_\perp^y
\end{aligned}\right\} \quad (1.21)$$

$$(g^o_{11})^{+(+)} = \frac{1}{\omega m_o}\left(1 + \nu^{+(+)}_{oy}\frac{d_y}{a_y}\right)$$

$$(g^o_{22})^{+(+)} = (g^o_{21})^{+(+)} = 0$$

$$(g^o_{12})^{+(+)} = -\frac{1}{\omega a_o m_y}(\nu^{+(+)}_{yo}a_y + d_y)$$

$$(g^y_{11})^{+(+)} = -\frac{1}{\omega a_y m_o}\left(\nu^{+(+)}_{yo}a_o + d_o\right)$$

$$(g^y_{22})^{+(+)} = (g^y_{21})^{+(+)} = 0$$

$$(g^y_{12})^{+(+)} = \frac{1}{\omega m_y}\left(1 + \nu^{+(+)}_{yo}\frac{d_o}{a_o}\right)$$

(1.22)

## 1.5 STRENGTH OF A FABRIC-REINFORCED PLASTIC UNDER UNIAXIAL TENSION

Let us consider a case when the load is applied in the warp direction. In this case, the weft monolayer is the first to undergo failure (Fig. 1.5 (a)) since it will be subjected to stresses $\langle \sigma^y_o \rangle = \langle \sigma^y_\perp \rangle$, $\langle \sigma^y_y \rangle$ and $\langle \tau^{y\beta}_{yz} \rangle$ as stated in Equation (1.19) with $\langle\langle \sigma_y \rangle\rangle = 0$. The first stage of failure of a fabric-reinforced plastic defining the moment of its continuity loss of the first type is due to stress $\langle \sigma^y_o \rangle$ when it reaches the strength level in transverse tension $R^+_\perp$ of the matrix-impregnated weft filament. Here the effect of stresses $\langle \sigma^y_y \rangle$ and $\langle \tau^{y\beta}_{yz} \rangle$ may be ignored in the first approximation. The strength $R^+_\perp$ is obtained according to the following formulas [55, 57, 81]:

- in the case of matrix failure

$$R^+_\perp = \frac{R^+_A}{\bar{\sigma}_r\sqrt{1 - \nu^2_A}} \quad (a)$$

- in the case of bond failure between the fibers and the matrix

$$R^+_\perp = \frac{R_b}{\bar{\sigma}_r} \quad (b)$$

*Structural Mechanics of a Fabric-Reinforced Plastic* 21

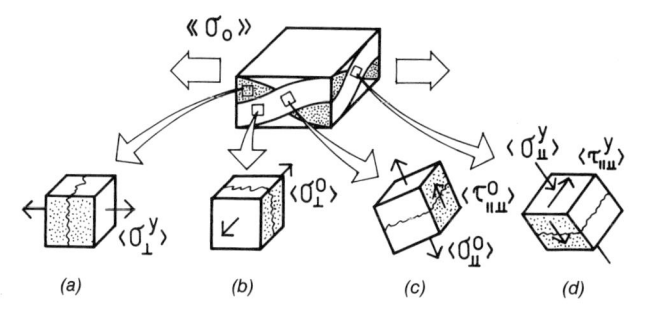

FIGURE 1.5. Schematic of the continuity loss of type 1(a), 2(b) and 3(c,d) in a fabric reinforced plastic under uniaxial tension.

- in the case of transverse rupture of the fibers

$$R_\perp^+ \approx R_{Br}^+ \qquad (c)$$

where $R_A^+$ is the polymer matrix strength, $R_b$ is the tensile strength of the bond, and $R_{Br}^+$ is the transverse tensile strength of the fibers.

Denoting the stress $\langle\langle \sigma_o \rangle\rangle$ at the instant of cracking of the weft monolayer by $\langle\langle \sigma_o^+ \rangle\rangle$ and considering that $\langle \sigma_o^y \rangle = R_\perp^+$ from Equation System (1.19) we obtain:

- for the matrix failure

$$\langle\langle \sigma_o^+ \rangle\rangle = \frac{R_A^+}{g_{21}^y \, \bar{\sigma}_r \sqrt{1 - v_A^2}} \qquad (1.23)$$

- for the failure of organic fibers under transverse tension

$$\langle\langle \sigma_o^+ \rangle\rangle = \frac{R_{Br}^+}{g_{21}^y \bar{\sigma}_r} = 0.91 \frac{R_{Br}^+}{g_{21}^y} \qquad (1.24)$$

- for the bond failure

$$\langle\langle \sigma_o^+ \rangle\rangle = \frac{R_b}{g_{21}^y \bar{\sigma}_r} \qquad (1.25)$$

It should be noted that the stress concentration factor $\bar{\sigma}_r$ in Equations (1.23)–(1.25) and further are valid for unidirectionally reinforced media inside the filament where the volume content of the fibers in the filament is constant and equals 0.7. The curves presented by Sih and Skudra [81] are sufficient to state that for a glass fiber reinforced plastic $\bar{\sigma}_r = 2.0$ and for an organic fiber reinforced plastic $\bar{\sigma}_r = 1.1$.

With a further increase of the load, two failure alternatives are possible [36]

(1) cracking of the warp monolayer due to the action of transverse stresses caused by Poisson's effect in the plane $oy$ (see Fig. 1.5 (b))
(2) the failure of warp and weft monolayers in the diagonally reinforced bands (see Fig. 1.5 (c))

The first case, when the cracks appear in the monolayer reinforced in the tension direction, will be called continuity loss of the second type. Denoting the stress $\langle\langle \sigma_o \rangle\rangle$ at the instant of cracking of the warp monolayer by the symbol $\langle\langle \sigma_o^{+(+)} \rangle\rangle$ we obtain the following description of conditions for the second type of continuity loss

$$\langle\langle \sigma_o^{+(+)} \rangle\rangle = \frac{R_{\perp o}^+}{(g_{21}^o)^+} \tag{1.26}$$

where $R_{\perp o}^+$ is obtained from formulas (a), (b), or (c).

The third type of continuity loss will be the second from the above, where the warp and the weft monolayers in diagonally reinforced bands fail under shear along the inclined planes (see Fig. 1.5 (c)). Note that due to the interaction between the interwoven warp and weft filaments (see Fig. 1.2) the failure of weft monolayers leads to the failure of warp monolayers as well (in the inclined planes) and vice versa. In compiling the conditions for the third-type continuity loss associated with the failure of, for instance, the warp monolayer, in the first approximation only the stresses $\langle \sigma_o^{o\beta} \rangle = \langle \sigma_o^o \rangle$ and $\langle \tau_{oz}^{o\beta} \rangle$ are taken into account. The stress $\langle \sigma_{y'}^{o\beta} \rangle = \langle \sigma_y^o \rangle$ is neglected as it acts parallel to the eventual crack plane (see Fig. 1.5 (c)). In determining conditions for the third-type continuity loss associated with the failure of the weft monolayers, the stresses $\langle \sigma_o^{y\beta} \rangle = \langle \sigma_o^y \rangle$ are omitted by analogy with the previous case.

The stresses in the elastic symmetry directions within the limits of bands $o\beta$ and $y\beta$ caused by the action of stresses $\langle \sigma_o^{o\beta} \rangle$, $\langle \tau_{oz}^{o\beta} \rangle$ and $\langle \sigma_{y'}^{y\beta} \rangle$, $\langle \tau_{yz}^{y\beta} \rangle$ are calculated according to the relationships

$$\langle \sigma_\parallel^i \rangle = \langle \sigma_i^{i\beta} \rangle \cos^2 \beta_i + 2 \langle \tau_{iz}^{i\beta} \rangle \sin \beta_i \cos \beta_i \tag{1.27}$$

$$\langle \sigma_\perp^i \rangle = \langle \sigma_i^{i\beta} \rangle \sin^2 \beta_i - 2 \langle \tau_{iz}^{i\beta} \rangle \sin \beta_i \cos \beta_i \tag{1.28}$$

Structural Mechanics of a Fabric-Reinforced Plastic   23

$$\langle \tau^i_{\perp \|} \rangle = \langle \sigma^{i\beta}_i \rangle \sin \beta_i \cos \beta_i - \langle \tau^{i\beta}_{iz} \rangle (\cos^2 \beta_i - \sin^2 \beta_i) \quad (1.29)$$

where $i = o, y$.

Stresses $\langle \sigma^o_\| \rangle$ and $\langle \sigma^y_\| \rangle$ insignificantly influence the shear strength of the monolayers. The monolayer strength is basically governed by the combined action of the stresses $\langle \sigma^o_\perp \rangle$, $\langle \tau^o_{\perp \|} \rangle$ and $\langle \sigma^y_\perp \rangle$, $\langle \tau^y_{\perp \|} \rangle$. In that case the strength criteria described in Bulaus & Radinsh [8] and Bichkov [12] have the form [36]:

- for shear failure of the matrix

$$\langle \sigma^i_\perp \rangle^2 (1 + v_A^2) + 2\langle \tau^i_{\perp \|} \rangle^2 (1 + v_A)$$
$$+ \langle \sigma^i_\perp \rangle (1 + v_A) \sqrt{\langle \sigma^i_\perp \rangle^2 (1 - v_A^2) + 4\langle \tau^i_{\perp \|} \rangle^2} = 2(R_A^+)^2 \quad (1.30)$$

- for transverse rupture of fibers

$$\frac{\langle \sigma^i_\perp \rangle}{R^+_{Br}} \bar{\sigma}_r + \left( \frac{\langle \tau^i_{\perp \|} \rangle \bar{\tau}_{rz}}{T_{Brz}} \right)^2 = 1 \quad (1.31)$$

- for bond failure

$$\frac{\langle \sigma^i_\perp \rangle}{R_b} \bar{\sigma}_r + \left( \frac{\langle \tau^i_{\perp \|} \rangle \bar{\tau}_{rz}}{T_b} \right)^2 = 1 \quad (1.32)$$

$$i = o, y$$

Letting stress $\langle\langle \sigma_o \rangle\rangle$ applied to the fabric-reinforced plastic and causing the monolayer failure in the diagonally reinforced bands to be represented by $\langle\langle \sigma_o^{+++} \rangle\rangle$ and with Equations (1.28), (1.29), and (1.19) taken into consideration, Equations (1.30)–(1.32) yield the following formulas for determining the instant of the third-type continuity loss when the warp or the weft monolayers undergo failure:

- shear failure of the matrix (if $v_A = 0.35$)

$$\langle\langle \sigma_o^{+++} \rangle\rangle = \sqrt{2} R_A^+ (u_i^2 + 2.7\, v_i^2 + u_i \sqrt{u_i^2 + 4v_i^2})^{-1/2} \quad (1.33)$$

- transverse failure of the fibers

$$\langle\langle \sigma_o^{+++} \rangle\rangle = \frac{1}{2} \left[ \sqrt{\left( \frac{\bar{\sigma}_r u_i}{R^+_{Br}} \right)^2 + 4\left( \frac{\bar{\tau}_{rz} v_i}{T_{Brz}} \right)^2} - \frac{\bar{\sigma}_r u_i}{R^+_{Br}} \right] \frac{T_{Brz}}{\bar{\tau}_{rz} v_i} \quad (1.34)$$

- bond failure

$$\langle\langle \sigma_o^{+++} \rangle\rangle = \frac{1}{2}\left[\sqrt{\left(\frac{\bar{\sigma}_r u_i}{R_b}\right)^2 + 4\left(\frac{\bar{\tau}_{rz} v_i}{T_b}\right)^2} - \frac{\bar{\sigma}_r u_i}{R_b}\right]\frac{T_b}{\bar{\tau}_{rz} v_i} \quad (1.35)$$

where, considering the previous continuity loss of the first and the second type, $u_i$ and $v_i$ are obtained from formulas

$$u_i = (g_{11}^i)^{+(+)} \sin^2 \beta_i - 2(g_{15}^i)^{+(+)} \sin \beta_i \cos \beta_i$$

$$v_i = (g_{11}^i)^{+(+)} \sin \beta_i \cos \beta_i - (g_{15}^i)^{+(+)} (\cos^2 \beta_i - \sin^2 \beta_i) \quad (1.36)$$

$$i = o, y$$

In fact, the third type of continuity loss occurs with the least value of stresses, $\langle\langle \sigma_o^{+++} \rangle\rangle$, which was obtained from Equations (1.33)–(1.35).

If the third type of continuity loss takes place before the second type of continuity loss, then coefficients $(g_{11}^i)^{+(+)}$ and $(g_{15}^i)^{+(+)}$ are used instead of $(g_{11}^i)^+$ and $(g_{15}^i)^+$ in the Equation System (1.36).

The third type of continuity loss is typical only for fabric-reinforced plastics. After it has taken place there is a crushing of the matrix across the whole bulk of the material which for practical purposes is equal to its complete failure, although the fabric-reinforced plastic will take up the tensile load before the warp fibers undergo rupture. With $R_o^+$ to denote the mean stress $\langle\langle \sigma_o \rangle\rangle$ at the moment of warp fiber rupture, a formula is obtained to approximate the ultimate strength of the fabric-reinforced plastic including the continuity loss of the third type:

$$R_o^+ \approx R_{Bz}^+ m_o \psi \cos \beta_o \quad (1.37)$$

Equation (1.37) enables us to determine the stress $R_o^+$ with an allowance because it contains factor $\cos \beta_o$. In reality a partial straightening of fibers takes place after continuity loss. If it is assumed that the fibers straighten out completely before they rupture, the following formula is substituted for (1.37)

$$R_o^+ \approx R_{Bz}^+ m_o \psi \quad (1.38)$$

It is also possible for the rupture of the warp fibers to occur before

*Structural Mechanics of a Fabric-Reinforced Plastic* 25

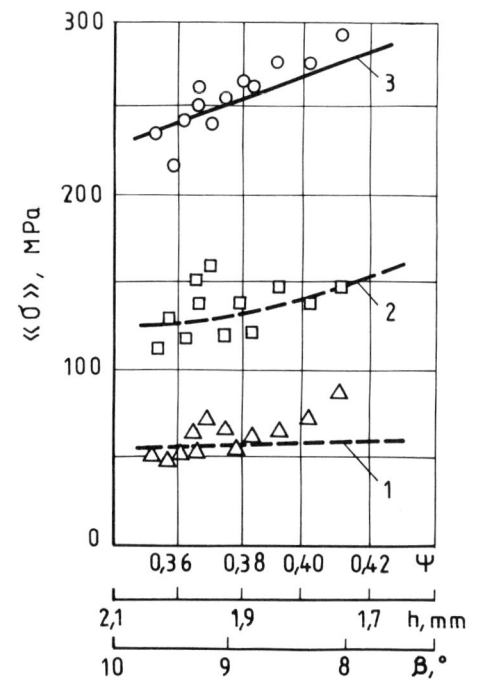

FIGURE 1.6. Results of experimental /90/ and theoretical investigations of the failure process of a glass fiber fabric reinforced plastic under uniaxial tension. The theoretical curves are based on the formulas: curve 1—(1.23); curve 2—(1.33); curve 3—(1.38); with the initial data: $m_o$ = 0.52; $R_A^+$ = 60 MPa; $R_{Bz}$ = 1.35 GPa.

the third-type continuity loss. Thus the ultimate strength of the fabric-reinforced plastic is expressed by

$$R_o^+ = \frac{R_{Bz}^+ E_o}{E_{Bz} S_o} \qquad (1.39)$$

where

$$S_o = (g_{11}^o)^{+(+)} \cos^2 \beta_o + 2(g_{15}^o)^{+(+)} \cos \beta_o \sin \beta_o$$

If the fiber rupture takes place before the second-type continuity loss, coefficients $(g_{11}^o)^+$ and $(g_{15}^o)^+$ are used instead of $(g_{11}^o)^{+(+)}$ and $(g_{15}^o)^{+(+)}$.

Fig. 1.6 presents experimental data from Owen and Rice [90] as well as the theoretical curves based on the relationships given in the above paragraph.

## 1.6 ALGORITHM FOR DETERMINING THE FAILURE STAGES IN A FABRIC-REINFORCED PLASTIC UNDER UNIAXIAL TENSION

The calculation of the failure stages of a fabric-reinforced plastic under uniaxial tension in the warp direction is made as follows.

(1) Matrix strength $R_A^+$, fiber strengths $R_{Bz}^+$, $R_{Br}^+$, $T_{Brz}$ and bond strengths $R_b$ and $T_b$ are established.

(2) Stress concentration factors $\bar{\sigma}_r$ and $\bar{\tau}_{rz}$ corresponding to the reinforcement factor $\psi = 0.7$ are found using the curves from Sih & Skudra [81] and Skudra & Bulvas [55].

(3) Coefficients $g_{ij}^o$ and $g_{ij}^y$ binding the external stresses and stresses in the structural components of the fabric-reinforced plastic both before and after continuity loss are found from Equations (1.19)–(1.22).

(4) The least limit of the first-type continuity loss is obtained from Equations (1.23)–(1.25).

(5) The limit for continuity loss of the second type is found from Equation (1.26).

(6) The least limit of the third-type continuity loss is set up according to Equations (1.33)–(1.35).

(7) The least final strength of a fabric-reinforced plastic under tension acting in the warp direction is established using Equations (1.37)–(1.39).

## 1.7 STRENGTH OF A FABRIC-REINFORCED PLASTIC UNDER UNIAXIAL COMPRESSION

Let us consider compression in a fabric-reinforced plastic in the warp direction. Under compression, preliminary cracking of the material is not observed. This leads to the supposition that the final failure of the fabric-reinforced plastic coincides with the third type of continuity loss, i.e., failure of the matrix in the impregnated filaments in the diagonally reinforced bands due to the combined action of the stresses $\langle \sigma_{\parallel}^o \rangle$ and $\langle \tau_{\parallel\perp}^o \rangle$ (see Fig. 1.5). Writing $R_o^-$ for the stress $\langle\langle \sigma_o \rangle\rangle$ at the instant of failure we determine for the case of the warp monolayer failure that

$$R_o^- = \langle\langle \sigma_o^{+++} \rangle\rangle \qquad (1.40)$$

where the least stress $\langle\langle \sigma_o^{+++} \rangle\rangle$ is found from Equations (1.33)–(1.35) by substituting coefficients $g_{ij}^o$ and $g_{ij}^y$ for $(g_{ij}^o)^{+(+)}$ and $(g_{ij}^y)^{+(+)}$ in Equation (1.36).

## 1.8 DEFORMATION DIAGRAM OF A FABRIC-REINFORCED PLASTIC. BIMODULARITY

The preliminary continuity loss under tension is seen in the deformation diagram of fabric-reinforced plastics. Before complete rupture of the samples the deformation diagrams for tension usually show two typical breaking points that correspond to continuity loss of the first and the third types. Under compression the deformation diagram remains linear until failure.

According to the initial assumptions, when the applied stress $\langle\langle \sigma_o \rangle\rangle$ reaches the value corresponding to the level of continuity loss, all the structural elements of the fabric-reinforced plastic that are of similar type and that find themselves in a disadvantageous state of stress undergo failure simultaneously, and the result is a stepwise change in the elastic properties of the material. But in reality the stress state of all these structural elements is not equal, owing to the scattered geometrical parameters (e.g., phase shift of filament undulations in some fabrics, scattered values of angles $\beta_o$ and $\beta_y$). In addition, there is the scattering of elastic and strength properties of the matrix and the fibers. That is why the stiffness variation at the instant of continuity loss is not stepwise but smooth, as shown by the points on the test diagram of a strained plastic reinforced by fabric T-42-36 where organic fiber SVM is included in the warp direction and glass fiber VMP is used in the weft (Fig. 1.7). In order to take into account the smooth character of the continuity loss it is necessary to use the stochastic model of failure. It should be remarked here that the determinate failure model used in the present paper allows us to take into consideration the most important features of the fabric-reinforced plastic failure with reasonable accuracy.

Consider, for example, the theoretical structure of a deformation diagram for a fabric-reinforced plastic (see Fig. 1.7) plotted with the following initial data:

- for SVM fibers—$E_{Bz}$ = 110 Gpa; $E_{Br}$ = 3.6 GPa; $\nu_{Bzr} = \nu_{Br\theta}$ = 0.16; $G_{Bzr}$ = 2.2 GPa; $G_{Br\theta}$ = 1.5 GPa; $R_{Bz}^+$ = 2.3 GPa; $R_{Br}^+$ = 70 MPa; $T_{Bzr}$ = 45 MPa;
- for VMP fibers—$E_B$ = 75 GPa; $\nu_B$ = 0.22; $G_B$ = 31 GPa; $R_{Bz}^+$ = 1.55 GPa

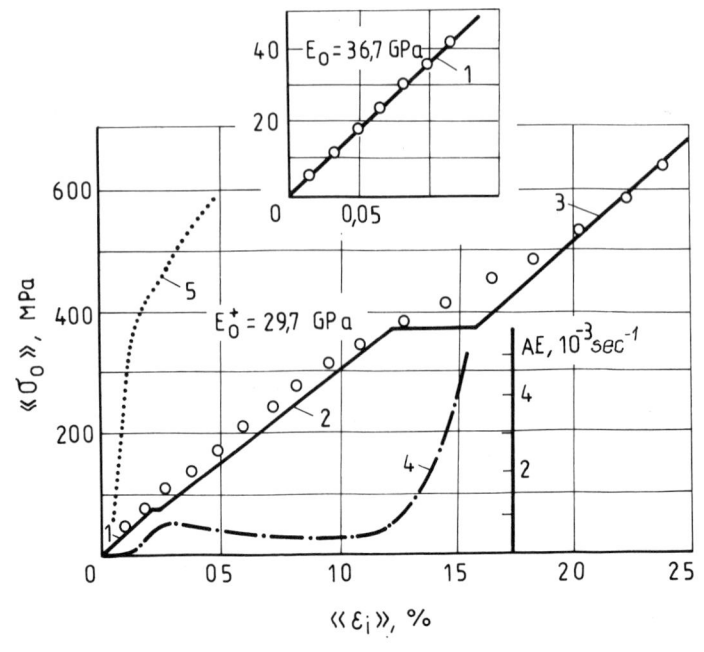

**FIGURE 1.7.** Longitudinal (1, 2, 3) and transversal deformation (5) diagrams for a fabric reinforced plastic and acoustic emissions (4) under uniaxial tension in the warp direction.

- for the matrix EHD-U—$E_A = 3$ GPa; $\nu_A = 0.35$; $G_A = 1.11$ GPa; $R_A^+ = 75$ MPa; $T_A = 65$ MPa;
- for the T-42-36 fabric—$m_o = 0.53$; $n_o = 0.35$; $n_y = 0.40$; $\beta_o = 12°$; $\beta_y = 11°$; $\psi = 0.55$.

Curve 1 based on Equation (1.11) illustrates the elasticity modulus of a fabric-reinforced plastic under compression and the initial modulus under tension. The stress $\langle\langle \sigma_o^+ \rangle\rangle$ which causes the first type of continuity loss, i.e., the first fracture, is obtained from Equation (1.23), and the respective deformation is obtained according to Hooke's law

$$\langle\langle \epsilon_o^+ \rangle\rangle = \frac{\langle\langle \sigma_o^+ \rangle\rangle}{E_o} \quad (1.41)$$

After the stress $\langle\langle \sigma_o \rangle\rangle$ has reached the value $\langle\langle \sigma_o^+ \rangle\rangle$ there is a stepwise transition to curve 2 based on Equation (1.14). Considering that $E_o^+ \approx E_o^{+(+)}$ the influence of the second-type continuity loss can be neglected when plotting the diagram. The stress corresponding to the instant of the third-type continuity loss $\langle\langle\sigma_o^{+++}\rangle\rangle$, i.e., the second fracture, is determined by Equation (1.33) whereas the respective ultimate strain is found according to the following formula

$$\langle\langle\epsilon_o^{+++}\rangle\rangle = \frac{\langle\langle\sigma_o^{+++}\rangle\rangle}{E_o^{+(+)}} \qquad (1.42)$$

where $E_o^{+(+)}$ is obtained from Equation (1.16).

According to initial assumptions, there is an instantaneous shear failure of the warp and the weft filaments in all the diagonally reinforced regions when $\langle\langle\sigma_o^{+++}\rangle\rangle$ is reached, causing the straightening of the warp filaments. Curve 3 in the deformation diagram is built according to the formula

$$\langle\langle\sigma_o\rangle\rangle = E_{Bz}m_o\psi(\langle\langle\epsilon_o\rangle\rangle - \Delta\langle\langle\epsilon_o\rangle\rangle) \qquad (1.43)$$

where $\Delta\langle\langle\epsilon_o\rangle\rangle$ is additional strain due to the straightening of the fibers and is calculated in the following way

$$\Delta\langle\langle\epsilon_o\rangle\rangle = n_o(\sec\beta_o - 1) \qquad (1.44)$$

The ultimate strength of a fabric-reinforced plastic is obtained from Equation (1.38) and the ultimate strain is obtained from the relationship

$$\langle\langle\epsilon_o^+\rangle\rangle = \epsilon_{Bz}^+ + \Delta\langle\langle\epsilon_o\rangle\rangle \qquad (1.45)$$

Fig. 1.7 illustrates also the measurements of longitudinal and transverse deformations in a fabric-reinforced plastic. In addition, curve 4 pictures the acoustic emission data during the loading. Intensified acoustic emission agrees fairly well with the deformation steps obtained by computation.

After a single loading exceeding the level of the first-type continuity loss the fabric-reinforced plastic becomes a bimodular material, as its elasticity modulus in compression is $E_o$ and in tension it is $E_o^+$.

Methods of establishing the mechanical properties of reinforced plastics, including fabric-reinforced plastics, are outlined in a detailed way in Tarnopolskii and Yu [67].

CHAPTER 2

# Structural Mechanics of Hybrid Composites

## 2.1 ELASTIC PROPERTIES OF A HYBRID COMPOSITE

At the present time, designers have a wide range of various reinforced plastics at their disposal which differ essentially in their cost and mechanical properties. A combination of two or more various types of fibers in a single plastic gives a hybrid composite, and its mechanical qualities may be varied in a wide range. There are two kinds of hybridity: laminated and dispersive (Fig. 2.1). In determining some elastic properties of the material it is important to consider the type of hybridity [55].

The elasticity modulus in the reinforcement direction of a unidirectionally reinforced hybrid does not practically depend upon the type of hybridity and is obtainable from the following formula

$$E_\| = (1 - \psi)E_A + (1 - \mu_C)\psi E_{Bz} + \mu_C \psi E_{Cz}$$
$$= \psi_A E_A + \psi_B E_{Bz} + \psi_C E_{Cz} \quad (2.1)$$

where $\psi_A$ and $E_A$ respectively are relative volume content and elasticity modulus of the polymer matrix; $\psi_B$ and $E_{Bz}$ are the same for the $B$ type fibers; $\psi_C$ and $E_{Cz}$ denote the same for the $C$-type fibers.

In Equation (2.1) and further, $z$ signifies the axial direction of fibers whereas $r$ stands for the radial direction.

The relative volumetric contents of the components are interrelated in the following way

$$\psi_A + \psi_B + \psi_C = 1$$

**32** STRUCTURAL MECHANICS OF A MATERIAL

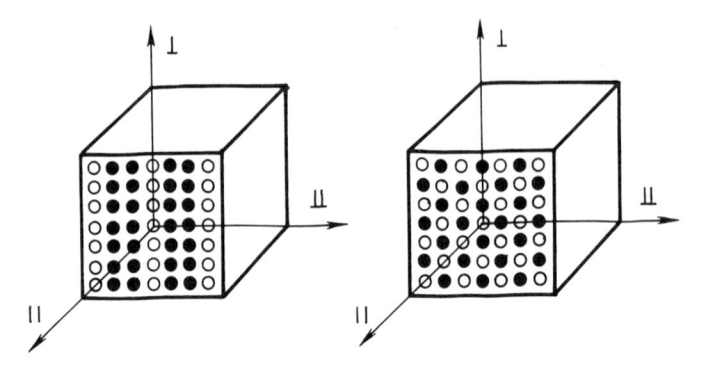

FIGURE 2.1. Structure of hybrid plastics: (a) laminated hybridity; (b) dispersive hybridity.

The total relative volumetric content of the fibers is

$$\psi = \frac{V_B + V_C}{V_A + V_B + V_C} = \frac{F_B + F_C}{F_A + F_B + F_C} = \psi_B + \psi_C$$

where $V$ and $F$ are the volume and the area of the cross-sections of the components respectively.

The ratio of the $C$-type fiber volume content to the total volume of fibers will be denoted by $\mu_C$

$$\mu_C = \frac{\psi_C}{\psi_B + \psi_C} = \frac{\psi_C}{\psi} \quad (2.2)$$

The transverse elasticity modulus $E_\perp$ for laminated hybridity is calculated according to the formula

$$E_\perp = E_\perp^B (1 - \mu_C) + E_\perp^C \mu_C \quad (2.3)$$

where $E_\perp^B$ and $E_\perp^C$ are the corresponding transverse elasticity moduli of the unidirectionally reinforced plastics where the reinforcement is by $B$- and $C$-type fibers.

For dispersive hybridity the transverse elasticity modulus $E_\perp$ is expressed as

$$E_\perp = \frac{E_\perp^B E_\perp^C}{E_\perp^C (1 - \mu_C) + E_\perp^B \mu_C} \quad (2.4)$$

The shear moduli are obtained in an analogous way [55]
- for laminated hybridity

$$G_{\|\perp} = G_{\|\perp}^B (1 - \mu_C) + G_{\|\perp}^C \mu_C \qquad (2.5)$$

- for dispersive hybridity

$$G_{\|\perp} = \frac{G_{\|\perp}^B G_{\|\perp}^C}{G_{\|\perp}^C (1 - \mu_C) + G_{\|\perp}^B \mu_C} \qquad (2.6)$$

It may be assumed in the first approximation that Poisson's ratio is practically independent from the type of hybridity and is determined according to the familiar summation law

$$\nu_{\perp\|} = \nu_{\underline{\perp}\|} = \nu_A \psi_A + \nu_{Brz} \psi_B + \nu_{Crz} \psi_C \qquad (2.7)$$

## 2.2 FAILURE PECULIARITIES OF A HYBRID COMPOSITE

The deformation law for a unidirectionally reinforced hybrid composite loaded in the reinforcement direction is written as

$$\langle \sigma_\| \rangle = (\psi_A E_A + \psi_B E_{Bz} + \psi_C E_{Cz}) \langle \epsilon_\| \rangle \qquad (2.8)$$

Equation (2.8) is based on the assumption that a complete bond exists between the fibers and the matrix, i.e., the fiber strains are equal to the hybrid composite strains

$$\epsilon_A = \epsilon_B = \epsilon_C = \langle \epsilon_\| \rangle$$

Suppose that there is the following relationship linking the ultimate strains of the components

$$\epsilon_{AR} > \epsilon_{BR} > \epsilon_{CR}$$

At the first stage of loading the hybrid composite is strained as stated by Equation (2.8). At the instant of loading when the $C$-type fibers reach their ultimate strain value $\epsilon_{CR}$, the first loading stage ends and the mean stress $\langle \sigma_\|' \rangle$ is determined as

$$\langle \sigma_\|' \rangle = (\psi_A E_A + \psi_B E_{Bz} + \psi_C E_{Cz}) \epsilon_{CR} \qquad (2.9)$$

After the failure of $C$-type fibers the stress-strain state of the hybrid composite undergoes abrupt changes. Depending on the loading regime the potential energy of the system may increase ($\Delta U > 0$), remain unchanged ($\Delta U = 0$) or decrease ($\Delta U < 0$) at the instant of failure.

If $\Delta U > 0$, practically the whole load after the failure of $C$-type fibers is taken by the $B$-type fibers. The mean stress $\langle \sigma'_\| \rangle$ applied at the instant of failure does not change, but strain $\langle \epsilon_\| \rangle$ increases to reach the value of

$$\langle \epsilon'_\| \rangle = \frac{(\psi_A E_A + \psi_B E_{Bz} + \psi_C E_{Cz})\epsilon_{CR}}{(\psi_A E_A + \psi_B E_{Bz})} \tag{2.10}$$

The deformation diagram for a glass/carbon plastic with $\Delta U > 0$ has been experimentally established by Kanlin [83] and presented here in Fig. 2.2.

Equation (2.10) may be rewritten as

$$\langle \epsilon'_\| \rangle = \frac{[(1-\psi)E_A + \mu_B \psi E_{Bz} + (1-\mu_B)\psi E_{Cz}]\epsilon_{CR}}{[(1-\psi)E_A + \mu_B \psi E_{Bz}]} \tag{2.11}$$

where

$$\mu_B = \frac{\psi_B}{\psi}$$

When $\langle \epsilon'_\| \rangle = \epsilon_{BR}$, immediately after the failure of $C$-type fibers an abrupt redistribution of stresses leads to the failure of $B$-type fibers as well. In this case Equation (2.11) has the form

$$\epsilon_{BR} = \frac{(1-\psi)E_A + \mu_{B\text{crit}}\psi E_{Bz} + (1-\mu_{B\text{crit}})\psi_{Cz}}{[(1-\psi)E_A + \mu_{B\text{crit}}\psi E_{Bz}]}\epsilon_{CR} \tag{2.12}$$

If the total relative volume content of fibers $\psi$ is constant, Equation (2.12) enables us to determine the critical relative volume of $B$-type fibers

$$\mu_{B\text{crit}} = \frac{(1-\psi)E_A(\epsilon_{CR} - \epsilon_{BR}) + \psi E_{Cz}\epsilon_{CR}}{\psi E_{Bz}(\epsilon_{BR} - \epsilon_{CR}) + \psi E_{Cz}\epsilon_{CR}}$$

$$\approx \frac{E_{Cz}\epsilon_{CR}}{E_{Bz}(\epsilon_{BR} - \epsilon_{CR}) + E_{Cz}\epsilon_{CR}} \tag{2.13}$$

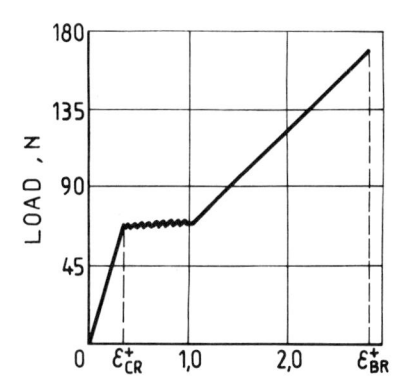

FIGURE 2.2. Experimentally established test diagram for a carbon-glass epoxy.

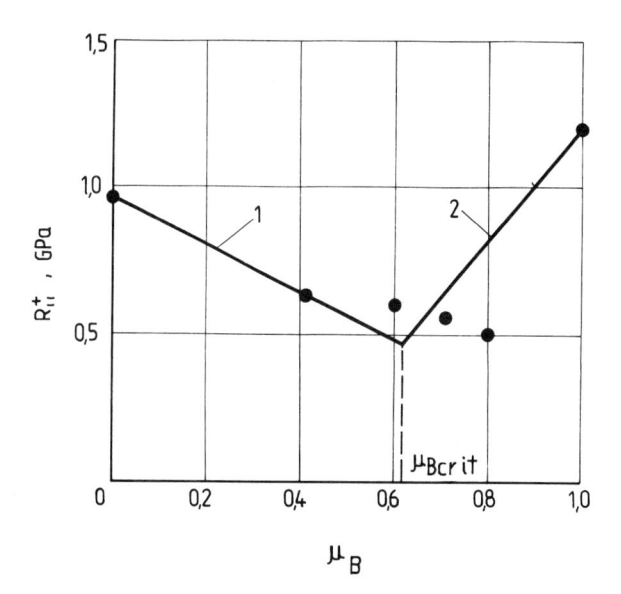

FIGURE 2.3. Dependence of tensile strength of a carbon-glass epoxy upon the relative fiber content with $\Delta U > 0$. Theoretical lines are plotted according to the formulas: curve 1—(2.14), curve 2—(2.15), with the initial data: $\psi = 0.62$; $E_{CR}^+ = 0.003$; $E_{BR}^+ = 0.036$; $E_A = 3500$ MPa; $E_{BZ} = 83{,}750$ MPa; $E_{CZ} = 502{,}500$ MPa.

**36** STRUCTURAL MECHANICS OF A MATERIAL

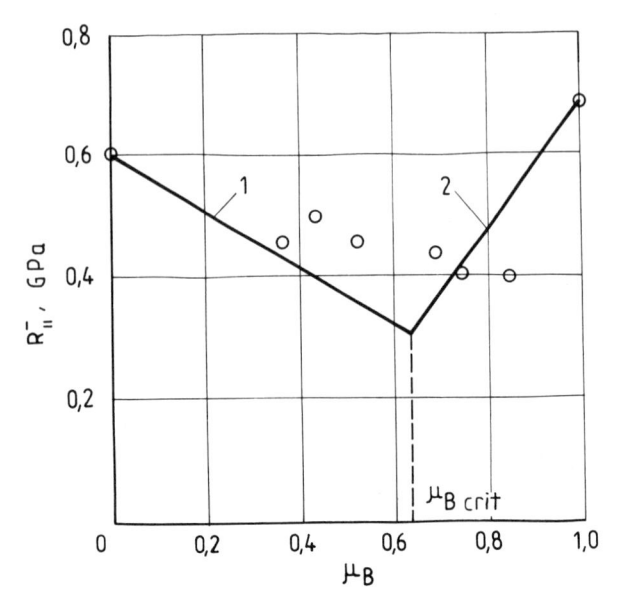

**FIGURE 2.4.** Dependence of compressive strength of a carbon-glass epoxy upon the relative fiber content with $\Delta U > 0$. Theoretical lines are plotted according to the formulas: curve 1—(2.14), curve 2—(2.15) with $\epsilon_{BR}^- = 0{,}021$ and $\epsilon_{CR}^- = 0{,}0019$. The other initial data are equivalent to those of Fig. 2.3.

With $\mu_B < \mu_{B\text{crit}}$ the strength of a three-component hybrid is obtained from the formula

$$R_{\parallel} = [(1 - \psi)E_A + \mu_B \psi E_{Bz} + (1 - \mu_B)\psi E_{Cz}]\epsilon_{CR} \quad (2.14)$$

whereas with $\mu_B > \mu_{B\text{crit}}$ it is obtained from

$$R_{\parallel} = [(1 - \psi)E_A + \mu_B \psi E_{Bz}]\epsilon_{BR} \quad (2.15)$$

Figs. 2.3 and 2.4 show the dependence of tensile strength ($R_{\parallel}^+$) and compressive strength ($R_{\parallel}^-$) of a carbon/glass plastic upon the relative content of glass fibers $\mu_B$ with a constant fiber volume content $\psi = 0.62$. The above mentioned figures also represent test data given by Kanlin [83].

When $\Delta U = 0$, the strain $\langle \epsilon'_{\parallel} \rangle$ may be determined from the change in the potential energy. At the instant of the failure of $C$-type fibers the energy accumulated in them is transferred to the undamaged part of the hybrid composite:

$$\tfrac{1}{2} E_{Cz}(1 - \mu_B)\psi \epsilon_{CR}^2 = \tfrac{1}{2}[E_{Bz}\mu_B\psi + E_A(1 - \psi)](\langle \epsilon'_{\parallel} \rangle^2 - \epsilon_{CR}^2)$$

The previous equation yields

$$\langle \epsilon'_{\|} \rangle = \epsilon_{CR} \sqrt{1 + \frac{E_{Cz}(1 - \mu_B)\psi}{E_A(1 - \psi) + E_{Bz}\mu_B\psi}} \quad (2.16)$$

At the instant of $C$-type fiber failure the mean strain of the material increases reaching value $\langle \epsilon'_{\|} \rangle$, and the mean stress $\langle \sigma_{\|} \rangle$ decreases from $\langle \sigma'_{\|} \rangle$ to $\langle \sigma''_{\|} \rangle$

$$\langle \sigma''_{\|} \rangle = [E_A(1 - \psi) + \mu_B\psi E_{Bz}]\epsilon_{CR}$$

$$\times \sqrt{1 + \frac{E_{Cz}(1 - \mu_B)\psi}{E_A(1 - \psi) + E_{Bz}\mu_B\psi}} \quad (2.17)$$

When $\langle \epsilon'_{\|} \rangle = \epsilon_{BR}$ the failure of $C$-type fibers is followed by the failure of $B$-type fibers. Then Equation (2.16) may be rewritten to have the following form

$$\epsilon_{BR} = \epsilon_{CR} \sqrt{1 + \frac{E_{Cz}(1 - \mu_{Bcrit})\psi}{E_A(1 - \psi) + E_{Bz}\mu_{Bcrit}}}$$

This relationship enables us to determine $\mu_{Bcrit}$ for the given loading regime when $\Delta U = 0$. We obtain

$$\mu_{Bcrit} = \frac{E_{Cz}\psi - \left[\left(\frac{\epsilon_{BR}}{\epsilon_{CR}}\right)^2 - 1\right](1 - \psi)E_A}{\left[\left(\frac{\epsilon_{BR}}{\epsilon_{CR}}\right)^2 - 1\right]\psi E_{Bz} + \psi E_{Cz}}$$

$$\approx \frac{E_{Cz}}{\left[\left(\frac{\epsilon_{BR}}{\epsilon_{CR}}\right)^2 - 1\right]E_{Bz} + E_{Cz}}$$

(2.18)

The strength of a hybrid composite is given by Equations (2.14) and (2.15).

Fig. 2.5 shows the dependence of the tensile strength of a glass/carbon plastic upon $\mu_B$. The theoretical curves are based on Equations (2.14) and (2.15) considering Equation (2.18).

**38** STRUCTURAL MECHANICS OF A MATERIAL

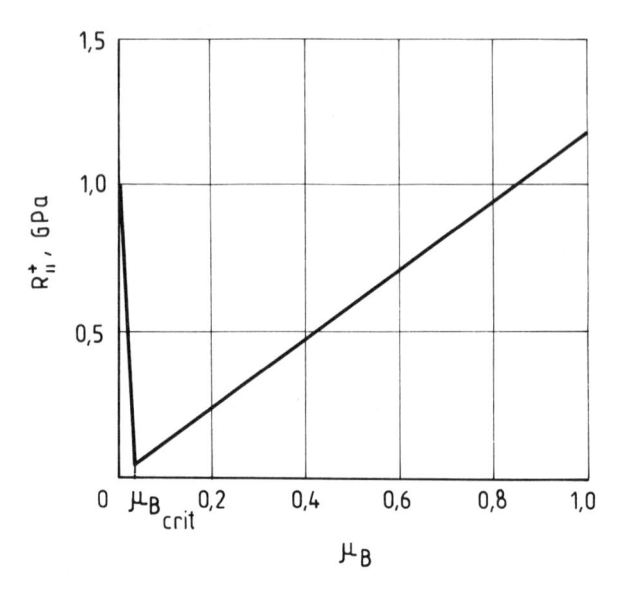

**FIGURE 2.5.** Dependence of tensile strength of a carbon-glass epoxy upon the relative volume content of fibers if $\Delta U = 0$. The theoretical curves are based on the initial data of Fig. 2.3.

It follows from Equation (2.18) that the value of parameter $\mu_{Bcrit}$ which corresponds to the minimum strength of the hybrid plastic depends to a great extent upon the ratio between the ultimate strains of $B$- and $C$-type fibers. This dependence is illustrated in Fig. 2.6 with the initial data taken for the carbon/glass plastic (see Fig. 2.3). Fig. 2.7 is a typical deformation diagram if $\Delta U = 0$.

If $\Delta U < 0$ at the instant of $C$-type fiber failure, the strain of the hybrid material does not change, but mean stress $\langle \sigma'_{\|} \rangle$ decreases abruptly, reaching the value $\langle \sigma''_{\|} \rangle$ which is given by the relationship

$$\langle \sigma''_{\|} \rangle = \epsilon_{CR}[(1 - \psi)E_A + \mu_B \psi E_{Bz}] \qquad (2.19)$$

When the $C$-type fibers have failed and $\langle \epsilon_{\|} \rangle < \epsilon_{BR}$, a further increase of the load makes the deformation law independent from the loading regime and is written as

$$\langle \sigma_{\|} \rangle = [(1 - \psi)E_A + \mu_B \psi E_{Bz}]\langle \epsilon_{\|} \rangle \qquad (2.20)$$

The dependence of the carbon/glass plastic strength upon $\mu_B$ in the loading case under consideration is illustrated by Fig. 2.8.

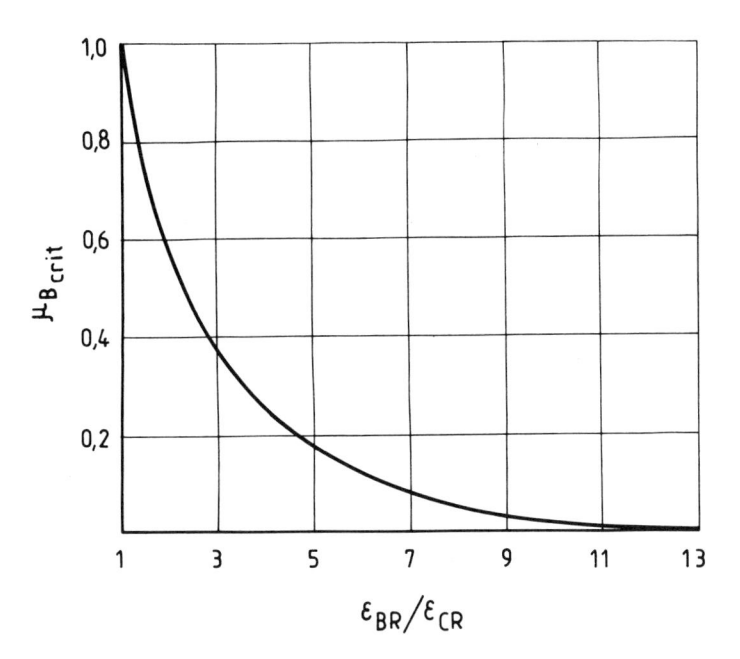

**FIGURE 2.6.** Dependence of parameter $\mu_{B_{crit}}$ upon the ratio between the ultimate strain of $B$ and C-type fibers.

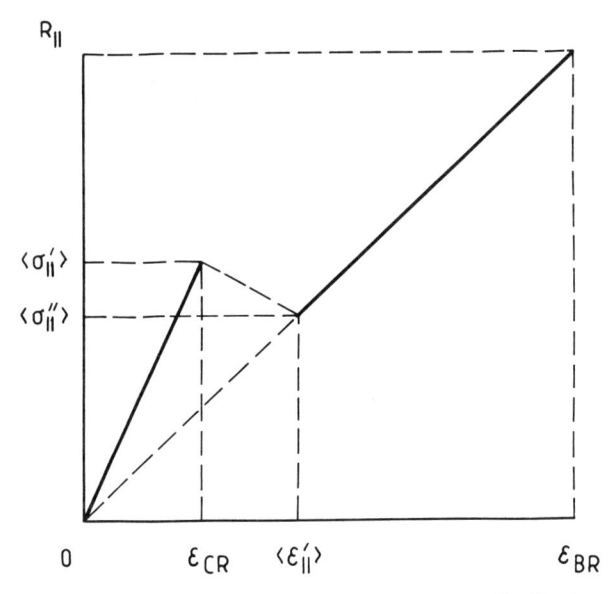

**FIGURE 2.7.** Deformation diagram of hybrid plastics if $\Delta U = 0$.

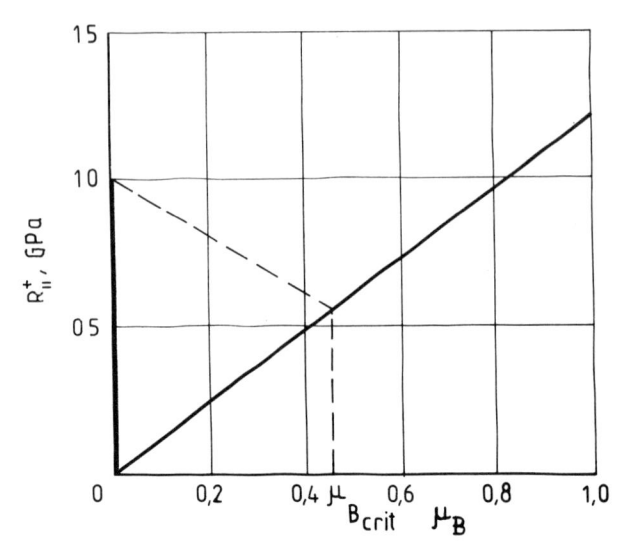

**FIGURE 2.8.** Dependence of carbon-glass reinforced plastic by extension of relative fiber content if $\Delta U < 0$.

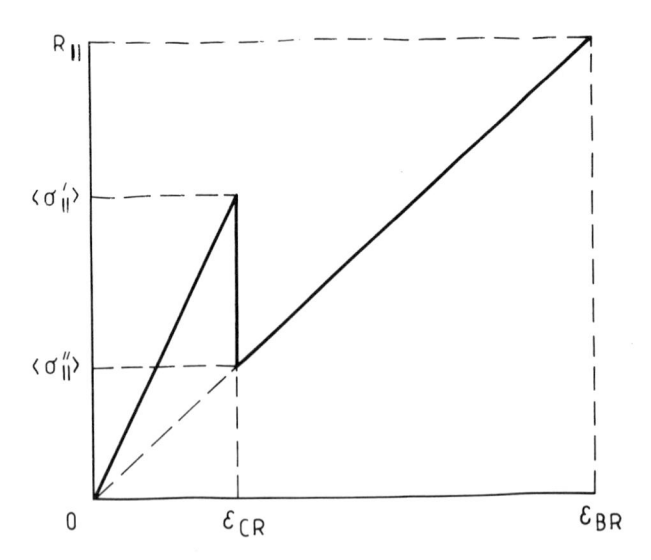

**FIGURE 2.9.** Deformation diagram of hybrid plastics if $\Delta U < 0$.

Mean stress $\langle \sigma'_\| \rangle$ at the instant of $C$-type fiber failure is found according to Equation (2.9) which can be rewritten as

$$\langle \sigma'_\| \rangle = [(1 - \psi)E_A + \mu_B \psi E_{Bz} + (1 - \mu_B)\psi E_{Bz}]\epsilon_{CR} \quad (2.21)$$

The dependence of $\langle \sigma'_\| \rangle$ upon $\mu_B$ for a carbon/glass plastic is shown with a dashed line in Fig. 2.8. It is evident from the diagram that with $\mu_B < \mu'_{B\text{crit}}$ mean stress $\langle \sigma''_\| \rangle$ responsible for the failure of the stiffer $C$-type fibers exceeds the strength of the hybrid composite $R_\|$ obtained from Equation (2.15). Therefore, if $\mu_B < \mu''_{B\text{crit}}$, it should be assumed that

$$R_\| = \langle \sigma'_\| \rangle = [(1 - \psi)E + \mu_B \psi E_{Bz} + (1 - \mu_B)\psi E_{Cz}]\epsilon_{CR} \quad (2.22)$$

When $\mu_B > \mu'_{B\text{crit}}$, strength $R_\|$ is given by Equation (2.15). The theoretical line plotted according to Equation (2.22) is shown by a solid line in Fig. 2.8. The qualitative characteristics of deformation are presented in Fig. 2.9.

CHAPTER 3
# Thermal Strains and Stresses

## 3.1 FUNCTION OF THERMAL EXPANSION

In order to determine thermal strains in reinforced plastics the thermal expansion coefficient, which is independent from temperature, has been used in various papers [13-17, 34, 59, 92, 94 and 95]. The coefficient of thermal expansion is a function of elastic characteristics and the thermal expansion coefficients of the polymer matrix and the fibers. It has been established experimentally [82, 85, 91, 94] that the elastic properties and the thermal expansion coefficients of the polymer matrix and the fibers are temperature-dependent. It follows that the concept of thermal expansion coefficient $\alpha$ as a material constant makes no sense for reinforced plastics and should be replaced by a new concept—the function of thermal expansion $\alpha(T)$ [60]. In that case the conventional relationship

$$\epsilon(T) = \alpha \Delta T$$

is replaced by

$$\epsilon(T_2 - T_1) = \int_{T_1}^{T_2} \alpha(T) dT$$

To simplify the determination of the thermal expansion function for a reinforced plastic $\alpha(T)$ it may be assumed in the first approximation that elastic characteristics of the fibers are not temperature dependent. The functions of thermal expansion of the matrix $\alpha_A(T)$ and of the fibers $\alpha_B(T)$ should be established by test. Fig. 3.1 shows characteristic

**44** STRUCTURAL MECHANICS OF A MATERIAL

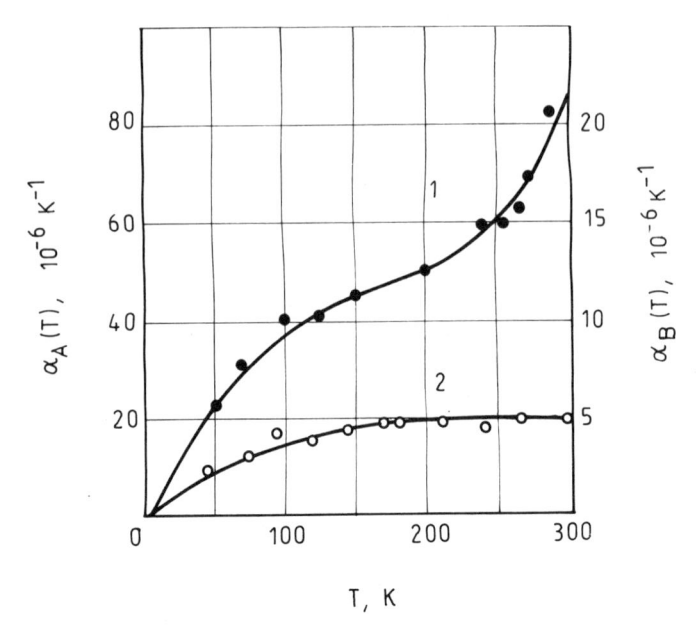

**FIGURE 3.1.** Thermal expansion functions for the epoxy resin $\alpha_A(T)$ and for glass fibers $\alpha_B(T)$.

types of thermal expansion functions for the epoxy resin and glass fibers as given by Hartwig & Puck [82]. It is convenient to approximate the experimental curves of the thermal expansion coefficient in the form of third-order polynomials.

The temperature dependence of the modulus $E_A(T)$ and of the Poisson's ratio of the matrix $\nu_A(T)$ should also be determined by test. Approximation by the third-order polynomial is suitable for these relationships as well.

When the temperature dependences of elastic characteristics of the matrix have been found, it is possible to proceed to the determination of the temperature dependence of elastic characteristics of a unidirectionally reinforced plastic. For this purpose $E_A(T)$, $\nu_A(T)$ and $G_A(T)$ have to be used in the known formulas for the calculation of elastic characteristics of a unidirectional plastic $E_\|$, $E_\perp$, $\nu_{\|\perp}$ and $G_{\|\perp}$ instead of $E_A$, $\nu_A$ and $E_A$. Thus, for instance, the application of equations presented in Skudra & Sbitnevs [59] yield

$$E_\|(T) = \psi E_{Bz} + (1 - \psi) E_A(T) \quad (3.1)$$

$$E_\perp(T) = \left[\left(1 - 2\sqrt{\frac{\psi}{\pi}}\right) \frac{1}{1 - \nu_A^2(T)} + 2\sqrt{\frac{\psi}{\pi}} J\right] E_A(T) \quad (3.2)$$

where

$$J = \frac{1}{bE_A(T)} \left[ \frac{\pi}{2} - \frac{2a}{\sqrt{a^2 - b^2}} \operatorname{arctg} \sqrt{\frac{a-b}{a+b}} \right]$$

$$a = \frac{1 - v_A^2(T)}{E_A(T)}$$

$$b = 2\sqrt{\frac{\psi}{\pi}} \left[ \frac{1 - v_{Brz} v_{Bzr}}{E_{Br}} - a \right] \quad (3.3)$$

$$v_{\|\perp}(T) = (1 - \psi) v_A(T) + \psi v_{Bzr}$$

$$v_{\perp\|}(T) = \frac{v_{\|\perp}(T) E_\perp(T)}{E_\|(T)}$$

$$G_{\|\perp}(T) = \left[ 2\sqrt{\frac{\psi}{\pi}} K \left( \frac{2K}{\sqrt{K^2 - 1}} \operatorname{arctg} \sqrt{\frac{K+1}{K-1}} - \frac{\pi}{2} \right) \right. \\ \left. + 1 - 2\sqrt{\frac{\psi}{\pi}} \right] G_A(T) \quad (3.4)$$

where

$$K = \frac{1}{\left(1 - \dfrac{G_A(T)}{G_{Brz}}\right) 2\sqrt{\dfrac{\psi}{\pi}}}$$

$$G_A(T) = \frac{E_A(T)}{2[1 + v_A(T)]}$$

In the above relationships $E_{Bz}$, $E_{Br}$, $G_{Brz}$, $v_{Brz}$ and $v_{Bzr}$ are elastic characteristics of the fibers, and $\psi$ is the relative volume content of the fibers.

As an example, let us discuss the determination of the elastic characteristics of an epoxy/carbon plastic with two types of reinforcement: high-modulus fibers and high-strength fibers (Table 3.1).

It is assumed that the strain characteristics of the fibers are not tem-

Table 3.1. Elastic Characteristics of Carbon Fibers.

| Type of Fibers | $E_{Bz}$, GPa | $E_{Br}$, GPa | $\nu_{Br z}$ | $\nu_{Bzr}$ | $G_{\|\|\perp}$, GPa | $\alpha_{Bz}$ $10^{-6}\,K^{-1}$ | $\alpha_{Br}$ $10^{-6}\,K^{-1}$ |
|---|---|---|---|---|---|---|---|
| High modulus | 411 | 6.6 | 0.006 | 0.35 | 20 | −1.2 | 27.3 |
| High strength | 262 | 13.0 | 0.013 | 0.25 | 20 | −1.2 | 27.3 |

Table 3.2. Elastic Characteristics of the Epoxy Resin

| $T, K$ | $E_A$, GPa | $\nu_A$ | $G_A$, GPa |
|---|---|---|---|
| 100 | 16.32 | 0.274 | 6.40 |
| 160 | 10.77 | 0.275 | 4.22 |
| 200 | 8.55 | 0.277 | 3.35 |
| 260 | 6.58 | 0.293 | 2.54 |
| 300 | 5.69 | 0.316 | 2.16 |
| 390 | 2.98 | 0.411 | 1.056 |

Table 3.3. Elastic Characteristics of a Plastic Reinforced with High-Modulus Carbon Fibers.

| $T, K$ | $E_A$, GPa | $E_\perp$, GPa | $\nu_{\|\|\perp}$ | $\nu_{\perp\|\|}$ | $G_{\|\|\perp}$, GPa |
|---|---|---|---|---|---|
| 100 | 213.4 | 11.78 | 0.3120 | 0.017 | 8.32 |
| 160 | 211.4 | 9.64 | 0.3125 | 0.014 | 5.83 |
| 200 | 210.3 | 8.49 | 0.3135 | 0.013 | 4.76 |
| 260 | 209.1 | 7.01 | 0.3215 | 0.011 | 3.700 |
| 300 | 208.6 | 6.30 | 0.3330 | 0.010 | 3.19 |
| 390 | 206.8 | 3.23 | 0.3805 | 0.006 | 1.63 |

Table 3.4. Elastic Characteristics of a Plastic Reinforced with High-Strength Carbon Fibers.

| $T, K$ | $E_{\|\|}$, GPa | $E_\perp$, GPa | $\nu_{\|\|\perp}$ | $\nu_{\perp\|\|}$ | $G_{\|\|\perp}$, GPa |
|---|---|---|---|---|---|
| 100 | 139.18 | 15.00 | 0.262 | 0.028 | 8.32 |
| 160 | 136.53 | 11.43 | 0.263 | 0.022 | 5.83 |
| 200 | 135.40 | 9.66 | 0.2635 | 0.0187 | 4.76 |
| 260 | 134.31 | 8.02 | 0.272 | 0.0166 | 3.70 |
| 300 | 133.70 | 7.16 | 0.283 | 0.0151 | 3.19 |
| 390 | 132.88 | 4.26 | 0.331 | 0.011 | 1.63 |

perature-dependent, whereas the temperature dependence of epoxy resin characteristics is determined from the following third-order polynomials

$$\alpha_A(T) = -4.68 + 11.00T \cdot 10^{-2} - 4.85T^2 \cdot 10^{-4}$$

$$+ 7.42T^3 \cdot 10^{-7} \ (10^{-5} \ K^{-1})$$

$$E_A(T) = 34.25 - 24.80T \cdot 10^{-2} + 77.10T^2 \cdot 10^{-5}$$

$$- 8.76T^3 \cdot 10^{-7} \ (GPa)$$

$$\nu_A(T) = 0.28 - 0.009T \cdot 10^{-3} - 0.12T^2 \cdot 10^{-5}$$

$$+ 0.53T^3 \cdot 10^{-8}$$

(3.5)

Temperature dependences of elastic characteristics for the epoxy resin and a unidirectional carbon/epoxy with $\psi = 0.50$ calculated according to Equations (3.1)–(3.4) are given in Tables 3.2 through 3.4.

The thermal expansion functions for a unidirectionally reinforced plastic will be obtainable by replacing the thermal expansion coefficients of the components $\alpha_A$ and $\alpha_B$ in the familiar formulas by the corresponding functions $\alpha_A(T)$ and $\alpha_B(T)$. The following formulas yielded the best agreement with the test data:

- in the reinforcement direction—the formula given by G. A. Vanin [13]

$$\langle \alpha_\parallel(T) \rangle = \alpha_A(T) - [\alpha_A(T) - \alpha_{Bz}] \frac{\psi E_{Bz}}{E_\parallel(T)} \quad (3.6)$$

- perpendicular to the reinforcement direction—the formula by A. Shapery [95]

$$\langle \alpha_\perp(T) \rangle = [1 + \nu_A(T)]\alpha_A(T)(1 - \psi) + \alpha_{Br}\psi(1 + \nu_{Bzr})$$
$$- \langle \alpha_\parallel(T) \rangle [\psi \nu_{Bzr} + (1 - \psi)\nu_A(T)]$$

(3.7)

The thermal expansion functions in the elastic symmetry directions based on Equations (3.6) and (3.7) for the two types of unidirectional carbon/epoxy are presented in Figs. 3.2 and 3.3. They also include test results of Rogers et al. [92].

**48** STRUCTURAL MECHANICS OF A MATERIAL

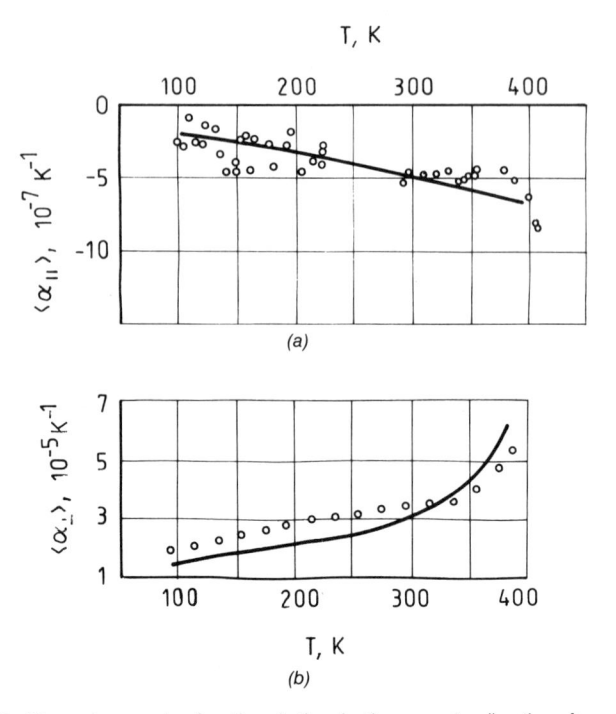

**FIGURE 3.2.** Thermal expansion functions in the elastic symmetry directions for a plastic containing high-modulus carbon fibers. The theoretical curves are based on formula (3.6) for $\langle \alpha_{\|}(T) \rangle$ and on (3.7) for $\langle \alpha_{\perp}(T) \rangle$.

The thermal expansion functions in the arbitrary $x$ and $y$ directions are given by

$$\langle \alpha_x(T) \rangle = \langle \alpha_{\|}(T) \rangle \cos^2 \beta + \langle \alpha_{\perp}(T) \rangle \sin^2 \beta \quad (3.8)$$

$$\langle \alpha_y(T) \rangle = \langle \alpha_{\|}(T) \rangle \sin^2 \beta + \langle \alpha_{\perp}(T) \rangle \cos^2 \beta \quad (3.9)$$

$$\langle \alpha_{xy}(T) \rangle = -2(\langle \alpha_{\perp}(T) \rangle - \langle \alpha_{\|}(T) \rangle) \sin \beta \cos \beta \quad (3.10)$$

where $\beta$ is the angle formed by $x$-axis and reinforcement direction.

Fig. 3.4 illustrates the thermal expansion function constricted according to Equation (3.8). The test data have been borrowed from Rogers et al. [92].

From generalized relationships given in Skudra & Sbitnevs [59] for calculating the thermal expansion coefficient of a laminate, the following

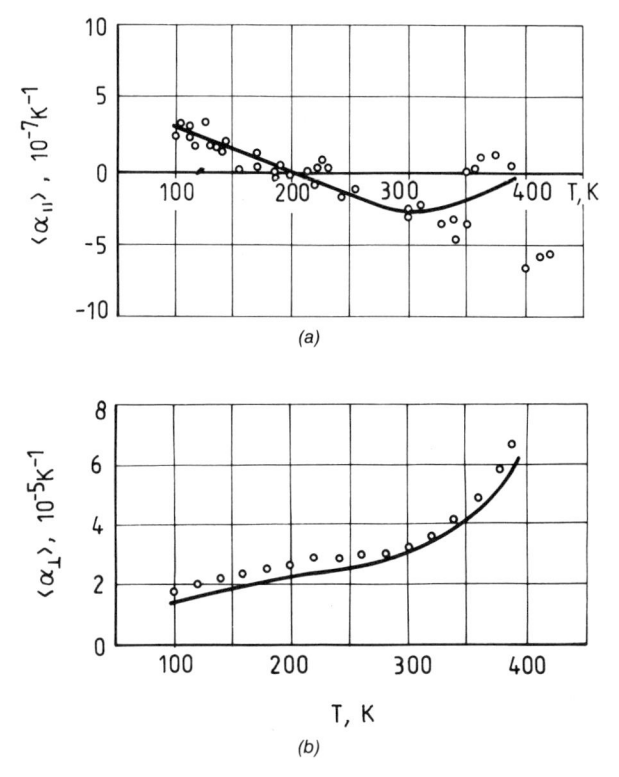

**FIGURE 3.3.** Thermal expansion functions in the elastic symmetry directions for a plastic having high strength carbon fiber reinforcement. The theoretical curves are based on formula (3.6) for $\langle \alpha_\parallel(T) \rangle$ and on formula (3.7) for $\langle \alpha_\perp(T) \rangle$.

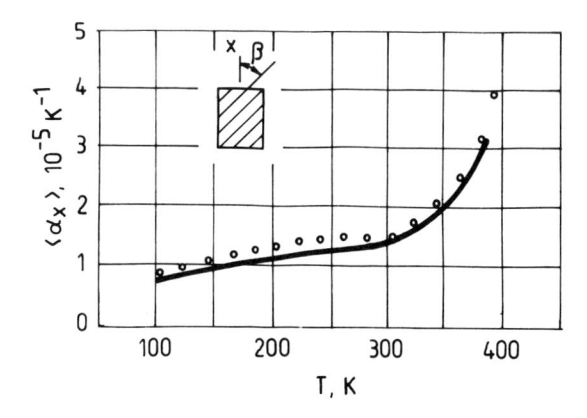

**FIGURE 3.4.** Thermal expansion function $\langle\langle \alpha_1(T) \rangle\rangle$ for a reinforced plastic containing high-strength carbon fibers, where $B = \pm 45°$.

generalized expressions of thermal expansion function for a laminate can be derived

$$\langle\langle\alpha_x(T)\rangle\rangle = A_{11}^{-1}(T) \sum_{k=1}^{n} [\bar{Q}_{11}(T)\langle\alpha_x(T)\rangle + \bar{Q}_{12}(T)$$

$$\times \langle\alpha_y(T)\rangle + \bar{Q}_{16}(T)\langle\alpha_{xy}(T)\rangle]_k(h_k - h_{k-1})$$

$$+ A_{12}^{-1}(T) \sum_{k=1}^{n} [\bar{Q}_{12}(T)\langle\alpha_x(T)\rangle$$

$$+ \bar{Q}_{22}(T)\langle\alpha_y(T)\rangle + \bar{Q}_{26}(T) \quad (3.11)$$

$$\times \langle\alpha_{xy}(T)\rangle]_k(h_k - h_{k-1}) + A_{16}^{-1}(T) \sum_{k=1}^{n} [\bar{Q}_{16}(T)\langle\alpha_x(T)\rangle$$

$$+ \bar{Q}_{26}(T)\langle\alpha_y(T)\rangle + \bar{Q}_{66}(T)\langle\alpha_{xy}(T)\rangle]_k(h_k - h_{k-1})$$

$$\langle\langle\alpha_y(T)\rangle\rangle = A_{12}^{-1}(T) \sum_{k=1}^{n} [\bar{Q}_{11}(T)\langle\alpha_x(T)\rangle + \bar{Q}_{12}(T)\langle\alpha_y(T)\rangle$$

$$+ \bar{Q}_{16}(T)\langle\alpha_{xy}(T)\rangle]_k(h_k - h_{k-1})$$

$$+ A_{22}^{-1}(T) \sum_{k=1}^{n} [\bar{Q}_{12}(T)\langle\alpha_x(T)\rangle$$

$$+ \bar{Q}_{22}(T)\langle\alpha_y(T)\rangle + \bar{Q}_{26}(T)\langle\alpha_{xy}(T)\rangle]_k(h_k - h_{k-1}) \quad (3.12)$$

$$+ A_{26}^{-1}(T) \sum_{k=1}^{n} [\bar{Q}_{16}(T)\langle\alpha_x(T)\rangle + \bar{Q}_{26}(T)\langle\alpha_y(T)\rangle$$

$$+ \bar{Q}_{66}(T)\langle\alpha_{xy}(T)\rangle](h_k - h_{k-1})$$

$$\langle\langle\alpha_{xy}(T)\rangle\rangle = A_{16}^{-1}(T) \sum_{k=1}^{n} [\bar{Q}_{11}(T)\langle\alpha_x(T)\rangle + \bar{Q}_{12}(T)$$

$$\times \langle\alpha_y(T)\rangle + \bar{Q}_{16}(T)\langle\alpha_{xy}(T)\rangle]_k(h_k - h_{k-1}) + A_{26}^{-1}(T)$$

$$\times \sum_{k=1}^{n} [\bar{Q}_{12}(T)\langle\alpha_x(T)\rangle + \bar{Q}_{22}(T)\langle\alpha_y(T)\rangle + \bar{Q}_{26}(T) \quad (3.13)$$

$$\times \langle\alpha_{xy}(T)\rangle]_k(h_k - h_{k-1}) + A_{66}^{-1}(T) \sum_{k=1}^{n} [\bar{Q}_{16}(T)\langle\alpha_x(T)\rangle$$

$$+ \bar{Q}_{26}(T)\langle\alpha_y(T)\rangle + \bar{Q}_{66}(T)\langle\alpha_{xy}(T)\rangle]_k(h_k - h_{k-1})$$

The following designations have been used in the above formulas

$$A_{11}^{-1}(T) = \frac{1}{\Delta} [A_{22}(T)A_{66}(T) - A_{26}^2(T)]$$

$$A_{12}^{-1}(T) = \frac{1}{\Delta} [A_{16}(T)A_{26}(T) - A_{12}(T)A_{66}(T)]$$

$$A_{22}^{-1}(T) = \frac{1}{\Delta} [A_{11}(T)A_{66}(T) - A_{16}^2(T)]$$

$$A_{16}^{-1}(T) = \frac{1}{\Delta} [A_{12}(T)A_{26}(T) - A_{22}(T)A_{16}(T)]$$

$$A_{26}^{-1}(T) = \frac{1}{\Delta} [A_{12}(T)A_{16}(T) - A_{11}(T)A_{26}(T)]$$

$$A_{66}^{-1}(T) = \frac{1}{\Delta} [A_{11}(T)A_{22}(T) - A_{12}^2(T)]$$

$$\Delta = A_{11}(T)[A_{22}(T)A_{66}(T) - A_{26}^2(T)] - A_{12}(T)[A_{66}(T)A_{12}(T)$$

$$- A_{16}(T)A_{26}(T)] + A_{16}(T)[A_{12}(T)A_{26}(T) - A_{22}(T)A_{16}(T)]$$

where

$$A_{ij}(T) = \sum_{k=1}^{n} [\bar{Q}_{ij}(T)]_k (h_k - h_{k-1})$$

In the particular case of the angle-ply laminate, the general Equations (3.11) through (3.13) yield the thermal expansion function [60] for a laminate

$$\langle\langle \alpha_1(T) \rangle\rangle = \langle \alpha_x(T) \rangle - \frac{\bar{Q}_{12}(T)\bar{Q}_{26}(T) - \bar{Q}_{22}(T)\bar{Q}_{16}(T)}{\bar{Q}_{11}(T)\bar{Q}_{22}(T) - \bar{Q}_{12}^2(T)} \langle \alpha_{xy}(T) \rangle \quad (3.14)$$

where

$$\bar{Q}_{11}(T) = Q_{11}(T) \cos^4 \beta + 2[Q_{12}(T) + 2Q_{66}(T)] \times \sin^2 \beta \cos^2 \beta + Q_{22}(T) \sin^4 \beta$$

$$\bar{Q}_{12}(T) = [Q_{11}(T) + Q_{22}(T) - 4Q_{66}(T)] \times \sin^2 \beta \cos^2 \beta + Q_{12}(T)(\sin^4 \beta + \cos^4 \beta)$$

$$\bar{Q}_{22}(T) = Q_{11}(T) \sin^4 \beta + 2[Q_{12}(T) + 2Q_{66}(T)] \times \sin^2 \beta \cos^2 \beta + Q_{22}(T) \cos^4 \beta$$

$$\bar{Q}_{16}(T) = [Q_{11}(T) - Q_{12}(T) - 2Q_{66}(T)] \sin \beta \cos^3 \beta + [Q_{12}(T) - Q_{22}(T) + 2Q_{66}(T)] \sin^3 \beta \cos \beta$$

$$\bar{Q}_{26}(T) = [Q_{11}(T) - Q_{12}(T) - 2Q_{66}(T)] \sin^3 \beta \cos \beta + [Q_{12}(T) - Q_{22}(T) + 2Q_{66}(T)] \sin \beta \cos^3 \beta$$

$$Q_{11}(T) = \frac{E_{11}(T)}{1 - \nu_{\|\perp}(T)\nu_{\perp\|}(T)}$$

$$Q_{22}(T) = \frac{E_\perp(T)}{1 - \nu_{\|\perp}(T)\nu_{\perp\|}(T)}$$

$$Q_{12}(T) = \frac{E_\|(T)\nu_{\|\perp}(T)}{1 - \nu_{\|\perp}(T)\nu_{\perp\|}(T)} = \frac{E_\perp(T)\nu_{\perp\|}(T)}{1 - \nu_{\|\perp}(T)\nu_{\perp\|}(T)}$$

$$Q_{66}(T) = G_{\|\perp}(T)$$

$Q_{ij}(T)$, $\langle \alpha_x(T) \rangle$ and $\langle \alpha_{xy}(T) \rangle$ are determined generally according to Equation (3.5) and the calculated results of elastic characteristics are presented in Tables 3.3 and 3.4. The theoretical curves constructed on the basis of Equation (3.14) for two types of carbon fiber reinforced plastics are plotted in Figs. 3.5 through 3.8. Test data have been borrowed from Owen & Rice [92].

The above relationships permit us to establish the thermal expansion functions if their components are known. If these latter functions are unknown but the test curve $\langle\langle \epsilon(T) \rangle\rangle$ is available, the curve is conveniently approximated by the third-order polynomial

$$\langle\langle \epsilon(T) \rangle\rangle = b_0 + b_1 T + b_2 T^2 + b_3 T^3 \qquad (3.15)$$

The thermal expansion function is given by the general relationship

$$\langle\langle \alpha(T) \rangle\rangle = \frac{\partial \langle\langle \epsilon(T) \rangle\rangle}{\partial T} \qquad (3.16)$$

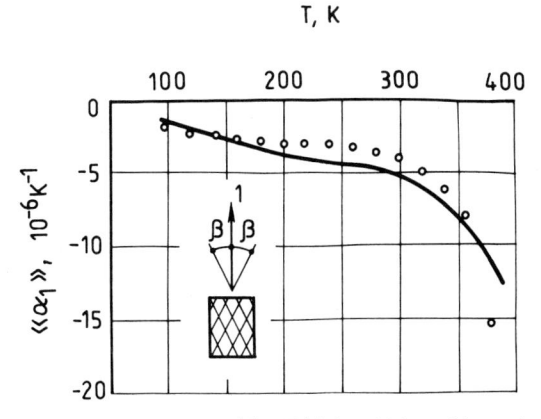

FIGURE 3.5. Thermal expansion function $\langle\langle \alpha_1(T) \rangle\rangle$ for a high-modulus carbon fiber reinforced epoxy with $B = \pm 30°$.

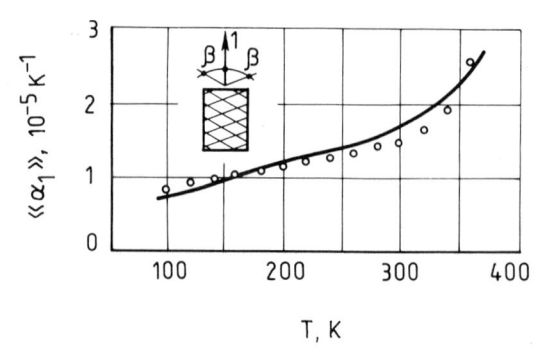

**FIGURE 3.6.** Thermal expansion function $\langle\langle\alpha_1(T)\rangle\rangle$ for a high-modulus carbon fiber reinforced epoxy with $B = \pm 60°$.

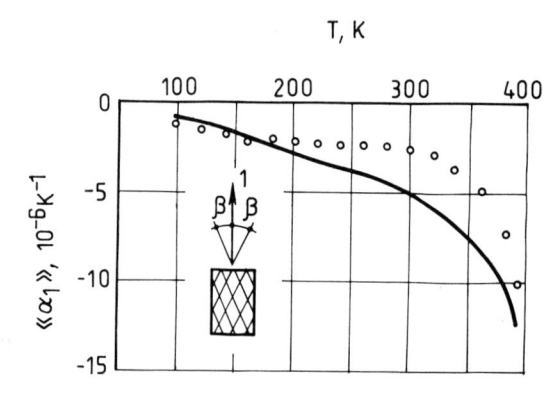

**FIGURE 3.7.** Thermal expansion function $\langle\langle\alpha_1(T)\rangle\rangle$ for a high-strength carbon fiber reinforced epoxy with $B = \pm 30°$.

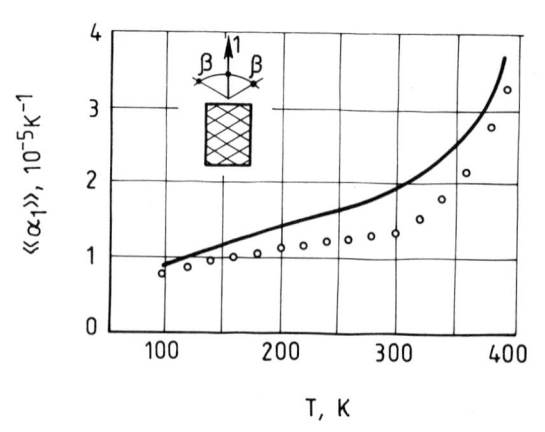

**FIGURE 3.8.** Thermal expansion function $\langle\langle\alpha_1(T)\rangle\rangle$ for a high-strength carbon fiber reinforced epoxy with $B = \pm 60°$.

After differentiating with respect to the polynomial in Equation (3.15), Equation (3.16) has the form

$$\langle\langle \alpha(T) \rangle\rangle = a_0 + a_1 T + a_2 T^2 \qquad (3.17)$$

## 3.2 THERMAL STRESSES IN PLIES OF A BALANCED REINFORCED PLASTIC

If the laminate is balanced with respect to its midplane, the thermal stresses in the elastic symmetry axes of a unidirectionally reinforced ply $k$ are given by the formula

$$\langle \sigma_\|(T_2 - T_1) \rangle_k$$
$$= [\bar{Q}_{11}(T_2)]_k \left[ \langle\langle \epsilon_1(T_2 - T_1) \rangle\rangle - \int_{T_1}^{T_2} \langle \alpha_1(T) \rangle_k dT \right]$$
$$+ [\bar{Q}_{12}(T_2)]_k \left[ \langle\langle \epsilon_2(T_2 - T_1) \rangle\rangle - \int_{T_1}^{T_2} \langle \alpha_2(T) \rangle_k dT \right] \qquad (3.18)$$
$$+ [\bar{Q}_{16}(T_2)]_k \left[ \langle\langle \gamma_{12}(T_2 - T_1) \rangle\rangle - \int_{T_1}^{T_2} \langle \alpha_{12}(T) \rangle_k dT \right]$$

$$\langle \sigma_\perp(T_2 - T_1) \rangle_k$$
$$= [\bar{Q}_{12}(T_2)]_k \left[ \langle\langle \epsilon_1(T_2 - T_1) \rangle\rangle - \int_{T_1}^{T_2} \langle \alpha_1(T) \rangle_k dT \right]$$
$$+ [\bar{Q}_{22}(T)]_k \left[ \langle\langle \epsilon_2(T_2 - T_1) \rangle\rangle - \int_{T_1}^{T_2} \langle \alpha_2(T) \rangle_k dT \right] \qquad (3.19)$$
$$+ [\bar{Q}_{66}(T)]_k \left[ \langle\langle \gamma_{12}(T_2 - T_1) \rangle\rangle - \int_{T_1}^{T_2} \langle \alpha_{12}(T) \rangle_k dT \right]$$

## 56  STRUCTURAL MECHANICS OF A MATERIAL

$\langle \tau_{\|\perp}(T_2 - T_1) \rangle_k$

$$= [\bar{Q}_{16}(T_2)]_k \left[ \langle\langle \epsilon_1(T_2 - T_1) \rangle\rangle - \int_{T_1}^{T_2} \langle \alpha_1(T) \rangle_k dT \right]$$

$$+ [\bar{Q}_{26}(T_2)]_k \left[ \langle\langle \epsilon_2(T_2 - T_1) \rangle\rangle - \int_{T_1}^{T_2} \langle \alpha_2(T) \rangle_k dT \right] \quad (3.20)$$

$$+ [\bar{Q}_{66}(T_2)]_k \left[ \langle\langle \gamma_{12}(T_2 - T_1) \rangle\rangle - \int_{T_1}^{T_2} \langle \alpha_{12}(T) \rangle_k dT \right]$$

where $\langle\langle \epsilon_1(T_2 - T_1) \rangle\rangle$, $\langle\langle \epsilon_2(T_2 - T_1) \rangle\rangle$ and $\langle\langle \gamma_{12}(T_2 - T_1) \rangle\rangle$ are mean strains of a laminated plastic resulting from temperature change from $T_1$ to $T_2$; $\langle \alpha_1(T) \rangle_k$, $\langle \alpha_2(T) \rangle_k$ and $\langle \alpha_{12}(T) \rangle_k$ are thermal expansion functions of a ply $k$ obtainable from Equations (3.8)–(3.10).

To determine the mean thermal strains of a laminated plastic, we use the results given in Skudra & Sbitnevs [59] substituting the thermal expansion function for the thermal expansion coefficient. This yields

$$\langle\langle \epsilon_1(T_2 - T_1) \rangle\rangle = A_{11}^{-1}(T_2) \sum_{k=1}^{n} \left[ \bar{Q}_{11}(T_2) \int_{T_1}^{T_2} \langle \alpha_1(T) \rangle dT \right.$$

$$+ \bar{Q}_{12}(T_2) \int_{T_1}^{T_2} \langle \alpha_2(T) \rangle dT + \bar{Q}_{16}(T_2) \int_{T_1}^{T_2} \langle \alpha_{12}(T) \rangle dT \Bigg]_k$$

$$\times (h_k - h_{k-1}) + A_{12}^{-1}(T_2) \sum_{k=1}^{n} \left[ \bar{Q}_{12}(T_2) \int_{T_1}^{T_2} \langle \alpha_1(T) \rangle dT \right.$$

$$+ \bar{Q}_{22}(T_2) \int_{T_1}^{T_2} \langle \alpha_2(T) \rangle dT + \bar{Q}_{26}(T_2) \int_{T_1}^{T_2} \langle \alpha_{12}(T) \rangle dT \Bigg]_k \quad (3.21)$$

$$\times (h_k - h_{k-1}) + A_{16}^{-1}(T_2) \sum_{k=1}^{n} \left[ \bar{Q}_{16}(T_2) \int_{T_1}^{T_2} \langle \alpha_1(T) \rangle dT \right.$$

$$+ \bar{Q}_{26}(T_2) \int_{T_1}^{T_2} \langle \alpha_2(T) \rangle dT + \bar{Q}_{66}(T_2)$$

$$\times \int_{T_1}^{T_2} \langle \alpha_{12}(T) \rangle dT \Bigg]_k (h_k - h_{k-1})$$

$$\langle\langle \epsilon_2(T_2 - T_1) \rangle\rangle = A_{12}^{-1}(T_2) \sum_{k=1}^{n} \left[ \bar{Q}_{11}(T_2) \int_{T_1}^{T_2} \langle \alpha_1(T) \rangle dT \right.$$

$$\left. + \bar{Q}_{12}(T_2) \int_{T_1}^{T_2} \langle \alpha_2(T) \rangle dT + \bar{Q}_{16}(T_2) \int_{T_1}^{T_2} \langle \alpha_{12}(T) \rangle dT \right]_k$$

$$\times (h_k - h_{k-1}) + A_{22}^{-1}(T_2) \sum_{k=1}^{n} \left[ \bar{Q}_{12}(T_2) \int_{T_1}^{T_2} \langle \alpha_1(T) \rangle dT \right.$$

$$\left. + \bar{Q}_{22}(T) \int_{T_1}^{T_2} \langle \alpha_2(T) \rangle dT + \bar{Q}_{26}(T) \int_{T_1}^{T_2} \langle \alpha_{12}(T) \rangle dT \right]_k \quad (3.22)$$

$$\times (h_k - h_{k-1}) + A_{26}^{-1}(T_2) \sum_{k=1}^{n} \left[ \bar{Q}_{16}(T_2) \int_{T_1}^{T_2} \langle \alpha_1(T) \rangle dT \right.$$

$$+ \bar{Q}_{26}(T_2) \int_{T_1}^{T_2} \langle \alpha_2(T) \rangle dT + \bar{Q}_{66}(T_2)$$

$$\left. \times \int_{T_1}^{T_2} \langle \alpha_{12}(T) \rangle dT \right]_k (h_k - h_{k-1})$$

$$\langle\langle \gamma_{12}(T_2 - T_1) \rangle\rangle = A_{16}^{-1}(T_2) \sum_{k=1}^{n} \left[ \bar{Q}_{11}(T_2) \int_{T_1}^{T_2} \langle \alpha_1(T) \rangle dT \right.$$

$$\left. + \bar{Q}_{12}(T_2) \int_{T_1}^{T_2} \langle \alpha_2(T) \rangle dT + \bar{Q}_{16}(T_2) \int_{T_1}^{T_2} \langle \alpha_{12}(T) \rangle dT \right]_k$$

$$\times (h_k - h_{k-1}) + A_{26}^{-1}(T_2) \sum_{k=1}^{n} \left[ \bar{Q}_{12}(T_2) \int_{T_1}^{T_2} \langle \alpha_1(T) \rangle dT \right. \quad (3.23)$$

$$\left. + \bar{Q}_{22}(T_2) \int_{T_1}^{T_2} \langle \alpha_2(T) \rangle dT + \bar{Q}_{26}(T_2) \int_{T_1}^{T_2} \langle \alpha_{12}(T) \rangle dT \right]_k$$

$$\times (h_k - h_{k-1}) + A_{66}^{-1}(T_2) \sum_{k=1}^{n} \left[ \bar{Q}_{16}(T_2) \int_{T_1}^{T_2} \langle \alpha_1(T) \rangle dT \right.$$

$$+ \bar{Q}_{26}(T_2) \int_{T_1}^{T_2} \langle \alpha_2(T) \rangle dT + \bar{Q}_{66}(T_2)$$

$$\left. \times \int_{T_1}^{T_2} \langle \alpha_{12}(T) \rangle dT \right]_k (h_k - h_{k-1})$$

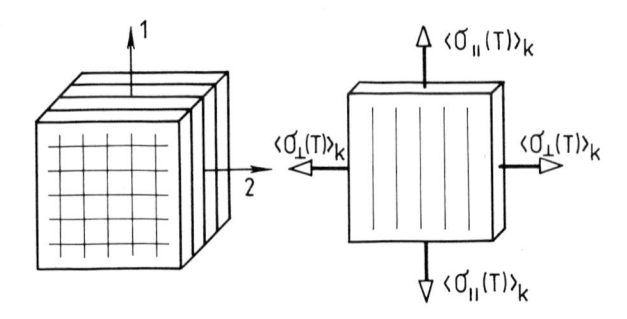

FIGURE 3.9. Calculation diagram for a cross-ply reinforced plastic.

As an example, consider the method of determining thermal stresses in a glass/epoxy ply. Let us discuss the most frequently occurring material consisting of unidirectionally reinforced plies oriented in two directions. This pertains to the case of orthogonal reinforcement as shown in Fig. 3.9. Considering that for the particular case $\bar{Q}_{16}(T_2) = \bar{Q}_{26}(T_2) = 0$ and $\langle\langle \gamma_{12}(T_2 - T_1) \rangle\rangle = 0$, Equations (3.18)–(3.20) for determining the stresses in the plies are simplified:

$$\langle \sigma_{\|}(T_2 - T_1) \rangle_k$$
$$= [\bar{Q}_{11}(T_2)]_k \left[ \langle\langle \epsilon_1(T_2 - T_1) \rangle\rangle - \int_{T_1}^{T_2} \langle \alpha_{\|}(T) \rangle_k dT \right] \quad (3.24)$$
$$+ [\bar{Q}_{12}(T_2)]_k \left[ \langle\langle \epsilon_2(T_2 - T_1) \rangle\rangle - \int_{T_1}^{T_2} \langle \alpha_{\perp}(T) \rangle_k dT \right]$$

$$\langle \sigma_{\perp}(T_2 - T_1) \rangle_k$$
$$= [\bar{Q}_{12}(T_2)]_k \left[ \langle\langle \epsilon_1(T_2 - T_1) \rangle\rangle - \int_{T_1}^{T_2} \langle \alpha_{\|}(T) \rangle_k dT \right] \quad (3.25)$$
$$+ [\bar{Q}_{22}(T_2)]_k \left[ \langle\langle \epsilon_2(T_2 - T_1) \rangle\rangle - \int_{T_1}^{T_2} \langle \alpha_{\perp}(T) \rangle_k dT \right]$$

$$\langle \tau_{\|\perp}(T_2 - T_1) \rangle_k = 0 \quad (3.26)$$

Equations (3.21) through (3.23), which are used to determine thermal strains, are also considerably simplified taking into account that $A_{16} = A_{26} = 0$.

$$\langle\langle\epsilon_1(T_2 - T_1)\rangle\rangle = \frac{A_{22}(T_2)}{A_{11}(T_2)A_{22}(T_2) - A_{12}^2(T_2)}$$

$$\times \sum_{k=1}^{n} \left[ \bar{Q}_{11}(T_2) \times \int_{T_1}^{T_2} \langle \alpha_\|(T) \rangle dT + \bar{Q}_{12}(T_2) \right.$$

$$\left. \times \int_{T_1}^{T_2} \langle \alpha_\perp(T) \rangle dT \right]_k (h_k - h_{k-1})$$

$$- \frac{A_{12}(T_2)}{A_{11}(T_2)A_{22}(T_2) - A_{12}^2(T_2)} \sum_{k=1}^{n} \left[ \bar{Q}_{12}(T_2) \right.$$

$$\times \int_{T_1}^{T_2} \langle \alpha_\|(T) \rangle dT + \bar{Q}_{12}(T_2)$$

$$\left. \times \int_{T_1}^{T_2} \langle \alpha_\perp(T) \rangle dT \right]_k (h_k - h_{k-1})$$

(3.27)

$$\langle\langle\epsilon_2(T_2 - T_1)\rangle\rangle = \frac{A_{12}(T_2)}{A_{11}(T_2)A_{22}(T_2) - A_{12}^2(T_2)}$$

$$\times \sum_{k=1}^{n} \left[ \bar{Q}_{11}(T_2) \times \int_{T_1}^{T_2} \langle \alpha_\|(T) \rangle dT + \bar{Q}_{12}(T_2) \right.$$

$$\left. \times \int_{T_1}^{T_2} \langle \alpha_\perp(T) \rangle dT \right]_k (h_k - h_{k-1})$$

$$- \frac{A_{11}(T_2)}{A_{11}(T_2)A_{22}(T_2) - A_{12}^2(T_2)} \sum_{k=1}^{n} \left[ \bar{Q}_{12}(T_2) \right.$$

$$\times \int_{T_1}^{T_2} \langle \alpha_\|(T) \rangle dT + \bar{Q}_{22}(T_2)$$

$$\left. \times \int_{T_1}^{T_2} \langle \alpha_\perp(T) \rangle dT \right]_k (h_k - h_{k-1})$$

(3.28)

$$\langle\langle \gamma_{12}(T_2 - T_1) \rangle\rangle = 0 \qquad (3.29)$$

Using the test results of Skudra & Sbitnevs [61] and Hartwig & Puck [82] provides thermal expansion functions $\langle \alpha_{\|}(T) \rangle$ and $\langle \alpha_{\perp}(T) \rangle$ for a unidirectional glass-epoxy. Transferring these results to cross-ply glass-epoxy it was possible to construct the thermal strain curve (Fig. 3.10) based on Equations (3.27) and (3.28).

After determining thermal strains in a cross-ply glass-epoxy it is possible to determine thermal stresses (Fig. 3.11) from Equations (3.24) and (3.25).

In the case of a balanced angle-ply ($\pm\beta$) plastic (Fig. 3.12) the thermal strains are calculated as follows

$$\langle\langle \epsilon_1(T_2 - T_1) \rangle\rangle = \int_{T_1}^{T_2} \langle \alpha_1(T) \rangle dT$$
$$+ \frac{\bar{Q}_{22}(T_2)\bar{Q}_{16}(T_2) - \bar{Q}_{12}(T_2)\bar{Q}_{26}(T_2)}{\bar{Q}_{11}(T_2)\bar{Q}_{22}(T_2) - \bar{Q}_{12}^2(T_2)} \int_{T_1}^{T_2} \langle \alpha_{12}(T) \rangle dT \quad (3.30)$$

$$\langle\langle \epsilon_2(T_2 - T_1) \rangle\rangle = \int_{T_1}^{T_2} \langle \alpha_2(T) \rangle dT$$
$$+ \frac{\bar{Q}_{11}(T_2)\bar{Q}_{26}(T_2) - \bar{Q}_{12}(T_2)\bar{Q}_{16}(T_2)}{\bar{Q}_{11}(T_2)\bar{Q}_{22}(T_2) - \bar{Q}_{12}^2(T_2)} \int_{T_1}^{T_2} \langle \alpha_{12}(T) \rangle dT \quad (3.31)$$

The theoretical curves constructed for various values of angles $\pm\beta$ and based on Equation (3.30) are given in Fig. 3.13. Test data have been borrowed from Hartwig & Puck [82].

Thermal stresses in the arbitrary ply $K$ are given by the following expressions:

$$\langle \sigma_1(T_2 - T_1) \rangle_k = \langle \sigma_2(T_2 - T_1) \rangle_k = 0$$

$$\langle \tau_{12}(T_2 - T_1) \rangle_k$$
$$= [\bar{Q}_{16}(T_2)]_k \left[ \langle\langle \epsilon_1(T_2 - T_1) \rangle\rangle - \int_{T_1}^{T_2} \langle \alpha_1(T) \rangle_k dT \right]$$
$$+ [\bar{Q}_{26}(T_2)]_k \times \left[ \langle\langle \epsilon_2(T_2 - T_1) \rangle\rangle - \int_{T_1}^{T_2} \langle \alpha_2(T) \rangle_k dT \right] \quad (3.32)$$
$$- [\bar{Q}_{66}(T_2)]_k \int_{T_1}^{T_2} \langle \alpha_{12}(T) \rangle_k dT$$

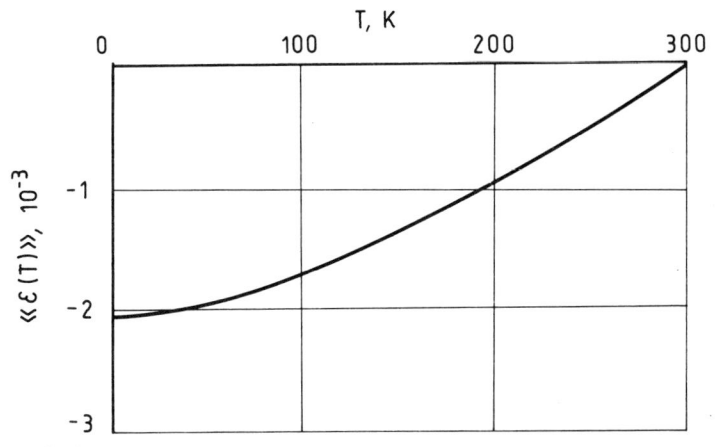

**FIGURE 3.10.** Temperature dependence of strains in the reinforcement direction for a cross-ply glass-epoxy with $\psi = 0.7$ resulting from temperature drop from 300 to 4.3 K.

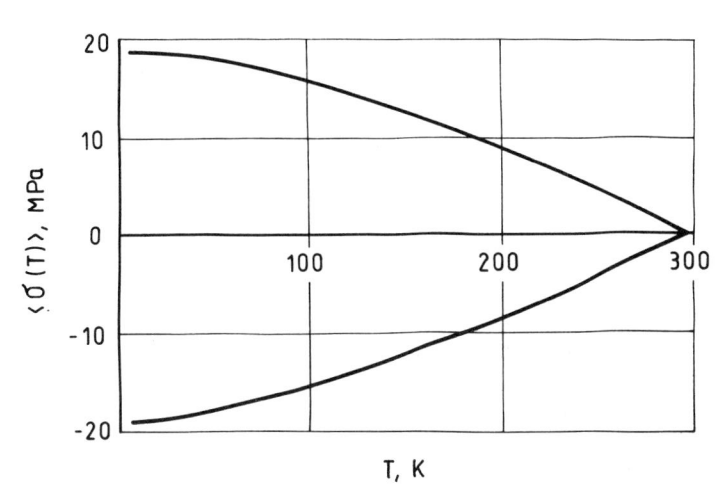

**FIGURE 3.11.** Temperature dependence of stresses in the elastic symmetry directions of an elementary ply in a cross-ply glass-epoxy with $\psi = 0.7$ as a result of temperature drop from 300 to 4.3 K: $1 - \langle \sigma_\perp(T) \rangle$; $2 - \langle \sigma_\parallel(T) \rangle$.

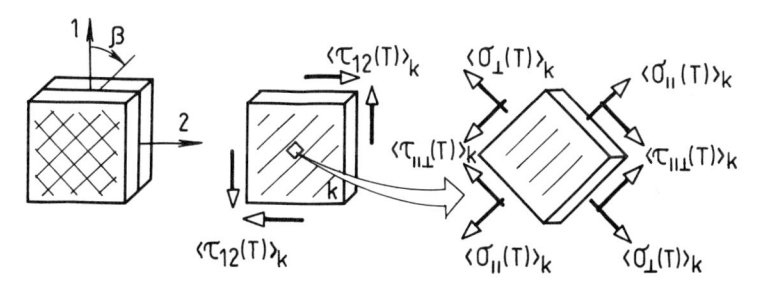

**FIGURE 3.12.** Calculation diagram for a cross-ply plastic.

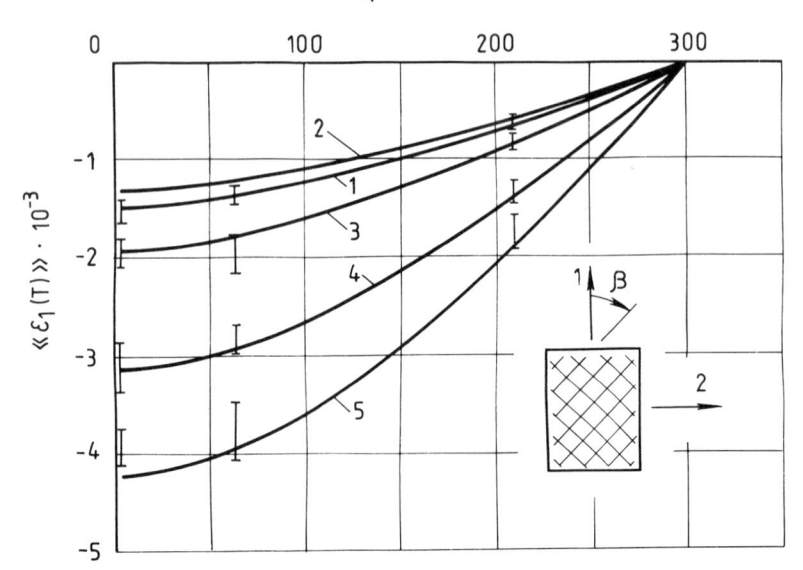

**FIGURE 3.13.** Temperature dependence of strains in a cross-ply glass-epoxy with $\psi = 0.7$ as a result of a temperature drop from 300 to 4.3 K: $B = 0°$ (1); $\pm 30$ (2); $\pm 45$ (3); $\pm 60$ (4); $90°$ (5).

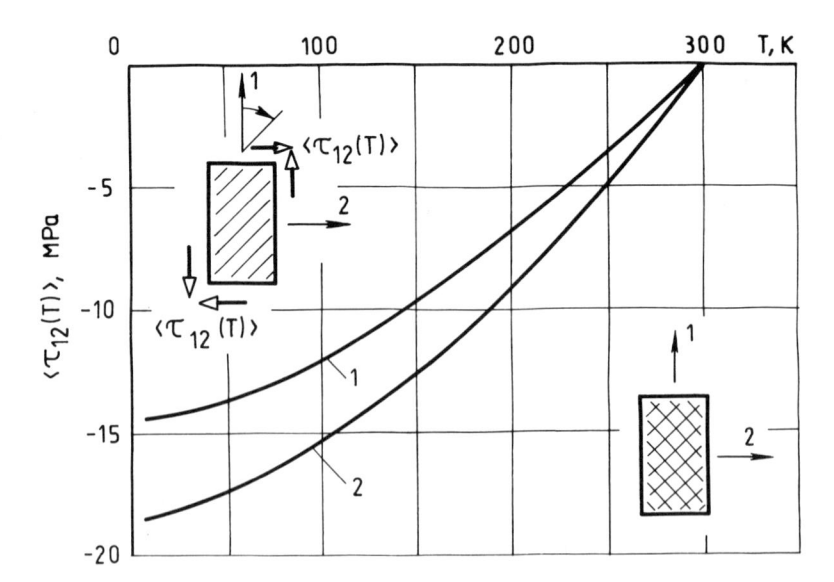

**FIGURE 3.14.** Temperature dependence of shear stresses in an elementary ply of a cross-ply glass-epoxy with $\psi = 0.7$ caused by a temperature drop from 300 to 4.3 K: $B = \pm 30$, $60°$ (1); $45°$ (2).

Fig. 3.14 represents the curves for a glass-epoxy based on Equation (3.32).

Thermal stresses in the elastic symmetry axes in a ply are determined by means of a transformation matrix

$$\left\{\begin{array}{c} \langle\sigma_{\parallel}(T_2 - T_1)\rangle \\ \langle\sigma_{\perp}(T_2 - T_1)\rangle \\ \langle\tau_{\parallel\perp}(T_2 - T_1)\rangle \end{array}\right\} = [T] \left\{\begin{array}{c} \langle\sigma_1(T_2 - T_1)\rangle \\ \langle\sigma_2(T_2 - T_1)\rangle \\ \langle\tau_{12}(T_2 - T_1)\rangle \end{array}\right\} \quad (3.33)$$

where

$$T = \begin{bmatrix} m^2 & n^2 & 2mn \\ n^2 & m^2 & -2mn \\ -mn & mn & m^2 - n^2 \end{bmatrix} \quad m = \cos\beta; \quad n = \sin\beta$$

CHAPTER 4

# Relative Damping

## 4.1 INTRODUCTORY REMARKS

An important feature of any structure subjected to vibrations is its power of relative damping (RD) which in its turn depends largely upon the RD of the materials involved. RD is expressed in a coefficient form

$$\Psi = \frac{\Delta U}{U} \qquad (4.1)$$

where $\Delta U$ is energy dissipated per cycle in a structure or a material sample; $U$ is the maximum strain energy during a cycle. In the present time an important place among structural materials is taken by reinforced plastic laminates (RPL). This necessitates the development of RD prediction methods with respect to RPL, taking into account their composition and the type of vibrations expected.

The works [26–29, 45, 74, 75, 88, 89] that discuss RD in reinforced plastics, contain the following approach that will be used in the present chapter. First, RD characteristics for a unidirectionally reinforced composite are determined by theoretical calculation [26] or by test [75]. Next, RD in a laminate is predicted proceeding from the microstructural approach and using the given characteristics. Here the frequency and amplitude [43, 64] dependence of RD may be ignored. Coefficients $\Psi_{\parallel}$, $\Psi_{\perp}$ and $\Psi_{\parallel\perp}$ have been introduced [74, 75, 88, 89] characterizing a unidirectionally reinforced composite, and they come to denote RD in a similar material subjected to corresponding types of vibrations. It is supposed in these papers that in the case of repeated combined stress state cycles, the energy of dissipation consists of several parts proportional to maximum energy parts of elastic deformation corresponding to the share of each stress. This supposition, however, leads to a small probability of RD equality—for instance, in cases when the material is in the state of cyclic uniaxial stress and in the state of cyclic uniaxial deformation. This shortcoming is not found in Zinovyev & Yermakov [26–29], which analyze RD in reinforced plastics, starting with an analysis of the isotropic components of a unidirectional glass-epoxy and finally analyzing a laminate under cyclic normal and shear forces in the lamina plane as well as bending and torsional moments. At the same time it is not quite evident why the number of RD characteristics for a unidirectionally reinforced material has been reduced. The effect of the interply shear on RD in an RPL has not been studied yet.

In this chapter the damping characteristics are discussed for an RPL subjected to a cyclic action of internal forces $N_x$, $N_y$, $N_{xy}$, $S_{xz}$, $S_{yz}$ (N/m), $M_x$, $M_y$, $M_{xy}$ (N·m/m) or $\{N\}$, $\{S\}$ and $\{M\}$ in the matrix form depending on the laminated structure of the material and the properties of its plies (Fig. 4.1). The most general case of a laminated structure is discussed where an arbitrary number of orthotropic plies are randomly oriented in an arbitrary lay-up plane. The principal assumptions are the following.

(1) Damping is caused by short-term viscoelastic properties of the component plies. The influence of possible interply flaws may be neglected.
(2) RD does not depend on the vibration amplitude.
(3) All the deformation laws are linear.

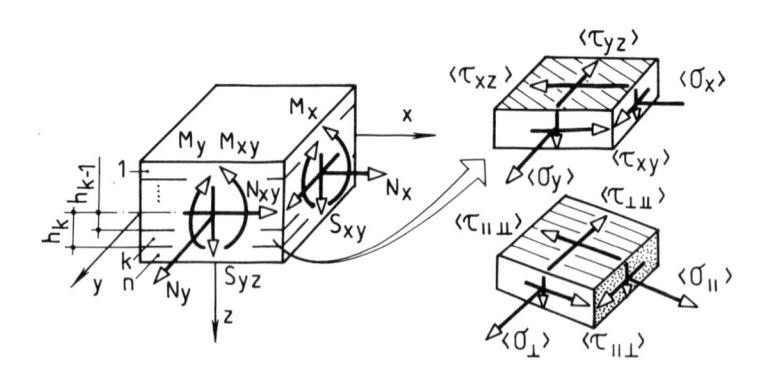

FIGURE 4.1. Calculation diagram for a reinforced plastic laminate.

(4) Deformation of RPL takes place according to the hypothesis of plane sections.
(5) Normal interply stresses may be neglected.
(6) RD may generally depend on the vibration frequency.

## 4.2 DAMPING IN A UNIDIRECTIONALLY REINFORCED PLY

In order to determine the damping in RPL we have to present damping characteristics of the component plies. With this aim in view the specific elastic strain energy in constituent plies should be determined in terms of stresses in the elastic symmetry axes

$$2W = \left\{ \begin{array}{c} \langle\sigma\rangle \\ \langle\tau\rangle \end{array} \right\}^T \left\{ \begin{array}{c} \langle\epsilon\rangle \\ \langle\gamma\rangle \end{array} \right\} = \left\{ \begin{array}{c} \langle\sigma\rangle \\ \langle\tau\rangle \end{array} \right\}^T [S] \left\{ \begin{array}{c} \langle\sigma\rangle \\ \langle\tau\rangle \end{array} \right\} \quad (4.2)$$

where

$$\{\langle\sigma\rangle\}^T = [\langle\sigma_\|\rangle; \langle\sigma_\perp\rangle; \langle\sigma_\Perp\rangle]$$

$$\{\langle\tau\rangle\}^T = [\langle\tau_{\|\Perp}\rangle; \langle\tau_{\perp\Perp}\rangle; \langle\tau_{\|\perp}\rangle]$$

$$\{\langle\epsilon\rangle\}^T = [\langle\epsilon_\|\rangle; \langle\epsilon_\perp\rangle; \langle\epsilon_\Perp\rangle]$$

$$\{\langle\gamma\rangle\}^T = [\langle\gamma_{\|\Perp}\rangle; \langle\gamma_{\perp\Perp}\rangle; \langle\gamma_{\|\perp}\rangle]$$

$[S]$ is a symmetric 6 by 6 square compliance matrix. If $\{\langle\sigma\rangle\}$ and $\{\langle\tau\rangle\}$ are amplitude value vectors of stresses with the material in a cyclic stress state, then the corresponding specific energy of dissipation in the stress form is

$$2\langle\Delta W\rangle = \left\{ \begin{array}{c} \langle\sigma\rangle \\ \langle\tau\rangle \end{array} \right\}^T [S^0] \left\{ \begin{array}{c} \langle\sigma\rangle \\ \langle\tau\rangle \end{array} \right\} \quad (4.3)$$

where $[S^0]$ is the matrix of inelastic characteristics of a ply analogous in form and number of constituents to the elastic compliance matrix $[S]$. Index "0" will be further used to denote other inelastic characteristics both of a ply and of the whole package. Due to the fact that interply normal stresses are absent in RPL and the elastic symmetry directions of the component plies can rotate only in the lay-up plane, it is reasonable to regard energies $\langle \Delta W \rangle$ and $\langle W \rangle$ as a sum of two independent members

$$\langle \Delta W \rangle = \langle \Delta W \rangle_p + \langle \Delta W \rangle_i$$
$$\langle W \rangle = \langle W \rangle_p + \langle W \rangle_i \qquad (4.4)$$

that appear respectively as a result of stresses in the plane of the plies (subscript "$p$") and as a result of interply shear stresses (subscript "$i$"). Specific dissipation energies $\langle \Delta W \rangle_p$ and $\langle \Delta W \rangle_i$ are expressed in the form of stresses acting in the elastic symmetry directions of unidirectionally reinforced component plies yielding the expressions

$$2\langle \Delta W \rangle_p = \{\langle \sigma \rangle\}_p^T [S^0]_p \{\langle \sigma \rangle\}_p$$
$$2\langle \Delta W \rangle_i = \{\langle \tau \rangle\}_i^T [S^0]_i \{\langle \tau \rangle\}_i \qquad (4.5)$$

where

$$\{\langle \sigma \rangle\}_p^T = [\langle \sigma_\| \rangle; \langle \sigma_\perp \rangle; \langle \tau_{\|\perp} \rangle]$$

$$\{\langle \tau \rangle\}_i^T = [\langle \tau_{\perp\perp} \rangle; \langle \tau_{\|\perp\perp} \rangle]$$

$$[S^0]_p = \begin{bmatrix} S_{11}^0 & S_{12}^0 & \\ S_{21}^0 & S_{22}^0 & \\ & & S_{66}^0 \end{bmatrix} = \begin{bmatrix} \varphi_{11} S_{11} & \varphi_{12} S_{12} & \\ \varphi_{21} S_{21} & \varphi_{22} S_{22} & \\ & & \varphi_{66} S_{66} \end{bmatrix}$$

$$[S^0]_i = \begin{bmatrix} S_{44}^0 & \\ & S_{55}^0 \end{bmatrix} = \begin{bmatrix} \varphi_{44} S_{44} & \\ & \varphi_{55} S_{55} \end{bmatrix}$$

The expression for $S_{ij}^0$ on account of the newly introduced coefficients $\varphi_{ij}$ ($S_{ij}^0 = \varphi_{ij} S_{ij}$) is justified because the diagonal coefficients $\varphi_{ii}$ ($i = 1$, 2, 4 through 6) are equal to the coefficients of the sample RD determined directly by test when cyclic stresses $\langle \sigma_\| \rangle$, $\langle \sigma_\perp \rangle$, $\langle \tau_{\perp\perp} \rangle$, $\langle \tau_{\|\perp\perp} \rangle$ and $\langle \tau_{\|\perp} \rangle$ are excited in the sample one by one and denoted respectively by $\Psi_i$. Thus, for example, according to the designation principle for unidirectionally reinforced material [74, 75, 88, 89] $\varphi_{11} = \Psi_\|$; $\varphi_{22} = \Psi_\perp$; $\varphi_{44} = \Psi_{\perp\perp}$; $\varphi_{55} = \Psi_{\|\perp\perp}$; $\varphi_{66} = \Psi_{\|\perp}$ (or $\Psi_1$; $\Psi_2$; $\Psi_4$; $\Psi_5$ and $\Psi_6$). Determination of $\Psi_\|$ and $\Psi_\perp$ by test is performed by applying longitudinal or bending vibrations longitudinally or transversely to the reinforced flat beam sample. When determining $\Psi_{\|\perp} = \Psi_{\|\perp\perp}$, the longitudinally reinforced sample has to be loaded by cyclic torque. Determination of $\Psi_{\parallel\perp}$ by cyclic torque requires the loading of sufficiently wide and relatively thin beams reinforced in the thickness direction. In the first approximation it may be assumed that $\Psi_{\parallel\perp} = \Psi_{\parallel\perp}$. An additional test is required in order to determine $\varphi_{12} = \varphi_{21}$, for instance, to study longitudinal vibrations of a bar sample reinforced at an angle of 45°. Here the RD coefficient represented by $\bar{\Psi}_1(45°)$ is measured, and $\varphi_{12}$ is given by

$$\varphi_{12} = \frac{1}{2S_{12}} [(S_{11} + 2S_{12} + S_{22} + S_{66})\bar{\Psi}_1(45°) \\ - \Psi_1 S_{11} - \Psi_2 S_{22} - \Psi_6 S_{66}] \quad (4.6)$$

The above formula is obtained by inserting Equations (4.2) and (4.3) (the load acting longitudinally at a 45° degree angle in the reinforcement direction) into the relationship $\Delta W/W$.

## 4.3 DAMPING IN A REINFORCED PLASTIC LAMINATE

In order to obtain damping characteristics for a laminate, the stresses acting in the elastic symmetry directions of the material, shown in Equation (4.5), will be expressed by the stresses acting along the beam axes $\{\langle\bar{\sigma}\rangle\}^T = [\langle\sigma_x\rangle, \langle\sigma_y\rangle, \langle\tau_{xy}\rangle]$ (see Fig. 4.1)

$$2\langle\Delta W\rangle_p = \{\langle\bar{\sigma}\rangle\}_p^T [T]^T [S^0]_p [T] \{\langle\bar{\sigma}\rangle\}_p \quad (4.7)$$

$$2\langle\Delta W\rangle_i = \{\langle\bar{\tau}\rangle\}_i^T [P]^T [S^0]_i [P] \{\langle\bar{\tau}\rangle\}_i \quad (4.8)$$

where the matrix $[T]$ is introduced in chapter 2 and the matrix $[P]$ is written as

$$[P] = \begin{bmatrix} \cos\beta & -\sin\beta \\ \sin\beta & \cos\beta \end{bmatrix}$$

Considering that the monolayers work within the lay-up composition, energy $\langle\Delta W\rangle_p$ is more conveniently expressed by strains $\langle\epsilon_x\rangle$, $\langle\epsilon_y\rangle$ and $\langle\gamma_{xy}\rangle$ (or $\{\langle\bar{\epsilon}\rangle\}$). Then a few matrix transformations yield the following instead of Equation (4.7):

$$2\langle\Delta W\rangle_p = \{\langle\bar{\epsilon}\rangle\}_p^T [\bar{Q}^0] \{\langle\bar{\epsilon}\rangle\}_p \quad (4.9)$$

where

$$[\bar{Q}^0] = [T]^{-1}[Q][S^0]_p[Q][T]^{-1T}$$

$$[Q] = \begin{bmatrix} \dfrac{E_1}{1 - \nu_{12}\nu_{21}} & \dfrac{\nu_{12}E_2}{1 - \nu_{12}\nu_{21}} \\ \dfrac{\nu_{21}E_1}{1 - \nu_{12}\nu_{21}} & \dfrac{E_2}{1 - \nu_{12}\nu_{21}} \\ & & G_{12} \end{bmatrix}$$

Expressing the strains of Equation (4.9) by strains and curvatures of the laminate midplane and the coordinate $z$, according to the hypothesis of plane sections, we determine that

$$\{\langle \bar{\epsilon} \rangle\}_p = \{\langle \bar{\epsilon}(z) \rangle\}_p = \{\langle\langle \epsilon^0 \rangle\rangle\} + z\{k\} \tag{4.10}$$

and expressing, in their turn, $\{\langle\langle \epsilon^0 \rangle\rangle\}$ and $\{k\}$ by internal forces acting in the laminate sections, and the laminate stiffnesses, we obtain

$$\begin{Bmatrix} \{\langle\langle \epsilon^0 \rangle\rangle\} \\ \{k\} \end{Bmatrix} = \begin{bmatrix} [A] & [B] \\ [B]^T & [D] \end{bmatrix}^{-1} \begin{Bmatrix} \{N\} \\ \{M\} \end{Bmatrix}$$

$$= \begin{bmatrix} [A'] & [B'] \\ [B']^T & [D'] \end{bmatrix} \begin{Bmatrix} \{N\} \\ \{M\} \end{Bmatrix} \tag{4.11}$$

Integration with respect to $\langle \Delta W \rangle_p$ across the laminate thickness yields the specific energy of dissipation per unit of surface area as a result of stresses acting in the ply plane

$$2\langle\langle \Delta W \rangle\rangle_p = \begin{Bmatrix} \{N\} \\ \{M\} \end{Bmatrix}^T \begin{bmatrix} [A^{0\prime}] & [B^{0\prime}] \\ [B^{0\prime}]^T & [D^{0\prime}] \end{bmatrix} \begin{Bmatrix} \{N\} \\ \{M\} \end{Bmatrix} \tag{4.12}$$

where

$$\begin{bmatrix} [A^{0\prime}] & [B^{0\prime}] \\ [B^{0\prime}]^T & [D^{0\prime}] \end{bmatrix} = \begin{bmatrix} [A'] & [B'] \\ [B']^T & [D'] \end{bmatrix} \begin{bmatrix} [A^0] & [B^0] \\ [B^0]^T & [D^0] \end{bmatrix} \begin{bmatrix} [A'] & [B'] \\ [B']^T & [D'] \end{bmatrix}$$

$$A_{ij}^0 = \sum_{k=1}^{n} (\bar{Q}_{ij}^0)_k (h_k - h_{k-1})$$

$$B_{ij}^0 = \frac{1}{2} \sum_{k=1}^{n} (\bar{Q}_{ij}^0)_k (h_k^2 - h_{k-1}^2)$$

$$D_{ij}^0 = \frac{1}{3} \sum_{k=1}^{n} (\bar{Q}_{ij}^0)_k (h_k^3 - h_{k-1}^3)$$

$$i, j = 1, 2, 6$$

The calculation diagram for $h_k$ is represented in Fig. 4.1 where $n$ is the number of plies.

The respective specific energy of elastic deformation is given by

$$2\langle\langle W \rangle\rangle_p = \begin{Bmatrix} \{N\} \\ \{M\} \end{Bmatrix}^T \begin{bmatrix} [A'] & [B'] \\ [B']^T & [D'] \end{bmatrix} \begin{Bmatrix} \{N\} \\ \{M\} \end{Bmatrix} \qquad (4.13)$$

The above Equation (4.12) represents the inelastic characteristics matrix for RPL in the case of cyclic action by internal forces $N_x$, $N_y$, $N_{xy}$, $M_x$, $M_y$ and $M_{xy}$. Rotation of the coordinate axes leads to the transformation of the inelastic characteristic matrix for RPL:

$$\begin{bmatrix} [\bar{A}^{0\prime}] & [\bar{B}^{0\prime}] \\ [\bar{B}^{0\prime}]^T & [\bar{D}^{0\prime}] \end{bmatrix} = \begin{bmatrix} [T] & \\ & [T] \end{bmatrix}^T \begin{bmatrix} [A^{0\prime}] & [B^{0\prime}] \\ [B^{0\prime}]^T & [D^{0\prime}] \end{bmatrix} \begin{bmatrix} [T] & \\ & [T] \end{bmatrix}$$

The dependence of the damping characteristics of RPL upon the interply shear is established using the investigation results (see Section 9.1) according to which two types of shear forces exist: $S_{yz}^b$, $S_{xz}^b$ and $S_{yz}^t$, $S_{xz}^t$ (or $\{S^b\}$ and $\{S^t\}$).

The distribution of interply shear stresses corresponding to all of the force factors are included in Equation (9.11). In the present discussion of the first approximation, the interply shear stresses are neglected, but they may eventually appear due to the presence of those derived membrane-bending forces and moments for which no corresponding shear forces exist. Thus by inserting Equation (9.11) into Equation (4.8) and integrating it with respect to $\langle \Delta W \rangle_i$ across the RPL thickness, we obtain the specific energy of dissipation per unit of RPL surface area resulting from the action of interply shear stresses

$$2\langle\langle\Delta W\rangle\rangle_i = \begin{Bmatrix}\{S^b\}\\\{S^t\}\end{Bmatrix}^T \begin{bmatrix}[u^0] & [g^0]\\ [g^0]^T & [p^0]\end{bmatrix}\begin{Bmatrix}\{S^b\}\\\{S^t\}\end{Bmatrix} \quad (4.14)$$

where

$$[u^0] = \int_{-h/2}^{h/2} [V^b(z)]^T [P]^T [S^0]_i [P][V^b(z)]dz$$

$$[p^0] = \int_{-h/2}^{h/2} [V^t(z)]^T [P]^T [S^0]_i [P][V^t(z)]dz$$

$$[g^0] = \int_{-h/2}^{h/2} [V^b(z)]^T [P]^T [S^0]_i [P][V^t(z)]dz$$

$[V^b(z)]$ and $[V^t(z)]$ will be explained in Equation (9.11).

The specific energy of elastic deformation caused by shear forces is given by

$$2\langle\langle W\rangle\rangle_i = \begin{Bmatrix}\{S^b\}\\\{S^t\}\end{Bmatrix}\begin{bmatrix}[u] & [g]\\ [g]^T & [p]\end{bmatrix}\begin{Bmatrix}\{S^b\}\\\{S^t\}\end{Bmatrix} \quad (4.15)$$

where $[u]$, $[g]$ and $[p]$ will be explained in Section 9.1.

Equation (4.14) includes an inelastic characteristics matrix for RPL in the cases when shear forces $S_{yz}^b$, $S_{xz}^b$ and $S_{yz}^t$, $S_{xz}^t$ act in a cyclic mode. The rotation of the coordinate axes transforms the above-mentioned matrix according to the formula

$$\begin{bmatrix}[\bar{u}^0] & [\bar{g}^0]\\ [\bar{g}^0]^T & [\bar{p}^0]\end{bmatrix} = \begin{bmatrix}[P] & \\ & [P]\end{bmatrix}^T \begin{bmatrix}[u^0] & [g^0]\\ [g^0]^T & [p^0]\end{bmatrix}\begin{bmatrix}[P] & \\ & [P]\end{bmatrix}$$

Summation of $\langle\langle\Delta W\rangle\rangle_p$ and $\langle\langle\Delta W\rangle\rangle_i$ as well as $\langle\langle W\rangle\rangle_p$ and $\langle\langle W\rangle\rangle_i$ yields total specific energies of dissipation and of elastic deformation per unit of RPL surface area

$$\langle\langle\Delta W\rangle\rangle = \langle\langle\Delta W\rangle\rangle_p + \langle\langle\Delta W\rangle\rangle_i$$
$$\langle\langle W\rangle\rangle = \langle\langle W\rangle\rangle_p + \langle\langle W\rangle\rangle_i \quad (4.16)$$

RD throughout the whole structural member is given by Equation (4.1), where total dissipation energies and total energies of elastic deformation are determined by integrating energies $\langle\langle \Delta W \rangle\rangle$ and $\langle\langle W \rangle\rangle$ across the whole area $A$ of the thin-wall structural member surface. Thus for a constant-structure RPL member we have

$$2\Delta U = \text{Tr}\left[\begin{bmatrix}[A^{0\prime}] & [B^{0\prime}]\\ [B^{0\prime}]^T & [D^{0\prime}]\end{bmatrix}\left[\int_A \begin{Bmatrix}\{N\}\\ \{M\}\end{Bmatrix} \begin{Bmatrix}\{N\}\\ \{M\}\end{Bmatrix}^T dA\right]\right]$$

$$+ \text{Tr}\left[\begin{bmatrix}[u^0] & [g^0]\\ [g^0]^T & [p^0]\end{bmatrix}\left[\int_A \begin{Bmatrix}\{S^b\}\\ \{S^t\}\end{Bmatrix} \begin{Bmatrix}\{S^b\}\\ \{S^t\}\end{Bmatrix}^T dA\right]\right]$$

$$2U = \text{Tr}\left[\begin{bmatrix}[A'] & [B']\\ [B']^T & [D']\end{bmatrix}\left[\int_A \begin{Bmatrix}\{N\}\\ \{M\}\end{Bmatrix} \begin{Bmatrix}\{N\}\\ \{M\}\end{Bmatrix}^T dA\right]\right] \quad (4.17)$$

$$+ \text{Tr}\left[\begin{bmatrix}[u] & [g]\\ [g]^T & [p]\end{bmatrix}\left[\int_A \begin{Bmatrix}\{S^b\}\\ \{S^t\}\end{Bmatrix} \begin{Bmatrix}\{S^b\}\\ \{S^t\}\end{Bmatrix}^T dA\right]\right]$$

where Tr [ ] is the sum of the diagonal matrix elements.

Thin-wall structural members are often subjected to cylindrical bending. If this bending occurs in plane $xz$, the Equations in (4.17) have the form

$$2\Delta U = \bar{\bar{A}}_{11}^{0\prime} \int_A N_x^2 dA + 2\bar{\bar{B}}_{11}^{0\prime} \int_A N_x M_x dA$$

$$+ \bar{\bar{D}}_{11}^{0\prime} \int_A M_x^2 dA + u_{55}^0 \int_A (S_{xz}^b)^2 dA$$

$$2U = \bar{\bar{A}}'_{11} \int_A N_x^2 dA + 2\bar{\bar{B}}'_{11} \int_A N_x M_x dA \quad (4.18)$$

$$+ \bar{\bar{D}}'_{11} \int_A M_x^2 dA + u_{55} \int_A (S_{xz}^b)^2 dA$$

where

$$\begin{bmatrix}\bar{\bar{A}}_{11}^{0\prime} & \bar{\bar{B}}_{11}^{0\prime}\\ \bar{\bar{B}}_{11}^{0\prime} & \bar{\bar{D}}_{11}^{0\prime}\end{bmatrix} = \begin{bmatrix}A_{11} & B_{11}\\ B_{11} & D_{11}\end{bmatrix}^{-1}\begin{bmatrix}A_{11}^0 & B_{11}^0\\ B_{11}^0 & D_{11}^0\end{bmatrix}\begin{bmatrix}A_{11} & B_{11}\\ B_{11} & D_{11}\end{bmatrix}^{-1}$$

$$\begin{bmatrix}\bar{\bar{A}}'_{11} & \bar{\bar{B}}'_{11}\\ \bar{\bar{B}}'_{11} & \bar{\bar{D}}'_{11}\end{bmatrix} = \begin{bmatrix}A_{11} & B_{11}\\ B_{11} & D_{11}\end{bmatrix}^{-1}$$

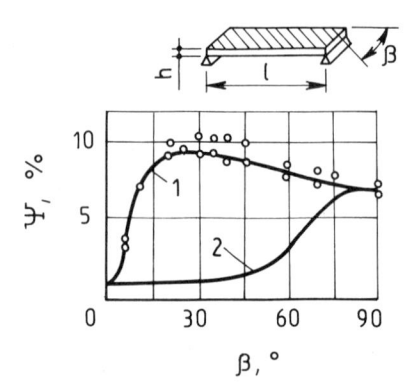

**FIGURE 4.2.** Dependence of RD in a unidirectional carbon-epoxy beam under three-point free (1) and cylindrical bending (2) upon the cutting angle of samples. The theoretical curves have been constructed according to the following formulas: curve 1—formula (4.17), curve 2—formula (4.18) with the initial data [75]: $l$ = 23 cm; $h$ = 0.23 cm; $E_\parallel$ = 103.6 GPa; $E_\perp$ = 7.59 GPa; $\gamma_{\parallel\perp}$ = 0.3; $G_{\parallel\perp} = G_{\perp\parallel}$ = 3.83 GPa; $\Psi_\parallel$ = 0.64%; $\Psi_\perp$ = 6.9%, $\Psi_\parallel(45°)$ = 9.1%; $\Psi_{\parallel\perp} = \Psi_{\perp\parallel}$ = 10%.

Fig. 4.2 illustrates RD coefficients from Adams & Bacon [75] obtained by testing unidirectional carbon-epoxy strips cut at various angles and loaded by three-point bending. It also includes the theoretical curves constructed by means of Equations (4.17) and (4.18) for cases of free and cylindrical bending. As is evident from the diagram, the free bending curve agrees satisfactorily with the test points. This is apparently explained by the appearance of side-strains in the areas of the sample close to the supports.

The component matrices of the inelastic characteristics for RPL as well as for a uniform material (see Equation (4.5)) may be expressed by means of the coefficients:

$$\begin{bmatrix} [A^{0\prime}] & [B^{0\prime}] \\ [B^{0\prime}]^T & [D^{0\prime}] \end{bmatrix} = \begin{bmatrix} [\varphi A'] & [\vartheta B'] \\ [\vartheta B']^T & [\delta D'] \end{bmatrix}$$

$$\begin{bmatrix} [u^0] & [g^0] \\ [g^0]^T & [p^0] \end{bmatrix} = \begin{bmatrix} [\gamma u] & [\kappa g] \\ [\kappa g]^T & [\eta p] \end{bmatrix}$$

(4.19)

The coefficients $\varphi$, $\delta$, $\vartheta$, $\gamma$, $\eta$ and $\kappa$ may also serve to characterize damping in RPL. In the most general case their number, i.e., the number of independent laminate characteristics for thin-wall products may reach 31—specifically, 21 characteristic of membrane-bending strains and 10 characteristics of interply shear. Therefore it would require 31 types of tests to determine all the RPL characteristics. This confirms the usefulness of the structural approach once again because it permits prediction of damping characteristics of RPL on the basis of its laminated structure and the properties of the component plies.

Of special significance are coefficients $\varphi_{11}$, $\varphi_{22}$, $\varphi_{66}$ as well as $\delta_{11}$, $\delta_{22}$ and $\delta_{66}$ corresponding to RD coefficients of the material under simple longitudinal ($\varphi_{11}$, $\varphi_{22}$), shearing ($\varphi_{66}$), bending ($\delta_{11}$, $\delta_{22}$) and torsional ($\delta_{66}$) vibrations. This relationship makes it possible to check easily (except $\varphi_{66}$) the truth of the methodology of theoretical prediction of RD as discussed above by means of a laboratory investigation of samples cut at various angles and subjected to the simple types of loading indicated. The following equations are to be solved at the same time:

$$\bar{\varphi}_{11} = \frac{\bar{A}_{11}^{0\prime}}{\bar{A}_{11}^{\prime}}; \quad \bar{\varphi}_{22} = \frac{\bar{A}_{22}^{0\prime}}{\bar{A}_{22}^{\prime}}; \quad \bar{\varphi}_{66} = \frac{\bar{A}_{66}^{0\prime}}{\bar{A}_{66}^{\prime}}$$
$$\bar{\delta}_{11} = \frac{\bar{D}_{11}^{0\prime}}{\bar{D}_{11}^{\prime}}; \quad \bar{\delta}_{22} = \frac{\bar{D}_{22}^{0\prime}}{\bar{D}_{22}^{\prime}}; \quad \bar{\delta}_{66} = \frac{\bar{D}_{66}^{0\prime}}{\bar{D}_{66}^{\prime}}$$
(4.20)

Fig. 4.3 represents the theoretical curves for coefficient variations pertaining to relative damping; the curves were constructed according to Equation (4.20) for RPL of various structures depending upon the cutting angle of the samples.

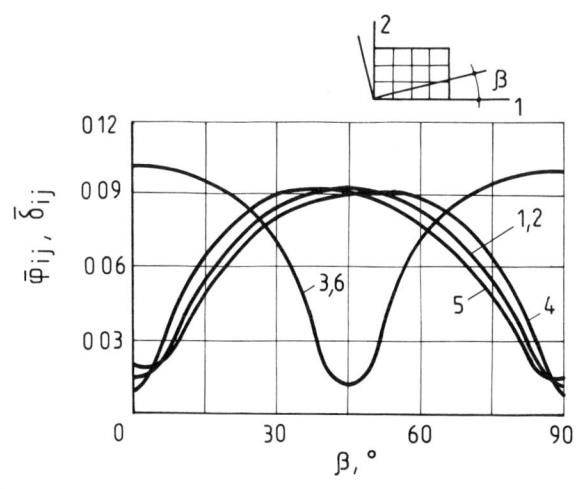

FIGURE 4.3. Dependence of RD characteristics of a carbon-epoxy laminate having [(0°/90°)₈]ₛ structure upon the sample cutting angle. The theoretical curves have been constructed according to formulas (4.20): curve 1—$\bar{\varphi}_{11}$, curve 2—$\bar{\varphi}_{22}$, curve 3—$\bar{\varphi}_{66}$, curve 4—$\bar{\delta}_{11}$, curve 5—$\bar{\delta}_{22}$, curve 6—$\bar{\delta}_{66}$. The initial data are those of Fig. 4.2.

## 4.4 DETERMINATION OF AN ALGORITHM FOR RELATIVE DAMPING

Determination of RD in a thin-walled structural member made of RPL taking into account the material structure and the type of cyclic load is carried out in the following sequence.

(1) Elastic and inelastic properties of the component plies are determined by laboratory testing. The description and the principles of obtaining the inelastic characteristics of plies are discussed in Section 4.2 from which it follows that the inelastic characteristics of a ply are components of matrices $[S^0]_p$ and $[S]_i$ obtained from Equation (4.5). The coefficients $\varphi_{11}$, $\varphi_{22}$, $\varphi_{66}$ and $\varphi_{12}$ required for compiling the matrix $[S^0]_p$ are determined from RD coefficients $\Psi_1$, $\Psi_2$, $\Psi_6$ and $\Psi_1$ (45°) obtained by test and by using Equation (4.6) taking into account equations $\varphi_1 = \Psi_1$, $\varphi_{22} = \Psi_2$ and $\varphi_{66} = \Psi_6$. The coefficients $\varphi_{44}$ and $\varphi_{55}$ required for compiling the matrix $[S^0]_i$ are determined from coefficients $\Psi_4$ and $\Psi_5$ obtained by test and using the equations $\varphi_{44} = \Psi_4$ and $\varphi_{55} = \Psi_5$.

(2) Elastic and inelastic characteristics of the component plies, the ply orientation in the lay-up and Equations (4.12) and (4.9) provide the basis for determining inelastic characteristics of RPL $A_{ij}^{0\prime}$, $B_{ij}^{0\prime}$ and $D_{ij}^{0\prime}$ which make it possible to take into account the cyclic action of forces and moments $N_x$, $N_y$, $N_{xy}$, $M_x$, $M_y$ and $M_{xy}$.

(3) Inelastic characteristics of RPL for including the effect of cyclic action of shear forces $S_{yz}^b$, $S_{xz}^b$, $S_{yz}^t$ and $S_{xz}^t$ are determined from Equation (4.14).

(4) RD across the whole structural member is calculated by inserting energies $\Delta U$ and $U$ into Equation (4.1) which were obtained from Equation (4.17). In order to do this, diagrams for all the membrane-bending forces and moments have to be constructed on member surface and to be multiplied in pairs, and the same has to be done separately for the diagrams of all the shear forces $S_{yz}^b$, $S_{xz}^b$ and $S_{yz}^t$, $S_{xz}^t$. If the structural member is working under cylindrical bending, Equation (4.18) is used instead of Equation (4.17).

CHAPTER 5

# Structural Theory of Creep

## 5.1 VISCOELASTIC PROPERTIES OF COMPONENTS

It has been proved by numerous tests that the polymer matrix possesses pronounced viscoelastic properties. Strains of the polymer matrix vary several times under constant long-term loading. Therefore time is an important variable in determining strains of the polymer matrix.

The deformation law for polymer matrices under uniaxial loading is given by

$$\epsilon(t) = \frac{1}{E_A}\left[\sigma(t) + \int_0^t K_A(t - \theta)\sigma(\theta)d\theta\right] \qquad (5.1)$$

The function $K$ has to be obtained by creep test of the material under constant loading.

When the load is constant ($\sigma(t)$ = const), Equation (5.1) is written as

$$\epsilon(t) = \frac{\sigma}{E_A}\left[1 + \int_0^t K_A(t - \theta)d\theta\right] = D(t)\sigma \qquad (5.2)$$

The creep function $D(t)$ is described in the form of diagrams, tables or analytical relationships.

Test data show that under constant loading the rate of strain at the first instant of time tends to infinity. Consequently, a function providing an infinitely high rate of strain at the instant of loading when $t = 0$ should be chosen as a creep kernel. This feature, however, should be moderate, otherwise the initial strains may also become infinite. The simplest type of such a function possessing a slight degree of this feature is Duffing's kernel [51]. With this kernel the creep curve corresponding to Equation (5.2) rises in an unrestricted manner, i.e., the long-term elasticity modulus is equal to zero. For materials possessing finite long-term elasticity modulus, A. R. Rzhanitsin [51] proposed a creep kernel of the form

$$K_A(t - \theta) = \frac{A_1 \exp[-\beta_A(t - \theta)]}{(t - \theta)^{\alpha_A}}$$

where $0 < \alpha_A < 1$. With this kernel and according to Equation (5.2) creep strains are determined by the incomplete $\gamma$-function of independent variables $\beta_A t$ and $\alpha_A$. The method for determining variables $\alpha_A$, $\beta_A$ and $A_1$ of Rzhanitsin's kernel are discussed by Kultunov [32]. Rabotnov [49] proposes well-behaved exponential functions $\xi_{\alpha_A}(-\beta_A, t - \theta)$ as singular creep kernels.

In the latter case, the creep functions of the polymer matrix have the form

$$D(t) = \frac{1}{E_A}\left[1 + \lambda_A \int_0^t \xi_{\alpha_A}(-\beta_A, t - \theta)d\theta\right] \quad (5.3)$$

The practical application of functions $\xi_{\alpha_A}$ is associated with the setting of the variables $E_A$, $\alpha_A$, $\beta_A$ and $\lambda_A$ which are assumed to be the rheological characteristics of the material. A method for determining these variables is proposed using the test creep curve and Laplace transformations [49]. A computer method for determining limited creep characteristics has been developed and implemented. If the computer program is not available, variables $\alpha_A$, $\beta_A$ and $\lambda_A$ may be determined by approximation methods that are quite unsophisticated, such as the graphical approach to creep curve approximation [47]. The underlying principle of the method is the large linear region in the semilogarithmic coordinates of the $\xi$-function graph inclining towards the abscissa at an angle proportional to the ratio of $\alpha_A$ to $\beta_A$. Another approach to determining variables $\alpha_A$, $\beta_A$ and $\lambda_A$ involves approximation of test creep curves of $\xi$-function tables and the function integral [50].

If variables $\alpha_A$, $\beta_A$ and $\lambda_A$ are known, the creep function $D(t)$ is completely determined according to (5.3). The integral of the $\xi_{\alpha_A}$ function, however, is determined by the sum of the power series, the practical finding of its values being possible by means of special tables [50].

Therefore, attempts were made to develop approximations of such series. The following approximation has been proposed by M. I. Rozovsky

$$\int_0^t \xi_{\alpha_A}(-\beta_A, t-\theta)d\theta \approx \frac{1}{\beta_A}[1 - \exp(-\beta_A \gamma t^{1+\alpha_A})] \qquad (5.4)$$

where

$$1 < \alpha_A < 0; \qquad \gamma = (1 + \alpha_A)^{1+\alpha_A}$$

In the case of Equation (5.4), Equation (5.2) yields

$$\epsilon_A(t) = \epsilon_0 \left\{ 1 + \frac{\lambda_A}{\beta_A}[1 - \exp(-\beta_A \gamma t^{1+\alpha_A})] \right\} \qquad (5.5)$$

where $\epsilon_0$ is elastic strain.

Equation (5.5) shows that if $t \to \infty$ the creep strain becomes asymptotic $\epsilon(\infty)$. Equation (5.5) enables us to write

$$\epsilon_A(\infty) = \epsilon_0 \left( 1 + \frac{\lambda_A}{\beta_A} \right)$$

Thus, knowing strains in instantaneous and infinitely long loading, the relationship $\lambda_A/\beta_A$ is uniquely given by

$$\frac{\lambda_A}{\beta_A} = \frac{\epsilon_A(\infty)}{\epsilon_0} - 1 \qquad (5.6)$$

Variables $\alpha_A$ and $\beta_A$ may be found using the relationships

$$\alpha_A = -1 + \frac{1}{\ln \frac{t_2}{t_1}} \ln \frac{\ln\left(1 - \frac{\beta_A}{\lambda_A} a_1\right)}{\ln\left(1 - \frac{\beta_A}{\lambda_A} a_2\right)}$$

$$\beta_A = -\frac{\ln\left(1 - \frac{\beta_A}{\lambda_A} a_1\right)}{[(1 + \alpha_A)t_1]^{1+\alpha_A}}$$

(5.7)

where

$$a_1 = \frac{\epsilon(t_1)}{\epsilon_0} - 1$$

$$a_2 = \frac{\epsilon(t_2)}{\epsilon_0} - 1$$

Here $t_1$ and $t_2$ denote two independently fixed values of loading time. Consequently according to Equation (5.7) the values of variables $\alpha_A$ and $\beta_A$ are ambiguous. The methods for determining creep parameters $\alpha_A$, $\beta_A$ and $\lambda_A$ according to Equations (5.6) and (5.7) may, however, be used with good results. Fig. 5.1 shows a creep test curve for phenol-formaldehyde resin under compression. The approximating curve has been plotted according to Equation (5.5) with the following creep parameter values: $\alpha_A = -0.35$; $\beta_A = 0.21$ days$^{-0.65}$; $\lambda_A = 0.17$ days$^{-0.65}$.

In this way the creep function for the polymer matrix, taking into account Equations (5.3) and (5.4), has the form

$$D(t) = \frac{1}{E_A}\left\{1 + \frac{\lambda_A}{\beta_A}[1 - \exp(-\beta_A \gamma t^{1+\alpha_A})]\right\} \qquad (5.8)$$

It should be noted that the sum exponential form of functions has a wide application as creep kernels. A disadvantage of these functions is

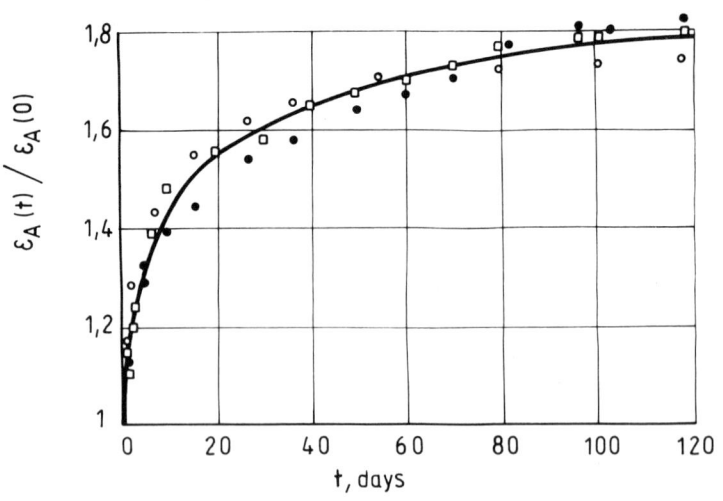

**FIGURE 5.1.** Relative creep deformation curve for a phenol-formaldehyde resin under compression. Three types of dots correspond to three various samples.

the large number of independent parameters, which increases with the number of summands. If the creep kernel is represented by one exponent, it is a specific case of Equation (5.4), provided $\alpha_A = 0$.

Equation (5.2) gives the polymer matrix creep for uniaxial stress state. In reinforced plastics, however, the polymer matrix is in a complex stress state even under the simplest kinds of loads. In determining the deformation law for the polymer matrix in a three-axial stress state a hypothesis of volume deformation elasticity will be used [2, 47], i.e., it is assumed that the polymer matrix volume is not subjected to time-dependent changes under static loading.

In such a case

$$\epsilon_1(t) = D(t)\left[\sigma_1 - \frac{\sigma_2 + \sigma_3}{2}\right] + \frac{1 - 2\nu_A}{E_A}(\sigma_2 + \sigma_3) \quad (5.9)$$

where $\sigma_1$, $\sigma_2$ and $\sigma_3$ are stress components; $\nu_A$ and $E_A$ are the Poisson's ratio and elasticity modulus of the polymer matrix; $D(t)$ is the creep function in the uniaxial stress state.

Expressions of strains $\epsilon_2(t)$ and $\epsilon_3(t)$ are written by analogy with Equation (5.9).

If the creep equation of the polymer matrix will be written symbolically as

$$\gamma(t) = J(t)\tau \quad (5.10)$$

then in terms of the hypothesis of elastic compressibility of the material the creep function in shear is expressed by the creep function under uniaxial loading as follows

$$J(t) = 3D(t) - \frac{1 - 2\nu_A}{E_A} \quad (5.11)$$

Consequently the viscoelastic properties of the polymer matrix in the case of linear dependence between stresses and strains are characterized by only one time function. Creep function $D(t)$ may serve the purpose; it is obtained by approximating the creep test curve under uniaxial loading. If the creep curve of the polymer matrix in shear is available, then the given data permit the definition of function $D(t)$ with the help of Equation (5.11).

Everything that has been said about viscoelastic properties is true with regard to polymer matrices where stresses and strains are found

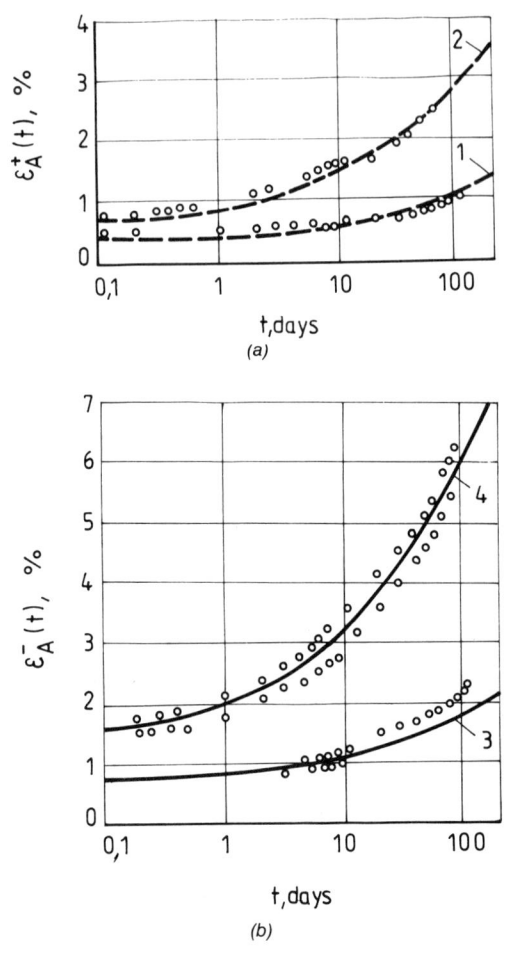

**FIGURE 5.2.** Creep curves for the epoxy resin EDT-10 under tension (a) and compression (b): $\sigma_A$ = 120 MPa (1), 200 (2), 240 (3), 500 MPa (4).

by a linear relationship. There are several cases, though (at high temperatures or high levels of loading, even at room temperatures for some polymer matrices) when polymer matrices are characterized by non-linear stress-strain relationships. This non-linearity in a number of polymer matrices is especially pronounced under sustained loads [30, 46].

Fig. 5.2 presents creep curves of polymer matrix EDT-10 under uni-axial tension and compression in isothermic conditions ($T = 22°C$). As seen in the diagrams, the degree of non-linearity changes with time.

As a result, no similarity condition is observed in creep curves or in isochronic curves. Creep curves of the epoxy resin EDT-10 under uniaxial loading will be best described by the rheological equation

$$\epsilon(t) = \frac{1}{E_A} \left\{ \sigma_A(t) + k_A \sigma_A^3(t) + \int_0^t K_A(t-\theta) \right.$$

(5.12)

$$\left. \times [\sigma_A(\theta) + (d_A + (k_A - d_A) \exp(-\kappa_A \theta)) \sigma_A^3(\theta)] d\theta \right\}$$

The creep kernel $K_A(t - \theta)$ characterizes creep in the linear region of deformation (usually at lower loading levels) whereas the three variables $k_A$, $d_A$ and $\kappa_A$ take care of the non-linear behaviour of the polymer matrix.

The exponent sum

$$K_A(t-\theta) = \sum_{j=1}^{n} C_{Aj} \exp[-\alpha_{Aj}(t-\theta)]$$

is a convenient form to be used for the creep kernel. By increasing the number of summands it is possible to guarantee the required accuracy of the approximation. A workable degree of accuracy under sustained loads is achieved when $n = 2$ or $n = 3$.

Provided $\sigma_A$ = const, Equation (5.12) is written as

$$\epsilon_A(t) = \frac{1}{E_A} \left\{ \sigma_A + k_A \sigma_A^3 + \sum_{j=1}^{n} \left[ \frac{C_{Aj}}{\alpha_{Aj}} (1 - \exp(-\alpha_{Aj} t)) \right. \right.$$

$$\times (\sigma_A + d_A \sigma_A^3) + \frac{C_{Aj}}{\alpha_{Aj} - \kappa_A} (\exp(-\kappa_A t)$$

(5.13)

$$\left. \left. - \exp(-\alpha_{Aj} t))(k_A - d_A) \sigma_A^3 \right] \right\}$$

Thus the description of creep curves of the polymer matrix with non-linear deformation characteristics requires, as seen from Equation (5.13), determination of $4 + 2n$ constants, $1 + 2n$ of them characterizing constants pertaining to non-linear behaviour of the material. Creep kernel parameters are obtained as a result of approximating the creep curve in the linear deformation area. For the resin EDT-10 under consideration (see Fig. 5.2(a)), approximation by the least-squares method yielded the following values of creep kernel parameters: $C_{A1} = 0.57$; $C_{A2} = 0.028$; $C_{A3} = 0.01056$; $\alpha_{A1} = 2$ days$^{-1}$; $\alpha_{A2} = 0.0833$ days$^{-1}$; $\alpha_{A3} = 0.007143$ days$^{-1}$.

In order to determine parameters $k_A$, $d_A$ and $\beta_A$ creep curves in the non-linear deformation area will be used. In the extreme case when $t = 0$, Equation (5.13) has the form

$$\epsilon_A(t) = \frac{1}{E_A}(\sigma_A + k_A \sigma_A^3) \tag{5.14}$$

Parameter $k_A$ is then obtained by approximating the non-linear deformation curve under short-term loading.

According to Equation (5.14) we have

$$k_A = \frac{E_A \epsilon_A(0) - \sigma_A}{\sigma_A^3} \tag{5.15}$$

By analogy, when $t \to \infty$ the relationship for determining parameter $d_A$ according to Equation (5.13) is

$$d_A = \frac{aE_A \epsilon_A(\infty) - (1+a)\sigma_A}{\sigma_A^3} - ak_A \tag{5.16}$$

where

$$a = \left( \sum_{i=1}^{n} \frac{C_{Aj}}{\alpha_{Aj}} \right)^{-1}$$

and $\epsilon_A(\infty)$ is steady-state creep strain at the stress level of $\sigma_A$.

Parameter $\kappa_A$ characterizes the rate of change in the non-linear time-dependent properties of the polymer matrix, and is obtained according to Equation (5.13) with fixed loading time $t = t_1$.

The following non-linearity parameter values were obtained for epoxy matrix EDT-10 from creep test curves of Fig. 5.2 according to Equations (5.15) and (5.16): $k_A = 350$ kPa$^{-2}$; $d_A^+ = 5400$ kPa$^{-2}$; $\kappa_A^+ = 0.4$ days$^{-1}$ under tension, and $k_A = 10.4$ kPa$^{-2}$; $d_A^- = 610$ kPa$^{-2}$; $\kappa_A^- = 0.55$ days$^{-1}$ under compression.

The agreement of test data with creep test curves constructed according to Equation (5.13) is fair.

So the linear-elastic and linear-viscoelastic properties of the polymer matrix EDT-10 appear to be practically equal in tension and in

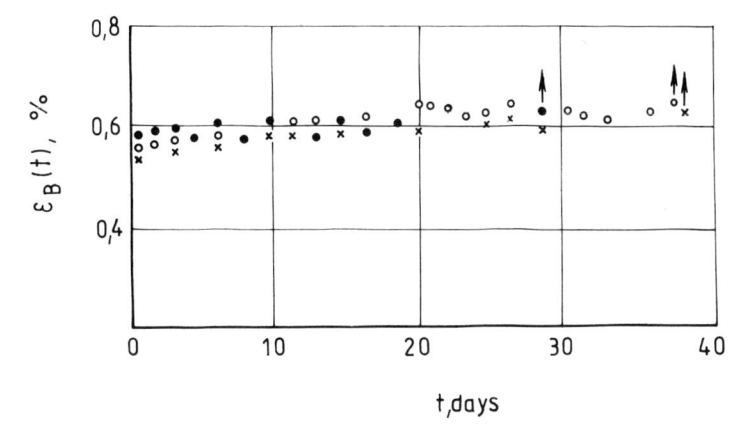

FIGURE 5.3. Creep curves for boron fibers at $\sigma_B$ = 2720 MPa (three configurations of dot marks correspond to three various samples).

compression but non-linear properties are more pronounced in tension. It should be remarked that Equation (5.13) permits a practicably accurate description of creep curves of a polymer matrix in simple stress state (uniaxial tension, compression or shear). It must be noted that there is practically no uniform theory of stress-deformation state in non-linear deformation field, even for isotropic materials.

Everything that has been said with respect to strain properties of the polymer matrix under sustained loading may refer to the reinforcing fibers. Although in difference from the polymer matrices the fibers used in structural plastics possess relatively small creep. Deformation properties of such fibers as boron, carbon and glass do not practically depend on loading time. Fig. 5.3 depicts creep test curves for boron fibers under constant tensile stress. The arrows indicate the instants of fiber failure.

Boron, carbon and glass fibers show practically no creep and they may be considered to be elastic but it may be an erroneous supposition with regard to organic fibers. Thus the results of Rogers et al. [92] testify that Kevlar-49 fibers possess creep properties (Fig. 5.4). Creep is a characteristic feature of high-strength organic filaments and of microplastics (polymer-impregnated threads after thermal processing) as seen in Figs. 5.5 and 5.6.

Specific creep curves (ratio of strain to initial strain) are smoothed and constructed on the basis of long-term test results [63] under stresses approaching 60 percent of the destructive values under short-term loading. Within the range of the stresses investigated the results show that the stress-strain relationship is linear at every instant of loading.

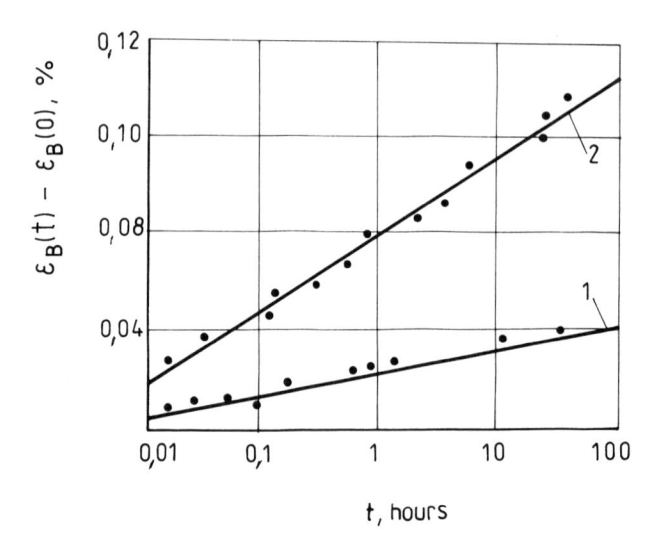

**FIGURE 5.4.** Creep strain curves for Kevlar-49 fibers: $\sigma_B$ = 240 MPa (1), 1940 MPa (2) [55].

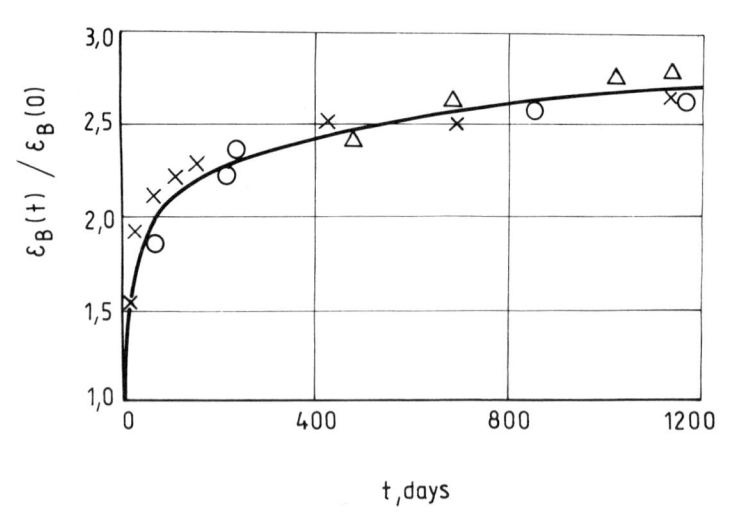

**FIGURE 5.5.** Relative creep deformation curve for an organic filament under constant loading levels. Three configurations of dot marks indicate three various samples.

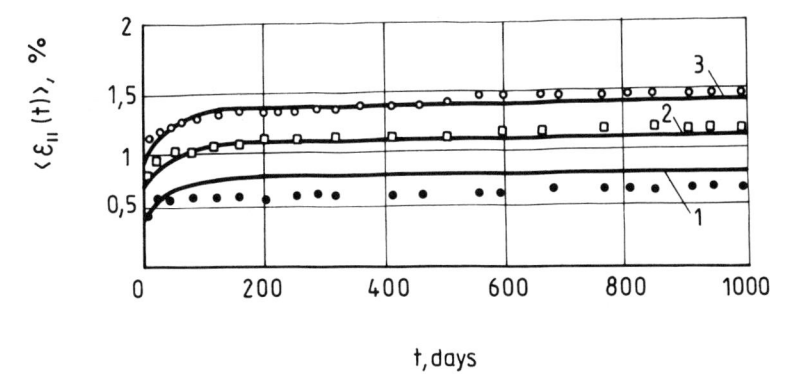

**FIGURE 5.6.** Creep curves of a plastic unidirectionally reinforced with organic fibers: $\langle\sigma_\|\rangle$ = 300 MPa (1), 450 (2), 600 MPa (3).

It may be concluded that the creep values of both organic filaments and microplastics obey the linear viscoelasticity theory and creep curves can be described by Equation (5.2).

## 5.2 CREEP IN A UNIDIRECTIONALLY REINFORCED PLY

### 5.2.1 Creep Under Longitudinally Applied Load

The principal structural element of reinforced plastic laminates is the unidirectionally reinforced ply. Therefore, creep in plastic laminates is governed by the viscoelastic properties and orientation geometry of the individual plies. Viscoelastic properties of unidirectionally reinforced plies in their turn are governed by the properties of their components. The inhomogeneous structure of reinforced plastics and the different deformation properties of their components explain the non-uniform distribution of stresses across the volume of the material. Stresses in the components vary and generally depend upon the relation between the viscoelastic properties and the volume content of the components. Under long-term loading there is time-dependent redistribution of stresses in the material components explained by a marked difference in viscoelastic properties of the polymer matrix and the reinforcing fibers. Let us discuss three independent loading modes of a unidirectionally reinforced ply: axial loading in the reinforcement direction, axial loading across reinforcement and longitudinal shear.

The calculation model for the axial loading of a unidirectionally reinforced plastic may be pictured as a bar of infinite length placed in a cylinder of finite thickness. In solving a boundary problem, expressions have been obtained for determining stresses in fibers and in the polymer matrix under short-term axial load [57]. Stresses acting transverse to the loading direction which are caused by the difference in Poisson's ratios of the polymer matrix and of the fibers do not exceed 10–12 percent of stresses acting in the fiber direction of glass and carbon fiber reinforced plastics. These stresses are generally neglected. Stresses in the reinforcement direction, however, are proportional to the stiffnesses of the polymer matrix and the fibers.

For structural reinforced plastics, the ratio of fiber stiffness in the longitudinal direction to the polymer matrix stiffness varies within the range of 20 to 120 even under a short-term load and depends on the fiber type. This ratio increases with the loading time. Boron, carbon, and glass fiber reinforced plastics suffer practically no creep under axial reinforcement direction loading. The process differs in unidirectional organic fiber reinforced plastics, which are characterized by creep. Complete strain in an organic fiber reinforced plastic, which consists of elastic strain and creep strain, may exceed elastic strain 1.6 times [63]. A feature of creep curves for organic fiber reinforced plastics (see Fig. 5.6) is the long time interval before they reach their maximum value.

Therefore, a crucial problem is the determination of creep strains in unidirectionally reinforced plastics under load applied in the reinforcement direction using viscoelastic property data of the material components. The problem consists of solving a system of linear integral equations of type (5.1), equations of strain compatibility and of stress equilibrium. This type of problem has been solved, for instance, for creep kernels of the polymer matrix and reinforcing fibers in the form of fraction-exponential or exponential functions [56]. We shall restrict our discussion to the simplest case, in which the exponential functions are satisfactory for the accuracy of approximating creep curves of the polymer matrix and the reinforcing fibers. Then the deformation law of the polymer matrix and the organic fibers under axial tension is described as

$$\epsilon_A(t) = \frac{1}{E_A} \left[ \sigma_A(t) + \lambda_A \int_0^t \exp\left[-\beta_A(t-\theta)\right] \sigma_A(\theta) d\theta \right] \quad (5.17)$$

$$\epsilon_B(t) = \frac{1}{E_B} \left[ \sigma_B(t) + \lambda_B \int_0^t \exp\left[-\beta_B(t-\theta)\right] \sigma_B(\theta) d\theta \right] \quad (5.18)$$

There should be an equilibrium between the mean stress $\langle \sigma_\| \rangle$ and the stresses in the polymer matrix $\sigma_A(t)$ and in the fibers $\sigma_B(t)$ at any time period

$$(1 - \psi)\sigma_A(t) + \psi\sigma_B(t) = \langle \sigma_\| \rangle \quad (5.19)$$

The case in which the creep kernels of the components are represented by fraction-exponential functions has been discussed in Skudra & Bulvas.

Omitting the details of solving Equations (5.17) through (5.19) if $\langle \epsilon_\|(t) \rangle = \epsilon_A(t) = \epsilon_B(t)$, the final result given will show that the creep curve for a unidirectionally reinforced plastic in the reinforcement direction is given by

$$\langle \epsilon_\|(t) \rangle = s_\| \left(1 + \frac{a_3}{b_3}\right)[1 + A_1 \exp(-z_1 t) \\ + A_2 \exp(-z_2 t)]\langle \sigma_\| \rangle \quad (5.20)$$

where

$$A_1 = \frac{a_3 - z_1(a_1 + a_2)}{(z_1 - z_2)(b_3 + a_3)} z_2$$

$$A_2 = \frac{a_3 - z_2(a_1 + a_2)}{(z_2 - z_1)(b_3 + a_3)} z_1$$

$$z_{1,2} = \frac{b_1 + b_2}{2}\left(1 \mp \sqrt{1 - 4\frac{b_3}{(b_1 + b_2)^2}}\right)$$

$$a_1 = \lambda_B[1 - (1 - \psi)E_A s_\|]$$

$$a_2 = \lambda_A[1 - \psi E_{Bz} s_\|]$$

$$a_3 = \lambda_A \lambda_B + \beta_A a_1 + \beta_B a_2$$

$$b_1 = \beta_B + (1 - \psi)\lambda_B E_A s_\|$$

$$b_2 = \beta_A + \psi \lambda_A E_{Bz} s_\|$$

$$s_\| = \frac{1}{E_\|}$$

$$b_3 = b_1\beta_A + b_2\beta_B - \beta_A\beta_B$$

(here $E_{\|}$ is the elasticity modulus of a unidirectionally reinforced plastic in the reinforcement direction).

According to Equation (5.20) the maximum creep strain with $t \to \infty$ is given by

$$\langle \epsilon_{\|}(\infty) \rangle = s_{\|}\left(1 + \frac{a_3}{b_3}\right)\langle \sigma_{\|} \rangle \qquad (5.21)$$

The ratio of maximum creep strain $\langle \epsilon_{\|}(\infty) \rangle$ to short-term elastic strain $\langle \epsilon_{\|}(0) \rangle$ characterizes the degree of creep in the material and also the rheonomic properties of the material under the prescribed type of load. As the deformation law of a unidirectionally reinforced plastic under axial loading applied in the reinforcement direction is given by Equation (5.20), the degree of creep is given by the relationship

$$\eta_{\|} = 1 + \frac{a_3}{b_3} c \eta_B \qquad (5.22)$$

where

$$c = E_{\|}\left[(1 - \psi)E_A \frac{\eta_B}{\eta_A} + \psi E_{Bz}\right]^{-1}$$

$$\eta_A = 1 + \frac{\lambda_A}{\beta_A}$$

$$\eta_B = 1 + \frac{\lambda_B}{\beta_B}$$

Analysing Equation (5.22), it becomes clear that the degree of creep in a unidirectionally reinforced ply in the reinforcement direction is proportional to the degree of creep in the fibers $\eta_B$, the proportionality factor $c$ with $\eta_B < \eta_A$ is larger than unity. This means that the degree of creep for a reinforced plastic in the reinforcement direction is always larger than the degree of fiber creep (it is supposed that the polymer matrix creep is larger than the creep in the reinforcing fibers).

### 5.2.2 Creep Under Longitudinal Shear

Longitudinal shear in a unidirectionally reinforced ply is a type of load that brings out very markedly the viscoelastic properties of the composite due to the viscoelasticity of the polymer matrix. In order to

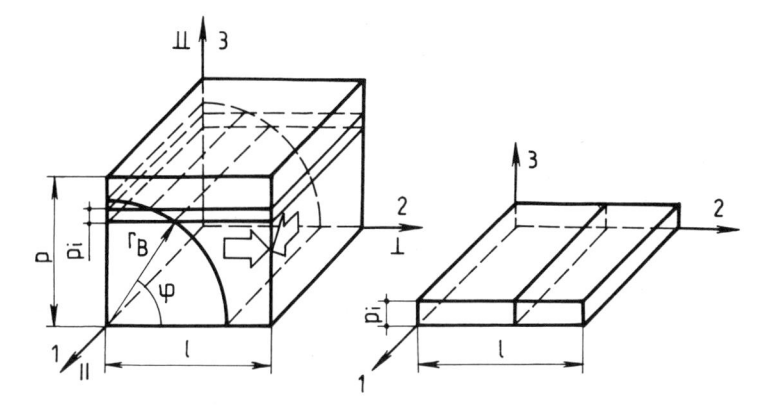

FIGURE 5.7. Calculation model for a unidirectionally reinforced plastic under longitudinal shear.

determine the viscoelastic properties of a unidirectionally reinforced ply judging from the viscoelastic properties of its components, the calculation model shown in Fig. 5.7 will be used. According to this model the material contains an unlimited number of plies of infinitely small thicknesses that are parallel to the loading plane. It is also assumed that each of the plies is in a uniform stress state and mean strains in all the plies are equal at any instant of loading. The shear strain in any ply consists of the polymer matrix strain and fiber strain. Considering that the polymer matrix strain is viscoelastic whereas the fibers strain is elastic, the stress-strain state of the components of the reinforced ply under longitudinal shear is determined by solving the system of equations

$$(1 - \psi_i)\gamma_{Ai}(t) + \psi_i\gamma_{Bi}(t) = \langle \gamma_{\|\perp}(t) \rangle$$

$$G_A\gamma_{Ai}(t) = \tau_{Ai}(t) + \int_0^t K_\tau(t - \theta)\tau_{Ai}(\theta)d\theta$$

$$G_{Brz}\gamma_{Bi}(t) = \tau_{Bi}(t)$$

$$\tau_{Ai}(t) = \tau_{Bi}(t) = \tau_{\|\perp i}(t)$$

$$\langle \tau_{\|\perp}(t) \rangle = \frac{r_B}{l}\int_0^{\pi/2} \tau_{\|\perp i}(\varphi)\cos\varphi\, d\varphi + \left(1 - \frac{r_B}{l}\right)\tau_A(t)$$

$$\tau_A(t) = G_A\left[\langle \gamma_{\|\perp}(t) \rangle - \int_0^t R_\tau(t - \theta)\langle \gamma_{\|\perp}(\theta) \rangle d\theta\right]$$

(5.23)

where $\tau_A(t)$ is the stress in these plies of the calculation model (Fig. 5.7) which do not incorporate fibers; $K_r(t - \theta)$ and $R_r(t - \theta)$ are the creep kernel and relaxion kernel of the polymer matrix under shear.

With $\langle \gamma_{\|\perp} \rangle$ = const the solution of the Equations in (5.23) makes it possible to establish the stress relaxion law (mean stress for a unidirectional ply and shear stress in its components).

Further, we shall discuss another problem—that of determining the creep strains of a unidirectional ply. It should be noted that stresses in the structural components of the ply vary during the creep process, i.e., there is a time-dependent stress redistribution. That is why the shear stress distribution diagrams are different at the instant of loading and at the end of the creep process. The character of shear stress distribution in a reinforced plastic is illustrated in Fig. 5.8. It is evident from the diagram that the shear stress does not vary in the plane coinciding with point $M$. The plane $m$ turns out to characterize the given reinforced plastic and the stress $\tau_{\|\perp m}$ is constant in this plane at any instant of loading. The position of plane $m$ is determined by angle $\varphi_M$

$$\varphi_M = \arccos \frac{l}{r_B} \left[ 1 + \frac{1 - \eta_{\|\perp}}{\eta_{\|\perp} - \eta_{\tau A}} \frac{G_A}{G_{Brz}} \right]^{-1} \quad (5.24)$$

where $\eta_{\tau A}$ is characteristic of the degree of creep for the polymer matrix under shear and is determined by the ratio of the values of the creep

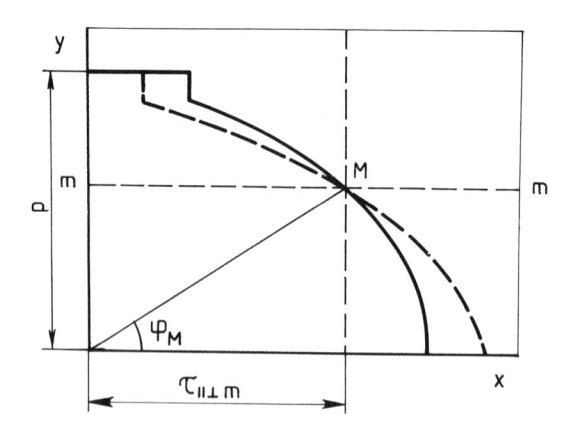

FIGURE 5.8. Characteristic shear stress distributions in the repeated element of a reinforced plastic at the loading instant (———) and after an unlimited continuous action of constant external stress (– – – –).

functions under shear $J(t)$ when $t \to \infty$ to the initial loading moment, i.e., when $t \to 0$. Equation (5.11) yields

$$\eta_{\tau A} = \frac{J(\infty)}{J(0)} = 1 + \frac{3}{2(1+\nu_A)} \frac{\lambda_A}{\beta_A} \quad (5.25)$$

The degree of creep for a unidirectionally reinforced plastic under longitudinal shear $\eta_{\|\perp}$ is determined to be the ratio of shear creep strain $\langle \gamma_{\|\perp}(\infty) \rangle$ to short-term loading elastic strain $\langle \gamma_{\|\perp}(0) \rangle$ for a fixed value of mean shear stress $\langle \tau_{\|\perp} \rangle$. Consequently this characteristic is expressed as the ratio of short time (at $t = 0$) longitudinal shear moduli to long-time (at $t = \infty$) shear moduli. The shear modulus for a composite with a prescribed structure, volume content, and elastic properties of its components is given by the expression

$$G_{\|\perp}(0) = \left[1 - \frac{r_B}{l} + \frac{r_B}{l} k\left(\frac{2k}{\sqrt{k^2-1}} \arctg \sqrt{\frac{k+1}{k-1}} - \frac{\pi}{2}\right)\right] G_A \quad (5.26)$$

whereas the long time shear modulus is expressed as

$$G_{\|\perp}(\infty) = \left[1 - \frac{r_B}{l}\right.$$

$$\left. + \frac{r_B}{l} k\left(\frac{2k}{\sqrt{k^2-1}} \arctg \sqrt{\frac{k+1}{k-1}} - \frac{\pi}{2}\right)\right] G_A(\infty) \quad (5.27)$$

where

$$G_A(\infty) = \frac{1}{\eta_{\tau A}} G_A$$

$$k = \left[\frac{r_B}{l}\left(1 - \frac{G_A}{G_{Brz}}\right)\right]^{-1}$$

According to the calculation model for the material the mean strain of the $m$-th ply is expressed as a sum of fiber strains and of the polymer matrix, and is equal to the mean strain of the composite

$$\gamma_{\|\perp m} = \langle \gamma_{\|\perp} \rangle = \frac{r_B}{l} \cos \varphi_M \gamma_{Bm} + \left(1 - \frac{r_B}{l} \cos \varphi_M\right) \gamma_{Am} \quad (5.28)$$

Equation (5.28), taking into account that the strain of any $i$-th ply (and consequently of $m$-th as well) is equal to the mean strain of the whole composite $\langle \gamma_{||\perp} \rangle$, enables us to determine stress in the $m$-th ply:

$$\tau_{||\perp m} = \frac{G_{Brz}}{G_{||\perp}} \left[ \frac{G_{Brz}}{G_A} \left( 1 - \frac{r_B}{l} \cos \varphi_M \right) + \frac{r_B}{l} \cos \varphi_M \right]^{-1} \langle \tau_{||\perp} \rangle \qquad (5.29)$$

Inserting the value of $\varphi_M$ from Equation (5.24) into Equation (5.29) we obtain

$$\tau_{||\perp m} = \frac{G_{Brz}}{G_{||\perp}} \left[ 1 - \frac{1 - \eta_{||\perp}}{1 - \eta_{\tau A}} \left( 1 - \frac{G_A}{G_{Brz}} \right) \right] \langle \tau_{||\perp} \rangle \qquad (5.30)$$

The creep curve for a unidirectionally reinforced plastic under longitudinal shear is determined by Equation (5.28) under the continuously acting constant stress $\langle \tau_{||\perp m} \rangle$. The resulting relationship between mean shear strain and the loading duration is

$$\langle \gamma_{||\perp}(t) \rangle = s_{66}(t) \langle \tau_{||\perp} \rangle \qquad (5.31)$$

Here $s_{66}(t)$ is the creep function of a unidirectionally reinforced plastic under longitudinal shear obtained according to the expression

$$s_{66}(t) = g_1 + g_2 I(t) \qquad (5.32)$$

where

$$g_1 = \frac{\eta_{||\perp} - \eta_{\tau A}}{1 - \eta_{\tau A}} \frac{1}{G_{||\perp}}$$

$$g_2 = \frac{1 - \eta_{||\perp}}{1 - \eta_{\tau A}} \frac{G_A}{G_{||\perp}}$$

In this way the creep of a unidirectionally reinforced plastic as seen from Equation (5.31) and taking into account Equation (5.32) is predictable from the viscoelastic properties of the polymer matrix. In Fig. 5.9 the solid line indicates the calculation curve for the creep of glass-phenol-formaldehyde under longitudinal shear. The calculation was

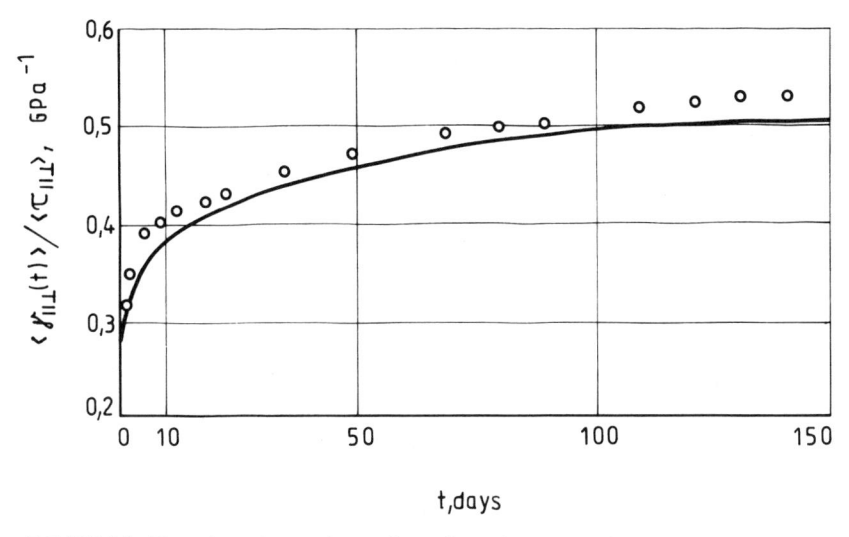

FIGURE 5.9. Time dependence of compliance for a glass-phenol-formaldehyde under longitudinal shear.

based on the following initial data: $\varphi = 0.52$; $G_{Brz} = 28$ GPa; $E_A = 3.6$ GPa; $\nu_A = 0.35$; $\alpha_A = -0.35$; $\beta_A = 0.21$ days$^{-0.65}$; $\lambda_A = 0.17$ days$^{-0.65}$. Fig. 5.9 also includes test data obtained under shear stress $\langle \tau_{\parallel\perp} \rangle = 22.5$ MPa. Fair agreement of test results and calculation data confirms the validity of the assumptions made in compiling the calculation formulas.

### 5.2.3 Creep Under Transverse Loading

In determining the stress-strain state of unidirectionally reinforced plastic components under long-term transverse loading, a volume boundary problem has to be solved for a non-homogeneous two-component medium. An accurate solution to this problem does not exist, however, at present. In order to arrive at the general and the most essential relationships characterizing distribution and redistribution of stresses and strains in composite structural elements under transverse loads, the unidirectionally reinforced plastic was analysed in Bulavs & Radinsh [9] and Skudra & Bulavs [96] as medium of dual periodicity where the repeating element selected as a calculation model is shown in Fig. 5.10. Applying the hypothetical section approach to the element to be calculated, a system of equations was set up for infinitely thin plies, which reflects the stress-strain state of the repeating element and of the whole reinforced plastic.

The solution to the problem of determining the transverse strain of unidirectionally reinforced plastic means solving a system of linear algebraic equations and also the equations of component deformation. Further it is assumed that fibers are transversely isotropic and elastic but the polymer matrix deforms according to Equation (5.1). As a result relationships [93, 98] were obtained for determining stresses in fibers and in the polymer matrix at any instant after a sustained static load is applied to a unidirectionally reinforced ply transversely to the fiber orientation.

A conclusion can be made on the basis of these results that the polymer matrix is in a non-uniform three-axial stress state, this stress state being practically independent of the loading duration in the case of glass- or boron-fiber reinforced plastics. Therefore the polymer matrix may be considered to be in a constant state of stress during the process of transversal creep in glass- and boron-fiber reinforced plastics. Consequently, Equation (5.9) may be viewed as a deformation law for the polymer matrix. According to the calculation diagram of Fig. 5.10, the creep of a reinforced plastic in the loading direction is equal to the creep of any of its plies.

Consider a ply containing the maximum of fibers or one that satisfies the condition $\varphi = 0$ (see Fig. 3.10). The strain of this ply is

$$\langle \epsilon_\perp(t) \rangle = \frac{r_B}{l} \epsilon_{Br} + \left(1 - \frac{r_B}{l}\right) \epsilon_A(t) \tag{5.33}$$

where $\epsilon_{Br}$ is elastic strain of the fibers in the direction of the mean external stress $\langle \sigma_\perp \rangle$ which is determined by Hooke's law; $\epsilon_A(t)$ is strain of the polymer matrix determined according to Equation (5.9).

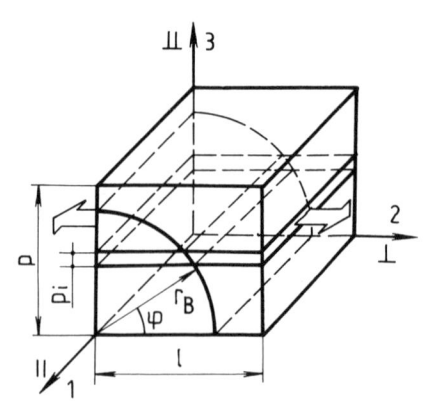

FIGURE 5.10. The repeating element in the model of a unidirectionally reinforced plastic.

Thus, for the creep strain of a unidirectionally reinforced plastic under transverse load we have the equation

$$\langle \epsilon_\perp(t) \rangle = s_{22}(t) \langle \sigma_\perp \rangle \tag{5.34}$$

where $s_{22}(t)$ is the creep function of a reinforced plastic in the transversal direction.

Taking into account the structure geometry of a unidirectionally reinforced plastic and the viscoelastic properties of its components an expression has been obtained for determining the creep function of a unidirectionally reinforced plastic in the transversal direction

$$s_{22}(t) = d_1 + d_2 D(t) \tag{5.35}$$

where

$$d_1 = \frac{r_B}{l} \frac{\bar{\sigma}_r}{E_{Br}} + \left(1 - \frac{r_B}{l}\right) \frac{1 - 2\nu_A}{2 E_A} (\bar{\sigma}_\theta + \bar{\sigma}_z)$$

$$d_2 = \left(1 - \frac{r_B}{l}\right) \left(\bar{\sigma}_r - \frac{\bar{\sigma}_\theta + \bar{\sigma}_z}{2}\right)$$

Here $\bar{\sigma}_r$, $\bar{\sigma}_\theta$ and $\bar{\sigma}_z$ denote coefficients characterizing stress concentration in the matrix of a reinforced plastic as a function of the elastic properties, volume ratio, and structural geometry of the components. Inserting the value of the creep function of the polymer matrix, which corresponds to $s_{22}(t)$ with $t \to \infty$, into Equation (5.35), an asymptote is obtained describing the maximum value of the transverse creep strain of a unidirectionally reinforced plastic. Taking into account Equation (5.8), Equation (5.35) yields

$$s_{22}(\infty) = d_1 + d_2\left(1 + \frac{\lambda_A}{\beta_A}\right) \frac{1}{E_A} \tag{5.36}$$

As $s_{22}(0)$ characterizes the stiffness of the reinforced plastic transversely to the reinforcement direction under short-term load, the degree of transverse creep is given by

$$\eta_\perp = \frac{s_{22}(\infty)}{s_{22}(0)} = \frac{d_1 E_A + d_2\left(1 + \frac{\lambda_A}{\beta_A}\right)}{d_1 E_A + d_2} \tag{5.37}$$

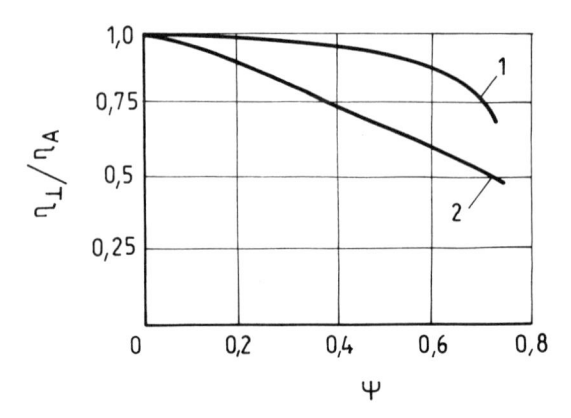

**FIGURE 5.11.** Dependence of the relative degree of creep for glass-epoxy (1) and carbon-epoxy (2) under transversal load on fiber volume content.

This relationship depends upon the viscoelastic properties of the polymer matrix as well as upon the type and volume of the reinforcement. Fig. 5.11 shows the dependence of coefficient $\eta_\perp$ upon the component volume content for glass-epoxy and carbon-epoxy. In Fig. 5.12 test data shown by dots, and the creep curve calculated according to Equation (5.34) refer to a unidirectionally glass fiber reinforced phenol-formaldehyde matrix with a fiber volume content of 0.52. A rectangular fiber distribution was assumed in the cross-section of the composite. The calculation was made with the following initial data: $E_{Bz} = E_{Br}$

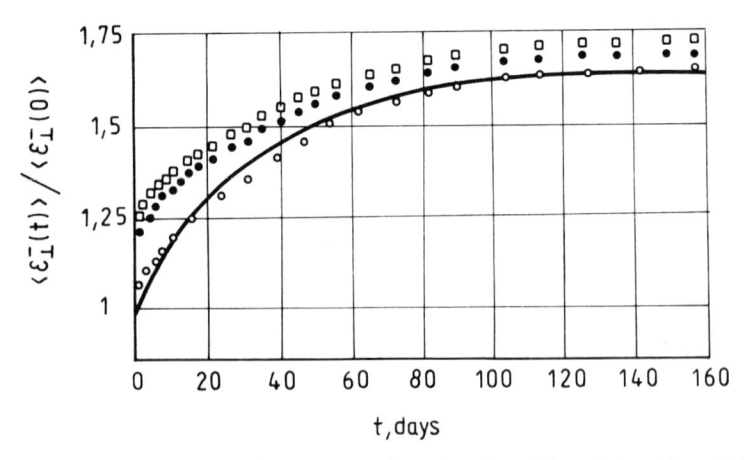

**FIGURE 5.12.** Relative creep deformation curve for a glass fiber reinforced phenol-formaldehyde resin under transverse compression: $\langle \sigma_\perp \rangle$ = 25.8 MPa (○); 57.3 (●); 56.5 (□) MPa.

= 70 GPa; $\nu_{Bzr} = \nu_{Br\theta} = 0.22$; $E_A = 3.6$ GPa; $\nu_A = 0.35$; $\alpha_A = -0.35$; $\beta_A = 0.21$ days$^{-0.65}$; $\lambda_A = 0.17$ days$^{-0.65}$; $\langle \sigma_\perp \rangle = 1$ MPa.

For reinforced plastics having pronounced fiber anisotropy (like carbon and organic fibers) the proposed method for determining the transverse creep properties is quite approximate.

Under constant long-term load acting transversely in carbon and organic fiber reinforced plastics the stress state in the components is time-dependent. Thus, for a carbon fiber reinforced plastic the maximum stress in the polymer matrix during creep rises by up to 30 percent compared with the initial value.

In Bulavs & Radinsh [9, 10] we find a solution to the problem of constructing transverse creep curves according to the viscoelastic properties of the fibers and the polymer matrix, taking into account the time-varying volume state of stress of components. We shall restrict ourselves to an approximated solution based on the hypothesis about the constant value of the maximum stress component in the $n$-th ply of the repeating element in the reinforced plastic calculation model. The orientation of the ply is described by angle $\varphi_N$ given by the expression

$$\varphi_N = \arccos \frac{l}{r_B} \left[ 1 + \frac{1 - \eta_\perp}{1 - \eta_A} \frac{E_A}{E_{Br}} \right]^{-1} \qquad (5.38)$$

where $\eta_A$ is the degree of creep of the polymer matrix under axial load; $\eta_\perp$ is the degree of creep of the reinforced plastic in the transverse direction.

Characteristic $\eta_\perp$ is governed by the ratio between strain values at the end of the creep process $\langle \epsilon_\perp(\infty) \rangle$ and short-term loading $\langle \epsilon_\perp(0) \rangle$ for a fixed mean stress $\langle \sigma_\perp \rangle$. Consequently $\eta_\perp$ is given by the relationship

$$\eta_\perp = \frac{E_\perp(\infty)}{E_\perp} \qquad (5.39)$$

The transverse elasticity modulus $E_\perp(\infty)$ corresponding to the elasticity modulus of the polymer matrix at the end of the creep process

$$E_A(\infty) = E_A \frac{\beta_A}{\lambda_A + \beta_A}$$

is calculated according to the structural theory using the viscoelastic properties of the components [8, 55]. As an approximated dependence, the following expression may be written

$$E_\perp = \left[\frac{r_B}{l} J + \left(1 - \frac{r_B}{l}\right)\frac{1}{1-v_A^2}\right] E_A$$

where

$$J = \frac{1}{bE_A}\left[\frac{\pi}{2} - \frac{2a}{\sqrt{a^2-b^2}} \text{arctg}\sqrt{\frac{a-b}{a+b}}\right]$$

$$a = \frac{1-v_A^2}{E_A}$$

$$b = \frac{r_B}{l}\left[\frac{1 - v_{Brz}v_{Bzr}}{E_{Br}} - a\right]$$

Expressing the strain of the $n$-th ply as a function of elastic and viscoelastic properties of the fibers and the polymer matrix we obtain a formula for determining time-invariable axial stress $\sigma_{\perp n}$ in that ply:

$$\sigma_{\perp n} = \frac{E_{Br}}{E_\perp}\left[1 - \frac{1-\eta_\perp}{1-\eta_A}\left(1 - \frac{E_A}{E_{Br}}\right)\right]\langle\sigma_\perp\rangle \quad (5.40)$$

Thus strain of the $n$-th ply and of the whole reinforced plastic under the long-term action of stress $\langle\sigma_\perp\rangle$ is given by

$$\langle\epsilon_\perp(t)\rangle = \frac{1}{E_\perp}\frac{1}{1-\eta_A}[\eta_\perp - \eta_A + (1-\eta_\perp)E_A D(t)]\langle\sigma_\perp\rangle \quad (5.41)$$

The creep function of the polymer matrix $D(t)$ is given by Equation (5.8).

## 5.3 VISCOELASTIC PROPERTIES OF LAMINATED REINFORCED PLASTICS UNDER SUSTAINED PLANE STRESS STATE

In practice, reinforced plastics are usually loaded in the reinforcement plane. Therefore, discussion of the plane stress state may be sufficient for determining the stress-strain state of a reinforced plastic laminate under sustained load. Provided there is compatibility between the deformation of the individual plies and the accuracy of stress equilibrium equations, the difference in the viscoelastic properties of these plies causes time-dependent redistribution of ply stresses. This phenomenon is not taken into consideration in the application of the widely known quasi-elastic method of solving viscoelastic problems for reinforced plastic laminates as well as in the methods of averaging the constituents of elastic or compliance matrices of individual plies in order to determine the strain properties of a laminate. Similar types of simplification may in a number of cases lead to essential deviations when determining strain characteristics of reinforced plastic laminates.

We may conclude that the problem of creep prediction for a reinforced plastic laminate judging from the viscoelastic properties of the individual plies is reduced to the determination of relationships according to which stresses change in the individual plies under prescribed external load and to finding strains of unidirectional plies under time-varying stresses.

### 5.3.1 Deformation Law for a Unidirectionally Reinforced Plastic Under a Sustained Plane Stress State

Let us consider the stress-strain state of an arbitrarily oriented unidirectionally reinforced ply loaded by time-varying stresses $\langle \sigma_{1'}(t) \rangle_k$, $\langle \sigma_{2'}(t) \rangle_k$ and $\langle \tau_{1'2'}(t) \rangle_k$ (Fig. 5.13). We shall use the matrix form introducing symbols $\langle \tau_{1'2'}(t) \rangle = \langle \sigma_{6'}(t) \rangle_k$, $\langle \gamma_{1'2'}(t) \rangle_k = \langle \epsilon_{6'}(t) \rangle_k$, $\langle \tau_{1'2'}(t) \rangle = \langle \sigma_{6'}(t) \rangle_k$ and $\langle \gamma_{1'2'}(t) \rangle_k = \langle \epsilon_{6'}(t) \rangle_k$. Stresses along the

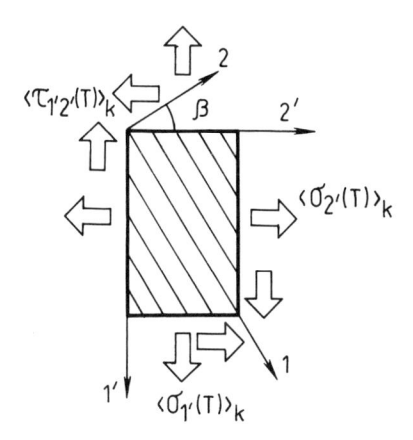

FIGURE 5.13. Stress state of an arbitrarily oriented unidirectionally reinforced ply.

elastic symmetry axes of a unidirectionally reinforced ply is then given by

$$[\langle \sigma_j(t) \rangle_k] = [T]_k [\langle \sigma_p(t) \rangle_k] \qquad (5.42)$$

$$j, p = 1, 2, 6.$$

The transformation matrix has the form

$$[T]_k = \begin{bmatrix} \cos^2 \beta_k & \sin^2 \beta_k & 2 \sin \beta_k \cos \beta_k \\ \sin^2 \beta_k & \cos^2 \beta_k & -2 \sin \beta_k \cos \beta_k \\ -\cos \beta_k \sin \beta_k & \cos \beta_k \sin \beta_k & \cos^2 \beta_k - \sin^2 \beta_k \end{bmatrix}$$

The deformation law for a unidirectionally reinforced ply in the case of plane stress state in the elastic symmetry axes is written as

$$[\langle \epsilon_i(t) \rangle_k] = [s_{ij}^{(k)}] [\langle \sigma_j(t) \rangle_k]$$

$$+ \int_0^t [s_{ij}^{(k)} K_{ij}^{(k)}(t - \theta)] [\langle \sigma_j(\theta) \rangle_k] d\theta \qquad (5.43)$$

$$i, j = 1, 2, 6$$

where $s_{ij}$ are compliance matrix components; $K_{ij}(t - \theta)$ are creep kernels characterizing viscoelastic properties of a unidirectionally reinforced ply in the elastic symmetry axes.

Considering the relationships for the axial rotation, the strains in plastics reinforced unidirectionally at an angle to the elastic symmetry axes are given by the relationship

$$[\langle \epsilon_q(t) \rangle_k] = [T]_k^{-1} [\langle \epsilon_i(t) \rangle_k] \qquad (5.44)$$

$$q = 1', 2', 6', \qquad i = 1, 2, 6.$$

A joint solution of Equations (5.42)–(5.44) yields the deformation law for a plastic reinforced unidirectionally at an angle to the elastic symmetry axis of the ply loaded by time-varying stresses and is written as

$$[\langle \epsilon_q(t) \rangle_k] = [s_{qp}^{(k)}][\langle \sigma_p(t) \rangle_k]$$

$$+ \int_0^t [s_{qp}^{(k)} K_{qp}^{(k)}(t - \theta)][\langle \sigma_p(\theta) \rangle_k] d\theta \quad (5.45)$$

$$i, j = 1, 2, 6; \quad q, p = 1', 2', 6'; \quad k = 1, \ldots, n,$$

where

$$[s_{qp}^{(k)} K_{qp}^{(k)}(t - \theta)] = [T]_k^{-1} [s_{ij}^{(k)} K_{ij}^{(k)}(t - \theta)][T]$$
$$[s_{qp}^{(k)}] = [T]_k^{-1} [s_{ij}^{(k)}][T] \quad (5.46)$$

The creep kernels for a unidirectionally reinforced plastic in the axes of elastic symmetry $K_{ij}(t - \theta)$ may be determined both by test and by calculation methods on the basis of component properties, as shown in the previous chapters.

Equations (5.45) may be integrated to obtain expressions for the calculation of strains in arbitrarily oriented unidirectionally reinforced plastic under a plane state of stress.

### 5.3.2 Deformation Law for a Reinforced Plastic Laminate Under Sustained Plane Stress State

In order to determine the stress-strain state of a reinforced plastic laminate under sustained plane stress states, we shall make the following conditions:

- strains in individual plies are equal to the strain in the whole laminate:

$$[\langle\langle \epsilon_q(t) \rangle\rangle] = [\langle \epsilon_q(t) \rangle_k]$$
$$q = 1', 2', 6', \quad k = 1, \ldots, n \quad (5.47)$$

- the mean stresses in the laminate are equal to the sum of the mean stresses in the individual plies:

$$[\langle\langle \sigma_p(t) \rangle\rangle] = \sum_{k=1}^{n} m_k [\langle \sigma_p(t) \rangle_k] \quad (5.48)$$

where $m_k$ is the specific volume of the $k$-th ply;

- the deformation law for arbitrarily oriented unidirectionally reinforced ply is assumed to be expressed by Equation (5.46).

The system of Equations (5.46) through (5.48) enables us to determine the stress-strain state of a laminate under sustained load applied in the plane of reinforcement, because for determining $6n + 3$ unknowns we have the same number of equations. It is convenient to apply the Laplace transformation for solving the given system of equations. With Equation (5.47) in account, Equations (5.46) and (5.48) enable us to write

$$[\langle\langle\bar{\epsilon}_q(p)\rangle\rangle] = [s^{(k)}_{qp}(1 + \bar{K}^{(k)}_{qp}(p))][\langle\bar{\sigma}_p(p)\rangle_k] \quad (5.49)$$

$$[\langle\langle\bar{\sigma}_p(p)\rangle\rangle] = \sum_{k=1}^{n} m_k[\langle\bar{\sigma}_p(p)\rangle_k] \quad q, p = 1', 2', 6' \quad (5.50)$$

where $\langle\langle\bar{\epsilon}_q(p)\rangle\rangle$, $\langle\langle\bar{\sigma}_p(p)\rangle\rangle$, $\bar{K}_{qp}(p)$ and $\langle\bar{\sigma}_p(p)\rangle$ are the Laplace representations of the respective functions.

The expressions determining the stress state of each unidirectional ply in a laminate under the prescribed variation of external strains are obtained from Equation (5.49) in the following way

$$[\langle\bar{\sigma}_p(p)\rangle_k] = [s^{(k)}_{qp}(1 + \bar{K}^{(k)}_{qp}(p))]^{-1}[\langle\langle\bar{\epsilon}_q(p)\rangle\rangle]$$
$$q, p = 1', 2', 6' \quad (5.51)$$

Alternatively, Equation (5.50) considering Equation (5.51) enables us to establish the functional link between the representations of external stresses and strains, this link determining creep and stress relaxation in a laminate in the general case

$$[\langle\langle\bar{\sigma}_p(p)\rangle\rangle] = \left[\sum_{k=1}^{n} m_k[s^{(k)}_{qp}(1 + \bar{K}^{(k)}_{qp}(p))]^{-1}[\langle\langle\bar{\epsilon}_q(p)\rangle\rangle]\right] \quad (5.52)$$

and hence

$$[\langle\langle\bar{\epsilon}_q(p)\rangle\rangle] = \left[\sum_{k=1}^{n} m_k[s^{(k)}_{qp}(1 + \bar{K}^{(k)}_{qp}(p))]^{-1}[\langle\langle\bar{\sigma}_p(p)\rangle\rangle]\right]$$
$$(5.53)$$
$$q, p = 1', 2', 6'$$

It should be remarked that the solution of Equations (5.52) and (5.53) is a complex problem involving large-scale computation even with the simplest (exponential) creep kernels, although for a number of materials and for specific cases of loading the problem may be considerably simplified with regard to component properties and ply packaging geometry in a laminate. So for boron-, carbon- and glass-fiber reinforced plastics, deformation along the reinforcement may be regarded as elastic with a high degree of accuracy, and it is sufficient to take into account viscoelastic properties under transversal loading and longitudinal shear.

### 5.3.3 Stress-Strain State of an Angle-Ply Laminate Under Axial Loading

As an example, consider an angle-ply laminate with its plies oriented at angles $\pm\beta$ with respect to the elastic symmetry axes ($m_1 = m_2 = 0.5$) and loaded along one of the symmetry axes, say, "1", by permanent stress $\langle\langle \sigma_{1'} \rangle\rangle$ (Fig. 5.14). Owing to the orthotropic symmetry of the material the condition $\langle\langle \epsilon_{6'} \rangle\rangle = 0$ holds true. By analysing Equation (5.51) in a developed form, it has been established that the following conditions are true for the plies oriented at angles $+\beta$ and $-\beta$ with respect to the elastic symmetry axes: $\langle \sigma_1(t) \rangle_1 = \langle \sigma_{1'}(t) \rangle_2$, $\langle \sigma_{2'}(t) \rangle_1 = \langle \sigma_{2'}(t) \rangle_2$, $\langle \tau_{1'2'}(t) \rangle_1 = -\langle \tau_{1'2'}(t) \rangle_2$. According to Equation (5.48) we obtain $\langle \sigma_{1'}(t) \rangle_1 = \langle \sigma_{1'}(t) \rangle_2 = \langle\langle \sigma_{1'} \rangle\rangle$; $\langle \sigma_{2'}(t) \rangle_1 = \langle \sigma_{2'}(t) \rangle_2 = 0$ and $\langle \tau_{1'2'} \rangle_1 + \langle \tau_{1'2'} \rangle_2 = 0$. It is evident that Equation (5.49) with the given conditions in consideration has to be solved in order to determine

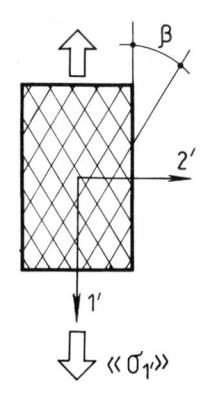

FIGURE 5.14. The loading scheme of an angle-ply laminate.

viscoelastic behavior of the material under discussion. The system is written as

$$\langle\langle \bar{\epsilon}_{1'}(p) \rangle\rangle = \bar{s}^{(k)}_{1'1'} \langle\langle \sigma_{1'} \rangle\rangle \frac{1}{p} + \bar{s}^{(k)}_{1'6'} \langle \tau_{1'2'}(p) \rangle_k$$

$$\langle\langle \bar{\epsilon}_{2'}(p) \rangle\rangle = \bar{s}^{(k)}_{1'2'} \langle\langle \sigma_{1'} \rangle\rangle \frac{1}{p} + \bar{s}^{(k)}_{2'6'} \langle \tau_{1'2'}(p) \rangle_k \qquad (5.54)$$

$$k = 1, 2$$

where

$$\bar{s}^{(k)}_{1'1'} = s^{(k)}_{11}[1 + \bar{K}^{(k)}_{11}(p)] \cos^4 \beta$$

$$+ s^{(k)}_{22}[1 + \bar{K}^{(k)}_{22}(p)] \sin^4 \beta + \{2s^{(k)}_{12}[1 + \bar{K}^{(k)}_{12}(p)]$$

$$+ s^{(k)}_{66}[1 + \bar{K}^{(k)}_{66}(p)]\} \sin^2 \beta \cos^2 \beta$$

$$\bar{s}^{(k)}_{1'2'} = \{s^{(k)}_{11}[1 + \bar{K}^{(k)}_{11}(p)] + s^{(k)}_{22}[1 + \bar{K}^{(k)}_{22}(p)]$$

$$- s^{(k)}_{66}[1 + \bar{K}^{(k)}_{66}(p)]\} \sin^2 \beta \cos^2 \beta$$

$$+ s^{(k)}_{12}[1 + \bar{K}^{(k)}_{12}(p)](\cos^4 \beta + \sin^4 \beta)$$

$$\bar{s}^{(k)}_{1'6'} = \{2[s^{(k)}_{11}[1 + \bar{K}^{(k)}_{11}(p)] \cos^2 \beta$$

$$+ s^{(k)}_{66}[1 + \bar{K}^{(k)}_{66}(p)] \sin^2 \beta] - [2s^{(k)}_{12}[1 + \bar{K}^{(k)}_{12}(p)] \qquad (5.55)$$

$$+ s^{(k)}_{66}[1 + \bar{K}^{(k)}_{66}(p)]](\cos^2 \beta - \sin^2 \beta)\} \sin \beta \cos \beta$$

$$\bar{s}^{(k)}_{2'6'} = \{2[s^{(k)}_{11}[1 + \bar{K}^{(k)}_{11}(p)] \sin^2 \beta$$

$$- s^{(k)}_{66}[1 + \bar{K}^{(k)}_{66}(p)] \cos^2 \beta] + [2s^{(k)}_{12}[1 + \bar{K}^{(k)}_{12}(p)]$$

$$+ s^{(k)}_{66}[1 + \bar{K}^{(k)}_{66}(p)]](\cos^2 \beta - \sin^2 \beta)\} \sin \beta \cos \beta$$

$$\bar{s}^{(k)}_{6'6'} = 4\{s^{(k)}_{11}[1 + \bar{K}^{(k)}_{11}(p)] - 2s^{(k)}_{12}[1 + \bar{K}^{(k)}_{12}(p)]$$

$$+ s^{(k)}_{22}[1 + \bar{K}^{(k)}_{22}(p)]\} \sin^2 \beta \cos^2 \beta$$

$$+ s^{(k)}_{66}[1 + \bar{K}^{(k)}_{66}(p)](\cos^2 \beta - \sin^2 \beta)^2$$

The last expression of Equation System (5.54) for the $k$-th ply yields

$$\langle \bar{\tau}_{1'2'}(p) \rangle = - \frac{\bar{s}^{(k)}_{1'6'}}{\bar{s}^{(k)}_{6'6'}} \frac{\langle\langle \sigma_{1'} \rangle\rangle}{p} \tag{5.56}$$

The strains of a laminate are given by two expressions of Equation System (5.54):

$$\langle\langle \bar{\epsilon}_1(p) \rangle\rangle = \left( \bar{s}^{(k)}_{1'1'} - \frac{(\bar{s}^{(k)}_{1'6'})^2}{\bar{s}^{(k)}_{6'6'}} \right) \frac{\langle\langle \sigma_{1'} \rangle\rangle}{p}$$

$$\langle\langle \bar{\epsilon}_2(p) \rangle\rangle = \left( \bar{s}^{(k)}_{1'2'} - \frac{\bar{s}^{(k)}_{1'6'} \bar{s}^{(k)}_{2'6'}}{\bar{s}^{(k)}_{6'6'}} \right) \frac{\langle\langle \sigma_{1'} \rangle\rangle}{p} \tag{5.57}$$

Let us analyse creep of an angle-ply glass-epoxy *AG-4s* ($\psi = 0.52$) and restrict ourselves to considering the viscoelastic properties of a unidirectionally reinforced ply under only transversal loading and longitudinal shear, namely, using the condition $K^{(k)}_{11}(t - \theta) = K^{(k)}_{12}(t - \theta) = 0$. Selecting the respective creep kernels in an exponential form $K^{(k)}_{22}(t - \theta) = C_{22} \exp[-\alpha_{22}(t - \theta)]$ and $K^{(k)}_{66}(t - \theta) = C_{66} \exp[-\alpha_{66}(t - \theta)]$ and performing the reverse Laplace transformation for Equation (5.56) we arrive at

$$\langle \tau_{1'2'}(t) \rangle_k = -\frac{\bar{s}^{(k)}_{1'6'}}{\bar{s}^{(k)}_{6'6'}} \left\{ 1 - \sum_{l=1}^{n} \frac{P_1(p_l)}{p_l P_2(p_l)} \right.$$

$$\left. \times [1 - \exp(p_l t)] \right\} \langle\langle \sigma_{1'} \rangle\rangle \tag{5.58}$$

where

$$P_1(p_l) = (a_1 - b_1) p_l + a_2 - b_2$$

$$P_2(p_l) = 2p_l + b_1$$

$$a_1 = \frac{1}{\bar{s}^{(k)}_{1'6'}} [\bar{s}^{(k)}_{1'6'}(\alpha_{22} + \alpha_{66}) - 2C_{22} \bar{s}^{(k)}_{22} \sin^3 \beta \cos \beta$$

$$- C_{66} \bar{s}^{(k)}_{66} (\cos^2 \beta - \sin^2 \beta) \sin \beta \cos \beta]$$

STRUCTURAL MECHANICS OF A MATERIAL

$$a_2 = \frac{1}{s_{1'6'}^{(k)}} [s_{1'6'}^{(k)} \alpha_{22}\alpha_{66} - 2C_{22}\alpha_{66}s_{22}^{(k)} \sin^3 \beta \cos \beta$$

$$- C_{66}\alpha_{22}s_{66}^{(k)}(\cos^2 \beta - \sin^2 \beta) \sin \beta \cos \beta]$$

$$b_1 = \frac{1}{s_{6'6'}^{(k)}} [s_{6'6'}^{(k)}(\alpha_{22} + \alpha_{66}) + 4C_{22}s_{22}^{(k)} \sin^2 \beta \cos^2 \beta$$

$$+ C_{66}s_{66}^{(k)}(\cos^2 \beta - \sin^2 \beta)^2]$$

$$b_2 = \frac{1}{s_{6'6'}^{(k)}} [s_{6'6'}^{(k)} \alpha_{22}\alpha_{66} + 4C_{22}\alpha_{66}s_{22}^{(k)} \sin^2 \beta \cos^2 \beta$$

$$+ C_{66}\alpha_{22}s_{66}^{(k)}(\cos^2 \beta - \sin^2 \beta)^2]$$

$p_l$ are the roots of equation $p^2 + b_1 p + b_2 = 0$.

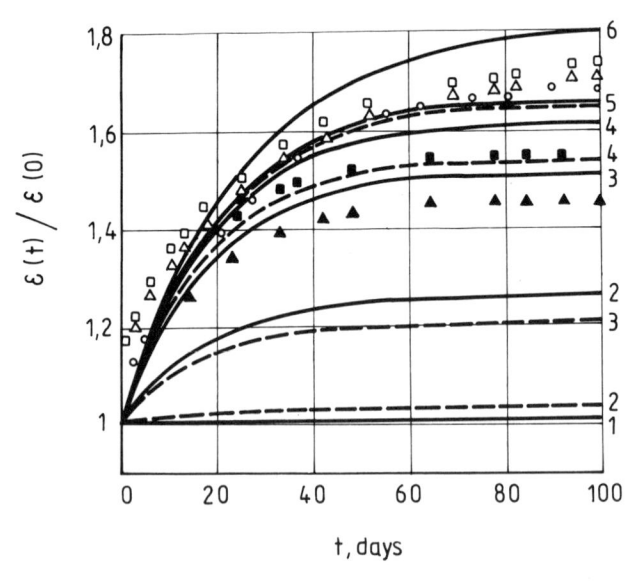

**FIGURE 5.15.** Relative creep deformation curves of unidirectionally (——, $B$) and two-directionally (- - -, $\pm B$) reinforced angle-ply laminates with various fiber orientations: $B = 0°$ (1), 15 (2), 30 (3), 45 (4), 90 (5). Test points were obtained for a unidirectional glass-epoxy AG-4S with $B = 90°$ (□, △, ○); 30° (■, ▲)

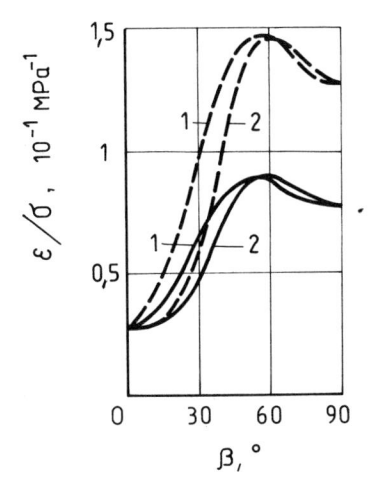

**FIGURE 5.16.** Compliance of unidirectionally (1)($B$) and two-directionally (2)($\pm B$) reinforced glass-epoxy angle-ply laminates with $t = 0$ (———) and $t \to \infty$ (– – –).

The relationships for determining the components $s_{q,p}^{(k)}$, $q, p = 1, 2, 6$ are obtained from Equation System (5.55) by inserting $K_{ij}^{(k)}(p) = 0$; $i, j = 1, 2, 6$.

Equations (5.57) and (5.58) permit us to determine strains of a laminate in directions $1'$ and $2'$:

$$\langle\langle \epsilon_{1'}(t) \rangle\rangle = \left\{ s_{1'1'}^{(k)} + \frac{C_{22}s_{22}^{(k)}}{\alpha_{22}} [1 - \exp(-\alpha_{22}t)] \sin^4 \beta \right.$$

$$+ \frac{C_{66}s_{66}^{(k)}}{\alpha_{66}} \sin^2 \beta \cos^2 \beta [1 - \exp(-\alpha_{66}t)]$$

$$+ s_{1'6'}^{(k)} \langle \tau_{1'2'}(t) \rangle_k \frac{1}{\langle\langle \sigma_{1'} \rangle\rangle} - 2 \frac{s_{1'6'}^{(k)}}{s_{6'6'}^{(k)}} s_{22}^{(k)}$$

$$\times \sin^3 \beta \cos \beta \Phi_{22}(t) - \frac{s_{1'6'}^{(k)}}{s_{6'6'}^{(k)}} s_{66}^{(k)}$$

$$\left. \times (\cos^2 \beta - \sin^2 \beta) \sin \beta \cos \beta \Phi_{66}(t) \right\} \langle\langle \sigma_{1'} \rangle\rangle$$

110  STRUCTURAL MECHANICS OF A MATERIAL

$$\langle\langle\epsilon_{2'}(t)\rangle\rangle = \left\{ s_{1'2'}^{(k)} + \frac{C_{22}s_{22}^{(k)}}{\alpha_{22}} \sin^2\beta \cos^2\beta \right.$$

$$\times [1 - \exp(-\alpha_{22}t)] - \frac{C_{66}s_{22}^{(k)}}{\alpha_{66}} \sin^2\beta \cos^2\beta$$

$$\times [1 - \exp(-\alpha_{66}t)] + s_{2'6'}^{(k)} \frac{\langle \tau_{1'6'}(t)\rangle_k}{\langle\langle\sigma_{1'}\rangle\rangle} \quad (5.59)$$

$$- 2 \frac{s_{1'6'}^{(k)}}{s_{6'6'}^{(k)}} s_{22}^{(k)} \sin\beta \cos^3\beta \Phi_{22}(t) + \frac{s_{1'6'}^{(k)}}{s_{6'6'}^{(k)}} s_{66}^{(k)}$$

$$\left. \times (\cos^2\beta - \sin^2\beta) \sin\beta \cos\beta \Phi_{66}(t) \right\} \langle\langle\sigma_{1'}\rangle\rangle$$

where

$$\Phi_{ss} = \frac{C_{ss}}{\alpha_{ss}} [\exp(-\alpha_{ss}t) - 1]$$

$$- \sum_{i=1}^{2} \frac{P_1(p_l)}{p_l P_2(p_l)} \left\{ \frac{C_{ss}}{\alpha_{ss}} [\exp(-\alpha_{ss}t) - 1] \right.$$

$$\left. - \frac{C_{ss}}{\alpha_{ss} + p_l} [\exp(-\alpha_{ss}t) - \exp(p_l t)] \right\}$$

$s = 2, 6$

Fig. 5.15 represents the specific creep curves of unidirectional and angle-ply glass-fiber reinforced plastics with phenol-formaldehyde resin and various fiber orientations; the curves were calculated from the Equations (5.45) and (5.59). The agreement between test data and calculated curves of creep is good. It should be observed that there is a significantly lower degree of creep in the anisotropic fiber (organic) reinforced plastics. As is seen from the diagram, the creep curves differ considerably within the range between 0 to 45° of angle $\beta$ value, i.e., between unidirectional and angle-ply laminates. Within this range the creep of an angle-ply laminate ($\pm\beta$) will be much lower than that of a unidirectional ($\beta$).

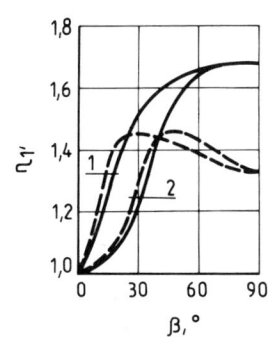

**FIGURE 5.17.** Dependence of the degree of creep for unidirectionally (1)(B) and two-directionally (2)(±B) glass- (———) and organic (- - -) fiber reinforced angle-ply laminates upon the fiber orientation.

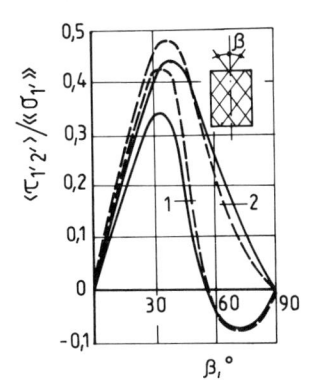

**FIGURE 5.18.** Dependence of specific shear stresses in angle-ply glass-epoxy (1) and carbon-epoxy (2) laminates upon fiber orientation with $t = 0$ (———) and $t \to \infty$ (- - -).

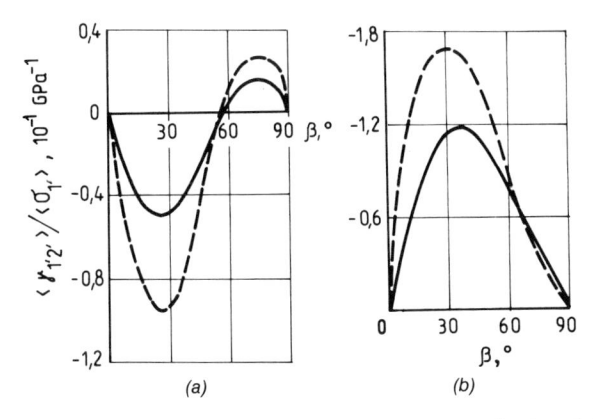

**FIGURE 5.19.** Dependence of specific shear strains of a unidirectionally glass- (a) and organic (b) fiber reinforced plastic upon fiber orientation with $t = 0$ (———) and $t \to \infty$ (- - -).

**112** STRUCTURAL MECHANICS OF A MATERIAL

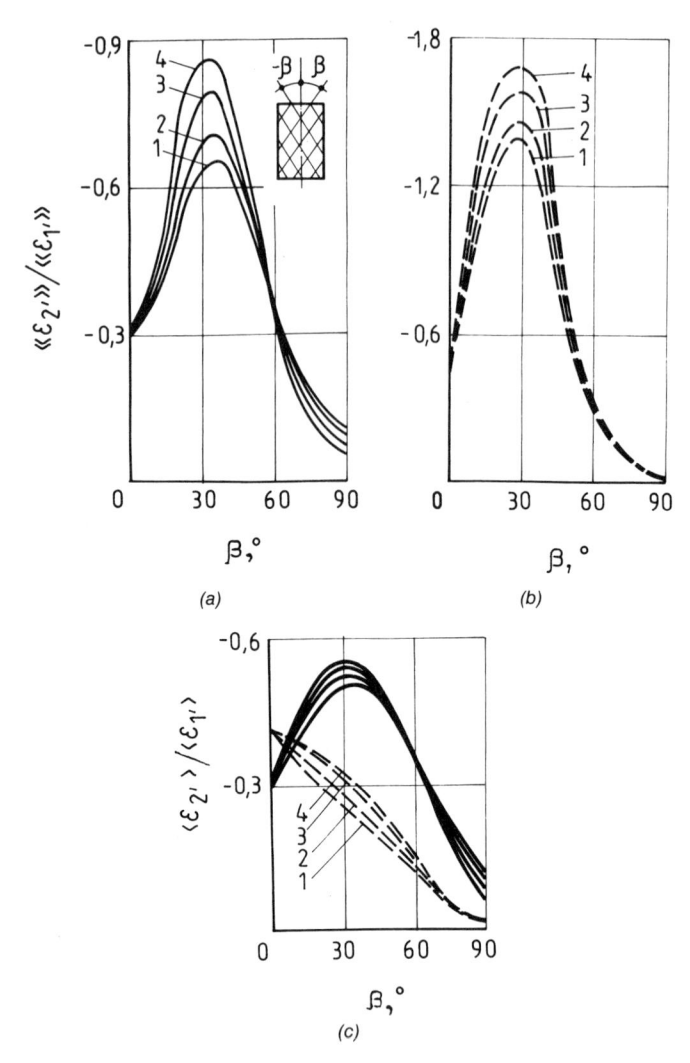

**FIGURE 5.20.** Dependence of the ratio of transverse strain to longitudinal strain for two-directionally (a, b) and a unidirectionally (c) glass- (———) and organic (- - -) fiber reinforced epoxy upon the fiber orientation: $t = 0$ (1), 5 days (2), 20 days (3), $t \to \infty$ (4).

The calculation results given in Figs. 5.16 and 5.17 permit us to evaluate the effects of ply interaction on the elastic and viscoelastic properties of a laminate depending upon its fiber orientation. Ignoring the ply interaction, which is present in a comparatively widely used method of averaging the components of elasticity and compliance matrices, may lead to considerable errors when determining elastic and viscoelastic properties of angle-ply laminates. As seen from Fig. 5.16 the maximum divergence in elastic strains of unidirectional and angle-ply laminates in the case of glass-epoxy reaches 35 percent, and with $t \to \infty$ it increases up to 70 percent; the respective divergences for organic fiber reinforced plastics amount to 125 and 155 percent ($\beta \approx 30°$).

Unidirectional and angle-ply laminates have widely differing degrees of creep (see Fig. 5.17). Angle-ply laminates if compared to the unidirectional, have a lower degree of creep for a glass-epoxy within the whole range of varying angle $\beta$ whereas for an organic fiber reinforced plastic the degree of creep is lower within the angle range of 0 to 40°. If $40° < \beta < 90°$ the degree of creep for angle-ply laminate is somewhat higher than that in unidirectional case.

As a result of shear strain crowding in variously oriented plies of an angle-ply laminate, shear stresses that are equal in magnitude, contrary in sign and increase in time appear in these directions (see Fig. 5.18). Shear stresses are essentially dependent on reinforcement orientation. For the glass-epoxy under discussion the shear stresses may be from 0.34 (with $t = 0$) to 0.43 (with $t \to \infty$) of the externally applied stress. Shear stress determination in the plies is of some interest in developing strength theory for a reinforced plastic laminate.

Shear strains appear when a unidirectional plastic is subjected to loading at an angle to the reinforcement direction. The most pronounced shear strains are observed in the angle $\beta$ variation range of 10 to 45° for a glass-epoxy and 10 to 70° for an organic fiber reinforced epoxy (Fig. 5.19). This phenomenon should be considered a disadvantage of unidirectionally reinforced plastics, because in some cases this may result in a loss of stability and consequently a loss of the bearing capacity of the structure.

Reinforced plastics loaded at an angle to the reinforcement direction exhibit significant transverse strains. The maximum value of Poisson's ratio for an angle-ply glass-epoxy laminate may vary from 0.65 (with $t = 0$) to 0.85 (with $t \to \infty$) whereas for an organic fiber reinforced plastic it will vary between 1.36 to 1.68 (Fig. 5.20). For a unidirectionally reinforced glass-epoxy the corresponding Poisson's ratios are equal to 0.5 and 0.55, for an organic fiber reinforced plastic the value is 0.42, which means that these values are lower in absolute terms as well as in the degree of time-dependent variation (see Fig. 5.20(b)).

CHAPTER 6

# Stress State of Reinforced Plastics Under Long-Term Loading

## 6.1 INTRODUCTORY REMARKS

The load applied to a reinforced plastic is distributed across its volume quite unevenly due to the highly inhomogeneous character of its structure. Under long-term loading there is stress redistribution in the components of the material and between them owing to the pronounced rheonomic properties of the components (the polymer matrix and the organic fibers). Even the simplest types of loading cause a volume stress state in the reinforced plastics components, which makes determination of the laws governing the time-dependent stress variations a complicated task in the general case and seems impossible to solve explicitly. Therefore, it will be useful to make reasonable assumptions about the stress state, the mechanical properties, and the component geometry.

All the assumptions simplifying the calculation of stress state in the components may be classified into two groups. The first group includes assumptions about the character of the stress redistribution among the components. Supposing that at any instant of loading the component stress distribution is proportional to the stiffness characteristics of these components at the given time, we arrive at quasi-elastic solutions. The other group of assumptions refers to the substitution of simpler stress states of the components for the complex ones. Thus, for example, the analysis of component stress state in a unidirectionally reinforced plastic under axial loading (supposing that the fibers are continuous, straight, and parallel, whereas the matrix is solid, i.e., without pores, initial cracks, or accidental flaws) has shown that it is sufficient to solve a uniaxial problem of two parallel simultaneously deforming members in order to evaluate time-dependent stress variations in the polymer matrix and the fibers.

The general methods for solving problems of simultaneously deforming parallel-connection members are given by Rzhanitsin [51]. Solutions to similar problems involving specific creep kernels have been obtained and included in Skudra et al. [56]. A joint solution of Equations (5.17) though (5.19) enables us to state that for the exponential function form of creep kernels the time-dependent stress variation in the components is described by the following relationships

$$\sigma_A(t) = E_A \Big\{ \langle \epsilon_{\|}(t) \rangle - \lambda_A \int_0^t \exp[-(\lambda_A + \beta_A)(t - \theta)] \langle \epsilon_{\|}(\theta) \rangle d\theta \Big\}$$

$$\sigma_B(t) = E_{Bz} \Big\{ \langle \epsilon_{\|}(t) \rangle - \lambda_B \int_0^t \exp[-(\lambda_B + \beta_B)(t - \theta)] \langle \epsilon_{\|}(\theta) \rangle d\theta \Big\}$$

(6.1)

The deformation law for a reinforced plastic deforming in the loading direction may be established by test or from Equation (5.20) using the strain characteristics of the components. In the latter case, Equation System (6.1) yields

$$\sigma_k(t) = \frac{E_k}{E_{\|}} \left(1 + \frac{a_3}{b_3}\right) \left[\lambda_k \left(\frac{1}{v_k} + \frac{1}{v_k - z_1} + \frac{1}{v_k - z_2}\right)\right.$$

$$\left. \times \exp(-v_k t) + \sum_{i=1}^{2} \frac{A_i(\beta_k - z_i)}{v_k - z_i} \exp(-z_i t) + \frac{\beta_k}{v_k}\right] \langle \sigma_{\|} \rangle \quad (6.2)$$

$$v_k = \beta_k + \lambda_k; \quad k = A, B$$

Fig. 6.1 illustrates stress variations in the organic fiber reinforced plastic components. Stresses in the polymer matrix relax with time from $0.049 \langle \sigma_{\|} \rangle$ to $0.021 \langle \sigma_{\|} \rangle$ whereas the stresses in the impregnated organic filament increase from $1.513 \langle \sigma_{\|} \rangle$ to $1.527 \langle \sigma_{\|} \rangle$.

When unidirectionally reinforced plastics are loaded in the fiber direction, the stress state of the fibers may be viewed as practically unchangeable with time regardless of the fiber type, and their stress state is determined by elastic solution. There is stress relaxation in the polymer matrix, provided the creep rate of the matrix is higher than that of the fibers.

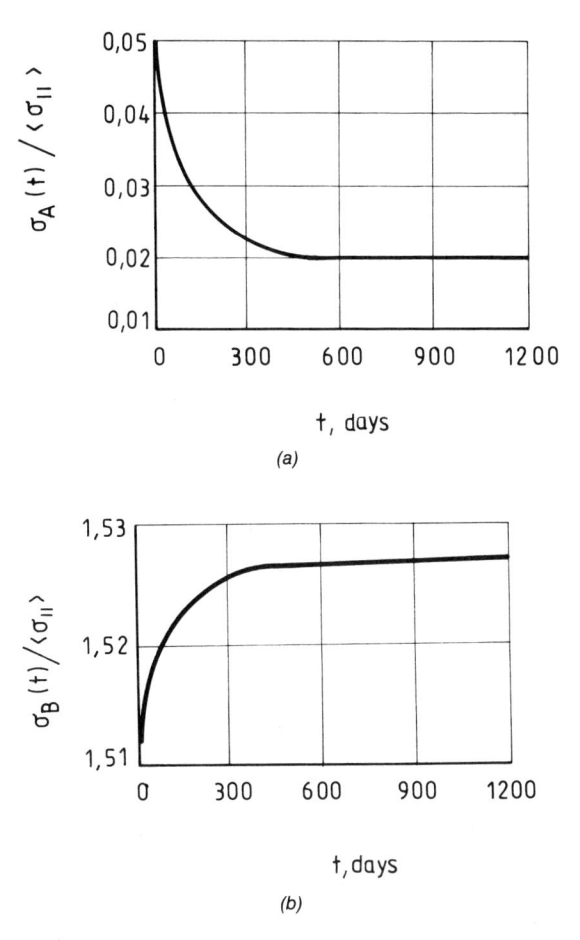

FIGURE 6.1. Dependence of specific stresses upon the duration of sustained longitudinal loading in the polymer matrix (a) and the organic fibers (b).

## 6.2 STRESS STATE OF UNIDIRECTIONALLY REINFORCED PLASTICS COMPONENTS IN LONGITUDINAL SHEAR

In order to determine the stress-strain state of unidirectionally reinforced plastics components under long-term static longitudinal shearing load we shall refer to a unidirectionally reinforced member selected as a calculation model (see Fig. 5.7).

In this case the interrelation between the geometrical variables of the material $p$, $l$, $r_B$ and the fiber volume content $\psi$ are dependent upon

the type of fiber distribution in the cross-section of the composite. With a square distribution of the fibers, $p = l$ and $r_B/l = 2\sqrt{\psi/\pi}$. With a hexagonal distribution of the fibers, $p = \dfrac{\sqrt{3}}{2} l$ and $r_B/l = 2\sqrt{\dfrac{\sqrt{3}}{2}\dfrac{\psi}{\pi}}$.
Using the method of hypothetic section of the calculation element into infinitely thin elementary layers by the planes that are parallel to the fiber packaging plane and to the reinforcement direction, a system of equations is set up reflecting the stress-strain state of the repeating element and of the whole composite material.

The solution of Equation System (5.23) with respect to stresses $\tau_{\|\perp i}(t)$ with the given law of shear strain $\langle \gamma_{\|\perp}(t) \rangle$, variations yield an integral equation

$$s_i \tau_{\|\perp i}(t) + (1 - \psi_i) \dfrac{1}{G_A} \int_0^t K_{A\tau}(t - \theta) \tau_{\|\perp i}(\theta) d\theta = \langle \gamma_{\|\perp}(t) \rangle \qquad (6.3)$$

where

$$s_i = \dfrac{1 - \psi_i}{G_A} + \dfrac{\psi_i}{G_{Brz}}$$

Equation (6.3) will be conveniently solved by the Laplace transformation. Then the stress is described by

$$\bar{\tau}_{\|\perp i}(p) = \dfrac{1}{1 + \dfrac{1 - \psi_i}{s_i G_A} \bar{K}_{A\tau}(p)} \dfrac{1}{s_i} \langle \bar{\gamma}_{\|\perp}(p) \rangle \qquad (6.4)$$

Making use of

$$\dfrac{1 - \psi_i}{s_i G_A} \bar{K}_{A\tau}(p) = \bar{K} \qquad (6.5)$$

and considering that

$$1 - \bar{R}_i = \dfrac{1}{1 + \bar{K}_i}$$

where $\bar{R}_i$ shows the relaxation kernel of the $i$-th layer, Equation (6.4) is written as

$$\bar{\tau}_{\|\perp i}(p) = \frac{1}{S_i}(1 - \bar{R}_i)\langle \bar{\gamma}_{\|\perp}(p)\rangle$$

Applying the reverse Laplace transformation we obtain

$$\tau_{\|\perp i}(t) = \frac{1}{S_i}\left[\langle \gamma_{\|\perp}(t)\rangle - \int_0^t R_i(t - \theta)\langle \gamma_{\|\perp}(\theta)\rangle d\theta\right] \quad (6.6)$$

The need to determine the resolvent of the creep kernel of the $i$-th ply is restricted by the class of functions used as creep kernels to describe the strain properties of the polymer matrix. Preference is given to the kernels in the form of fraction-exponential or exponential functions. A sufficiently accurate description of the creep curves of the polymer matrix is assured by the selection of creep kernel as a sum of exponents $K_{A\tau}(t - \theta) = \sum_{i=1}^{n} C_{A\tau i} \exp[-\alpha_{A\tau i}(t - \theta)]$. In such a case, $\bar{K}_{A\tau i}(p) = \sum_{i=1}^{n} \frac{C_{A\tau i}}{p + \alpha_{A\tau i}}$ and Equation (6.4) has the form

$$\bar{\tau}_{\|\perp i}(p) = \frac{P_1^{(i)}(p)}{P_2^{(i)}(p)}\langle \bar{\gamma}_{\|\perp}(p)\rangle \quad (6.7)$$

for which performing the reverse Laplace transformation, i.e., determining the resolvent for the creep kernel of the $i$-th layer, does not cause difficulty ($P_1^{(i)}(p)$ and $P_2^{(i)}(p)$ are $n$-th order polynomials with respect to $p$).

For example, if creep in the polymer matrix is described by a single exponent, like $K_{A\tau}(t - \theta) = C_{A\tau} \exp[-\alpha_{A\tau}(t - \theta)]$, then the stress in any layer and consequently in any point of the composite according to the familiar law of time-dependent variation of the mean strain $\langle \gamma_{\|\perp}(t)\rangle$ is given by the relationship

$$\tau_{\|\perp i}(t) = \frac{\langle \gamma_{\|\perp}(t)\rangle}{S_i}$$

$$- \frac{b_i}{S_i^2}\int_0^t \exp\left[-\left(\alpha_{A\tau} + \frac{b_i}{S_i}\right)(t - \theta)\right]\langle \gamma_{\|\perp}(\theta)\rangle d\theta \quad (6.8)$$

where

$$b_i = C_{A\tau} S_{A66}(1 - \psi_i)$$

The law of deformation $\langle \gamma_{\|\perp}(t) \rangle$ may be determined either by test or theoretically from Equation (5.31).

The analysis of the component stress state under longitudinal shear has shown that the nonuniform distribution of stresses in the polymer matrix is such that the maximum value corresponds to the plane $\varphi = 0$ (see Fig. 5.7). This enables us, according to Equation (6.8), to write the law of the maximum shear stress variation

$$\max \tau_{\|\perp}(t) = \left\{ \frac{s_{66}(t)}{s_0} - \frac{b_0}{s_0^2} \int_0^t \right. \tag{6.9}$$

$$\left. \times \exp\left[-\left(\alpha_{A\tau} + \frac{b_0}{s_0}\right)(t-\theta)\right] s_{66}(\theta) d\theta \right\} \langle \tau_{\|\perp} \rangle$$

where

$$s_0 = \frac{1 - r_B/l}{G_A} + \frac{r_B/l}{G_{Brz}}$$

$$b_0 = C_{A\tau} \frac{1 - r_B/l}{G_A}$$

The ratio of the maximum stress value to the mean stress $\langle \tau_{\|\perp} \rangle$ is an important structural characteristic of the material. For the exponential creep kernel of the polymer matrix under shear the function of time-dependent stress concentration variation is given by

$$\bar{\tau}_{\|\perp}(t) = \frac{\max \tau_{\|\perp}(t)}{\langle \tau_{\|\perp} \rangle} = A \frac{\alpha_{A\tau}}{\alpha_{A\tau} + b_0}$$

$$- \frac{g_2}{s_0} \frac{C_{A\tau}}{\alpha_{A\tau}} \left(1 + \frac{1}{\alpha_{A\tau}}\right) \exp(-\alpha_{A\tau} t) \tag{6.10}$$

$$+ \left(\frac{Ab_0}{\alpha_{A\tau} + b_0} + \frac{g_2}{s_0} \frac{C_{A\tau}}{\alpha_{A\tau}}\right) \exp\left[-\left(\alpha_{A\tau} + \frac{b_0}{s_0}\right)t\right]$$

where

$$A = \frac{1}{s_0}\left[g_1 + g_2 \frac{C_{A\tau}}{\alpha_{A\tau}}\left(1 + \frac{1}{\alpha_{A\tau}}\right)\right]$$

If the strain $\langle \gamma_{\|\perp}(t) \rangle$ is taken as constant in time, Equation (6.6) describes the relationship for determining stress relaxation in any point of the reinforced plastic. According to the equilibrium condition between the internal and the external stresses, Equation System (5.23) leads to the determination of the stress relaxation law for the composite as a whole

$$\tau_{\|\perp i}(t) = \frac{\langle \gamma_{\|\perp} \rangle}{S_i}\left[1 - \int_0^t R_i(t-\theta)d\theta\right] \qquad (6.11)$$

$$\langle \tau_{\|\perp}(t) \rangle = \left\{\frac{r_B}{l}\int_0^{\pi/2} \frac{1}{S_i}\left[1 - \int_0^t R_i(t-\theta)d\theta\right]\cos\varphi\, d\varphi\right.$$

$$\left. + \left(1 - \frac{r_B}{l}\right)G_A\left[1 - \int_0^t R_{A\tau}(t-\theta)d\theta\right]\right\}\langle \gamma_{\|\perp} \rangle \qquad (6.12)$$

Alteration in the stress distribution with time under longitudinal shear according to Equation (6.8) for glass- and organic fiber reinforced plastics is represented in Fig. 6.2. The calculation was made for matrices having the degree of creep of $\eta_{A\tau} = \gamma_A(\infty)/\gamma_A(0) = 3$ ($G_A = 1.3$ GPa, $C_{A\tau} = 0.0872$ days$^{-1}$, $\alpha_{A\tau} = 0.0435$ days$^{-1}$) and $\eta_{A\tau} = 10$ ($G_A = 1.3$ GPa, $C_{A\tau} = 0.11765$ days$^{-1}$, $\alpha_{A\tau} = 0.01307$ days$^{-1}$), with the ratio of fiber-matrix shear moduli $G_{Brz}/G_A = 20$ (glass-fiber reinforced plastic) and $G_{Brz}/G_A = 2$ (organic fiber reinforced plastic), the fiber volume content being $\psi = 0.6$.

It should be observed that the stress distribution across the planes parallel to the loading plane is not uniform. The shear stresses reach their maximum values in the planes having the highest fiber volume content. The stress concentration factor under longitudinal shear, i.e., the ratio of the maximum to the mean stress values, depends both upon the component volume ratio and upon their shear moduli ratio. Fig. 6.3 illustrates the dependence of stress concentration factor $\bar{\tau}$ with $t = 0$ upon the fiber-matrix shear moduli ratio. The stress concentration factor increases with an increase in the shear moduli ratio $G_{Brz}/G_A$ and the fiber volume content $\psi$. The highest stress concentration is observed in boron-epoxy ($G_{Brz}/G_A \approx 100$) with high fiber volume contents

(a)

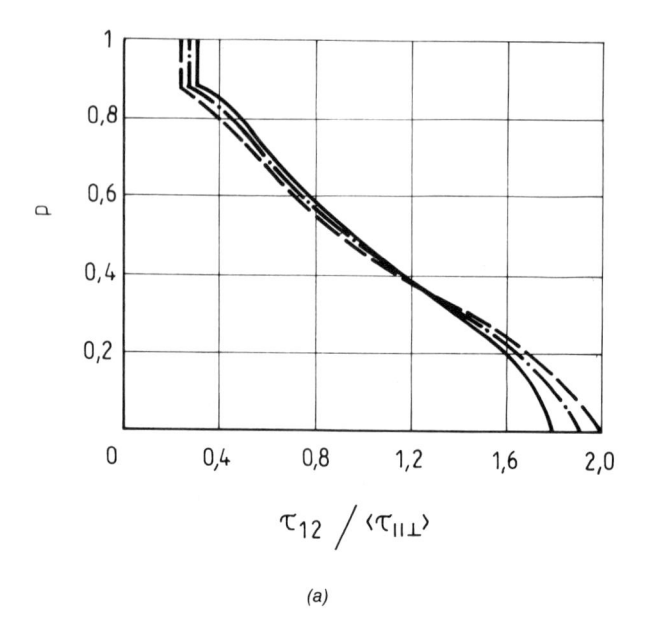

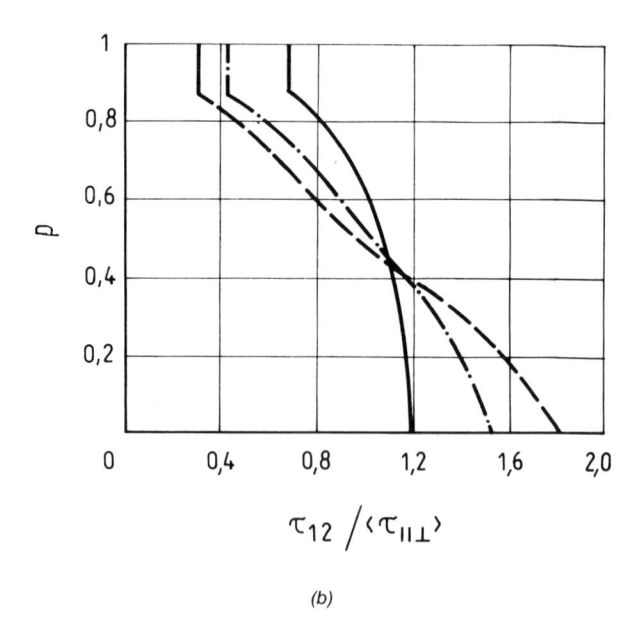

(b)

**FIGURE 6.2.** Dependence of stress distribution in unidirectionally glass- (a) and organic (b) fiber reinforced epoxy under longitudinal shear upon the loading duration (fiber volume content equal to 60 percent), $t = 0$ (———), $t \to \infty$, $\eta_A = 3$ (— · — · —), $t \to \infty$, $\eta_A = 10$ (– – –).

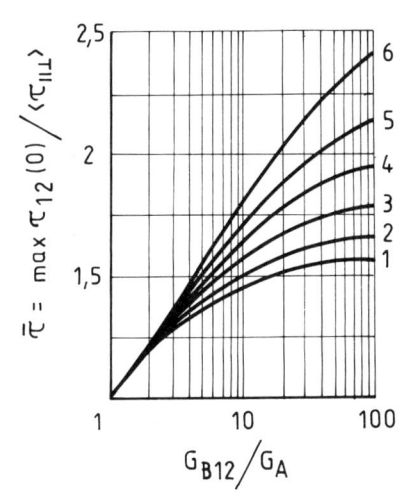

FIGURE 6.3. Dependence of stress concentration factor under longitudinal shear upon the fiber-matrix shear moduli ratio and the fiber volume content: $\psi$ = 0.3 (1), 0.4 (2), 0.5 (3), 0.6 (4), 0.65 (5), 0.7 (6).

whereas the lowest stress concentration occurs in organic fiber reinforced plastics ($G_{Brz}/G_A = 2$).

The analysis of Equation (6.6) made it evident that the nonuniform character of the stress distribution (see Fig. 6.2), specifically in the unidirectionally reinforced layer under longitudinal shear, increases with time. Fig. 6.4 shows how the stress concentration factor for glass- and organic fiber reinforced plastics changes with time. The more pronounced variation is observed with lower fiber-matrix shear moduli ratios, i.e., for organic fiber reinforced plastics.

Stress redistribution in the unidirectionally reinforced ply increases with the increase in the degree of creep of the polymer matrix. The dependence of the stress concentration factor's variation with time upon the degree of creep of the polymer matrix for glass- and organic fiber reinforced plastics is shown in Fig. 6.5. The increase in the stress concentration factor with time is more pronounced with lower fiber-matrix shear moduli ratio (in organic fiber reinforced plastics) and with high fiber volume contents.

## 6.3 STRESS STATE OF UNIDIRECTIONALLY REINFORCED PLASTICS COMPONENTS UNDER TRANSVERSAL LOADING

As has already been stated, the determination of the stress-strain state in unidirectionally reinforced plastics components is in the general

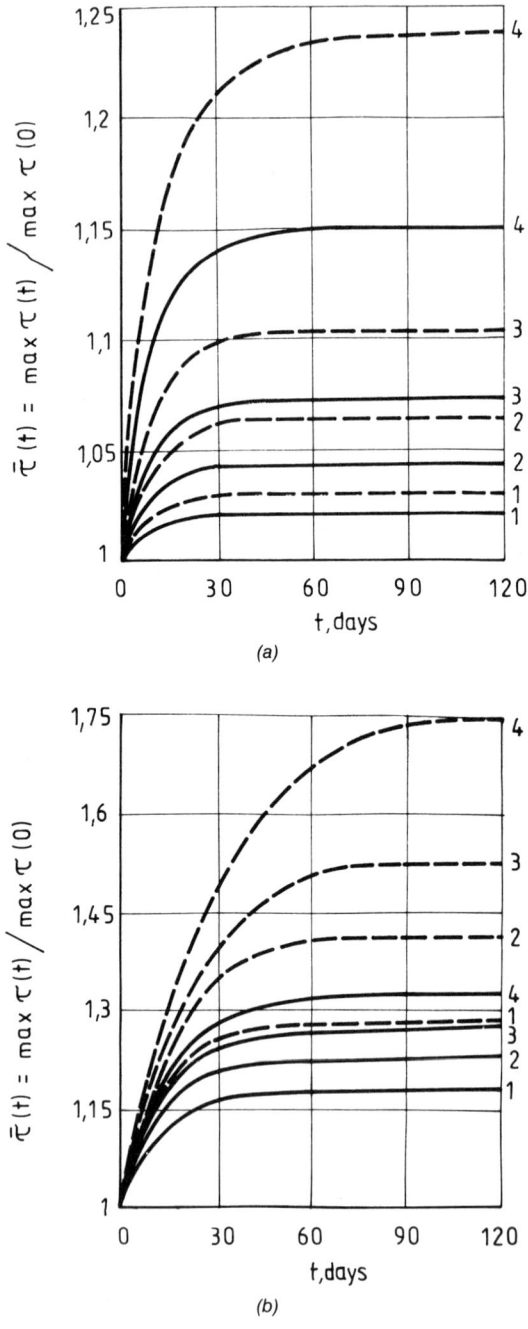

**FIGURE 6.4.** Dependence of stress concentration factor in glass- (1) and organic fiber reinforced epoxy (b) under longitudinal shear upon the loading time: $\eta_A = 3$ (———), 10 (- - -) $\psi = 0.3$ (1), 0.5 (2), 0.7 (3).

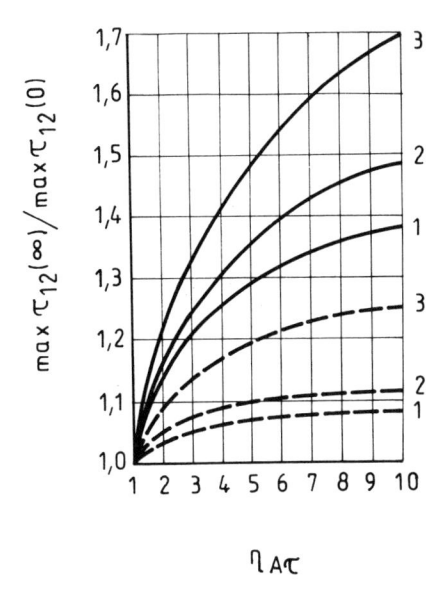

FIGURE 6.5. Dependence of maximum stress concentrations in plastics with anisotropic (———) and isotropic (– – –) fiber reinforcement under long-term longitudinal shear upon the degree of creep of the polymer matrix: $\psi = 0.5$ (1), 0.6 (2), 97 (3).

case reduced to the solution of the three-dimensional boundary problem of elasticity theory. In an effort to obtain simpler and more convenient solutions, the approximation methods were used in the calculation. One of the approximation methods permitting sufficiently accurate determination of the basic laws of stress distribution and redistribution in the components under long-term transverse loading is the method used in Section 5.2—the hypothetic section of the calculation member into thin layers that are parallel to the reinforcement plane and to the transverse load direction.

The proposed calculation model should be considered approximated as it can only take normal stresses appearing in the plastic components without considering shear stresses. Another assumption of the calculation model is the one about the constancy of stresses in the layer. Nevertheless, the proposed model makes it possible to make a qualitative and correct evaluation of the stress field in the reinforced plastic. As to the quantitative truth validity of the results, the maximum stresses calculated according to the model proposed do not practically differ from the stresses calculated by the finite element method, as will be shown further.

The following initial preconditions were used before compiling the system of equations for determining the stress-strain state of the unidirectionally reinforced plastic components:

(1) All the elementary layers of the repeating element deform simultaneously, i.e., strains are equal in the loading direction

$$(1 - \psi_i)\epsilon_{A2}^{(i)}(t) + \psi_i\epsilon_{B2}^{(i)}(t) = \langle \epsilon_\perp(t) \rangle \qquad (6.13)$$

where $\epsilon_A^{(i)}(t)$, $\epsilon_B^{(i)}(t)$ are strains of the polymer matrix and the fibers in the $i$-th layer respectively.

(2) Strains of the fibers and the polymer matrix in the reinforcement direction are equal at any time

$$\epsilon_{A1}^{(i)}(t) = \epsilon_{B1}^{(i)}(t) = \epsilon_{A1}(t) = \nu_{\perp\parallel}\langle \epsilon_\perp(t) \rangle \qquad (6.14)$$

and in a number of cases, such as the case of high-modulus anisotropic fibers, it is useful to make a condition that

$$\nu_{\perp\parallel} = 0$$

(3) The section planes of the repeating element in the process of deformation continue to be planes, i.e.,

$$\epsilon_{A3}^{(i)}(t) = \epsilon_{B3}^{(i)}(t) \qquad (6.15)$$

(4) The polymer matrix and the fibers deform jointly and as a result there is an equation of equilibrium in any elementary layer

$$(1 - \psi_i)\sigma_{A3}^{(i)}(t) + \psi_i\sigma_{A3}^{(i)}(t) = 0 \qquad (6.16)$$

(5) Stresses in the loading direction are equal for any elementary layer in the polymer matrix and for the fibers, but they vary from layer to layer and on the average are equal to the externally applied stress

$$\sigma_{A2}^{(i)}(t) = \sigma_{B2}^{(i)}(t) = \sigma_2^{(i)}(t) \qquad (6.17)$$

$$\frac{r_B}{l}\int_0^{\pi/2} \sigma_2^{(i)}(t)\cos\varphi\, d\varphi + \left(1 - \frac{r_B}{l}\right)\sigma_{A2}(t) = \langle \sigma_\perp(t) \rangle \qquad (6.18)$$

where $\sigma_{A2}(t)$ is the stress in the unreinforced part of the calculation element.

(6) The polymer matrix is a linear isotropic viscoelastic material but the fibers are transversely isotropic

$$\epsilon_{Aj}^{(i)}(t) = S_{Ajm}\left[\sigma_{Am}^{(i)}(t) + \int_0^t K_A(t-\theta)\sigma_{Am}^{(i)}(\theta)d\theta\right] \quad (6.19)$$

$$\epsilon_{Bj}^{(i)}(t) = S_{Bjm}\sigma_{Bm}^{(i)}(t)$$

$$j, m = 1, 2, 3 \quad (6.20)$$

where $S_{Ajm}$, $S_{Bjm}$ are components of compliance matrices of the polymer matrix and of the fibers; $\epsilon_{Aj}^{(i)}(t)$, $\epsilon_B^{(i)}(t)$, $\sigma_{Am}^{(i)}(t)$, $\sigma_{Bm}^{(i)}(t)$ are strains and stresses in the polymer matrix and in the fibers in the $i$-th layer.

Application of the Laplace transformation to the system of equations obtained reduces to the solution of the elastic problem, and is the image of the respective unknown functions. The stress image $\bar{\sigma}^{(i)}$ and the strain image $\langle\bar{\epsilon}_\perp\rangle$ of the unidirectionally reinforced plastic in the direction of the externally applied load are linked in the following way:

$$\bar{\sigma}_{A1}^{(i)} = -\frac{1}{1-\psi_i}\left\{\frac{S_{A12}}{S_{A11}}\left(\psi_i\frac{S_{B11}}{S_{B12}}\bar{\sigma}_{B1}^{(i)} + \bar{\sigma}_2^{(i)}\right)\right.$$

$$\left. + \frac{1}{S_{A11}}\left[\frac{1}{1+\bar{K}_A} + \psi_i\left(\frac{S_{A12}}{S_{B12}} - \frac{1}{1+\bar{K}_A}\right)\right]\nu_{\perp\|}\langle\bar{\epsilon}_\perp\rangle\right\}$$

$$\bar{\sigma}_{B1}^{(i)} = \frac{1}{A_4 + B_4 + \bar{K}_A}$$

$$\times [(A_6 + B_6\bar{K}_A)\nu_{\perp\|}\langle\bar{\epsilon}_\perp\rangle - (A_5 + B_5\bar{K}_A)\bar{\sigma}_2^{(i)}] \quad (6.21)$$

$$\bar{\sigma}_2^{(i)} = \frac{F_1\bar{K}_A^2 + F_2\bar{K}_A + F_3}{D_1\bar{K}_A + D_2\bar{K}_A + D_3}\langle\bar{\epsilon}_\perp\rangle$$

$$\bar{\sigma}_{A3}^{(i)} = -\frac{1}{S_{A12}}\left(\frac{\nu_{\perp\|}}{1+\bar{K}_A}\langle\bar{\epsilon}_\perp\rangle + S_{A11}\bar{\sigma}_{A1}^{(i)}\right) - \bar{\sigma}_2^{(i)}$$

$$\bar{\sigma}_{B3}^{(i)} = -\frac{1}{S_{B12}}(\nu_{\perp\|}\langle\bar{\epsilon}_\perp\rangle + S_{B11}\bar{\sigma}_{B1}^{(i)}) - \bar{\sigma}_2^{(i)}$$

## STRUCTURAL MECHANICS OF A MATERIAL

where

$$F_1 = (B_3 B_4 - B_1 B_6) v_{\perp\|}$$

$$F_2 = B_4 + (A_3 B_4 + A_4 B_3 - A_1 B_6 - A_6 B_1) v_{\perp\|}$$

$$F_3 = A_4 + (A_3 A_4 - A_1 A_6) v_{\perp\|}$$

$$D_1 = B_2 B_4 - B_1 B_5$$

$$D_2 = A_2 B_4 + A_4 B_2 - A_1 B_5 - A_5 B_1$$

$$D_3 = A_2 A_4 - A_1 A_5$$

$$A_1 = \psi_i \left( S_{B12} - \frac{S_{B11} S_{B23}}{S_{B12}} \right) + B_1$$

$$A_2 = \psi_i (S_{B22} - S_{B23}) + B_2$$

$$A_3 = \frac{S_{A12}}{S_{A11}} + \psi_i \left( \frac{S_{B23}}{S_{B12}} - \frac{S_{A12}}{S_{A11}} \right) + B_3$$

$$A_4 = \frac{S_{A11}}{S_{A12}} \left( \frac{S_{B11} S_{B12}}{S_{B12}} - S_{B12} \right)(1 - \psi_i) + B_4$$

$$A_5 = \frac{S_{A11}}{S_{A12}} (S_{B22} - S_{B23})(1 - \psi_i) + B_5$$

$$A_6 = \left( 1 - \frac{S_{A11} S_{B22}}{S_{A12} S_{B12}} \right)(1 - \psi_i) + B_6$$

$$B_1 = \psi_i \frac{S_{A12} S_{B11}}{S_{A11} S_{B12}} (S_{A12} - S_{A11})$$

$$B_2 = S_{A11} - \frac{S_{A12}^2}{S_{A11}} + \psi_i (S_{A12} - S_{A11})$$

$$B_3 = \psi_i \frac{S_{A12}}{S_{B12}} \left( \frac{S_{A12}}{S_{A11}} - 1 \right)$$

$$B_4 = -s_{B11} B_6$$

$$B_5 = s_{A11} - s_{A12} + \psi_i \left( \frac{s_{A11}^2}{s_{A12}} - s_{A11} \right)$$

$$B_6 = \psi_i \left( \frac{s_{A12}}{s_{B12}} - \frac{s_{A11}^2}{s_{A12} s_{B12}} \right)$$

$\bar{K}_A$, $\bar{\sigma}_{Aj}^{(i)}$, $\langle \bar{\epsilon}_\perp \rangle$ are Laplace transformations of the respective functions.

By applying the reverse Laplace transformation to Equation System (6.21), we can find the relation between the mean strain in the loading direction and stress components in the polymer matrix and the fibers.

Thus, with the deformation law of the reinforced plastics in the loading direction designated, the stresses in any material point can be determined. The deformation law may be established by test or by calculation.

If $\bar{K}_A(t - \theta) = C_A \exp[-\alpha_A(t - \theta)]$, then $\bar{K}_A = \dfrac{C_A}{p + \alpha_A}$ and Equation System (6.21) yields

$$\sigma_2^{(i)}(t) = \frac{F_3}{D_3} \Big\{ \langle \epsilon_\perp(t) \rangle$$

$$+ \int_0^t \sum_{n=1}^{2} \frac{P_1(p_n)}{P_2(p_n)} \exp[p_n(t - \theta)] \langle \epsilon_\perp(\theta) \rangle d\theta \Big\} \quad (6.22)$$

where $P_1$ and $P_2$ are the roots of the equation $p^2 + d_1 p + d_2 = 0$

$$P_1(p) = C_A \left( \frac{F_2}{F_3} - \frac{D_2}{D_3} \right)(p + \alpha_A) + C_A^2 \left( \frac{F_1}{F_3} - \frac{D_1}{D_3} \right)$$

$$P_2(p) = 2p + C_A \frac{D_2}{D_3} + 2\alpha_A$$

$$d_1 = C_A \frac{G_2}{G_3} + 2\alpha_A$$

$$d_2 = \frac{C_A}{D_3}(C_A D_1 + \alpha_A D_2) + \alpha_A^2$$

## 130  STRUCTURAL MECHANICS OF A MATERIAL

By analogy we find the relationships for other stress components

$$\sigma_{B1}^{(i)}(t) = -\frac{A_5}{A_4}\left\{\sigma_2^{(i)}(t) + C_A\left(\frac{B_5}{A_5} - \frac{B_4}{A_4}\right)\right.$$

$$\left. \times \int_0^t \exp\left[-\left(\alpha_A + C_A\frac{B_4}{A_4}\right)(t-\theta)\right]\sigma_2^{(i)}(\theta)d\theta\right\}$$

$$+ \frac{A_6}{A_4}\left\{\nu_{\perp\|}\langle\epsilon_\perp(t)\rangle + C_A\nu_{\perp\|}\left(\frac{B_6}{B_4} - \frac{B_4}{A_4}\right)\right.$$

$$\left. \times \int_0^t \exp\left[-\left(\alpha_A + C_A\frac{B_4}{A_4}\right)(t-\theta)\right]\langle\epsilon_\perp(\theta)\rangle d\theta\right\}$$

$$\sigma_{A1}^{(i)}(t) = -\frac{\psi_i}{1-\psi_i}\frac{S_{A12}S_{B11}}{S_{A11}S_{B12}}\sigma_{B1}^{(i)}(t) - \frac{1}{1-\psi_i}\frac{S_{A12}}{S_{A11}}\sigma_2^{(i)}(t)$$

$$-\frac{\psi_i}{1-\psi_i}\frac{S_{A12}}{S_{A11}S_{B12}}\nu_{\perp\|}\langle\epsilon_\perp(t)\rangle - \frac{\nu_{\perp\|}}{S_{A11}}\left\{\langle\epsilon_\perp(t)\rangle\right.$$

$$\left. - C_A \int_0^t \exp\left[-(\alpha_A + C_A)(t-\theta)\right]\langle\epsilon_\perp(\theta)\rangle d\theta\right\}$$

(6.23)

$$\sigma_{A3}^{(i)}(t) = -\frac{\nu_{\perp\|}}{S_{A12}}\left\{\langle\epsilon_\perp(t)\rangle - C_A \int_0^t \exp\left[-(\alpha_A + C_A)(t-\theta)\right]\right.$$

$$\left. \times \langle\epsilon_\perp(\theta)\rangle d\theta\right\} - \frac{S_{A11}}{S_{A12}}\sigma_{A1}^{(i)}(t) - \sigma_2^{(i)}(t)$$

$$\sigma_{B3}^{(i)}(t) = \frac{\nu_{\perp\|}}{S_{B12}}\langle\epsilon_\perp(t)\rangle - \frac{S_{B11}}{S_{B12}}\sigma_{B1}^{(i)}(t) - \sigma_2^{(i)}(t)$$

The above relationships make it possible to determine all the stress components at any time instant in an arbitrary point of the reinforced plastic under transverse loading.

The stress state analysis of the reinforced plastic components, the stress state being given by Equation Systems (6.22) and (6.23), has shown that the maximum stresses are reached in the layers satisfying the condition $\varphi = 0$ (see Fig. 5.10). Therefore, a significant characteristic of the stress state of a unidirectionally reinforced plastic under long-term transverse loading is the function of the principal stress variations in their maximum concentration points. Equation Systems (6.22) and (6.23) are representative of such functions, provided $\varphi = 0$.

The stress state and the character of its variation in the maximum stress concentration points under constant mean stress $\langle \sigma_\perp \rangle$ are given by

$$\sigma_{Ar}(t) = \bar{\sigma}_r(t)\langle \sigma_\perp \rangle$$

$$\sigma_{A\theta}(t) = \bar{\sigma}_\theta(t)\langle \sigma_\perp \rangle \qquad (6.24)$$

$$\sigma_{Az}(t) = \bar{\sigma}_z(t)\langle \sigma_\perp \rangle$$

where

$$\bar{\sigma}_r(t) = \left.\frac{\sigma_{A2}(t)}{\langle \sigma_\perp \rangle}\right|_{\varphi=0}$$

$$\bar{\sigma}_\theta(t) = \left.\frac{\sigma_{A3}(t)}{\langle \sigma_\perp \rangle}\right|_{\varphi=0}$$

$$\bar{\sigma}_z(t) = \left.\frac{\sigma_{A1}(t)}{\langle \sigma_\perp \rangle}\right|_{\varphi=0}$$

For determining the solutions to the Equations in (6.24), Equation (5.34) may be used as the creep law of a unidirectionally reinforced plastic under transversal loading. The analysis of Equation System (6.24) has shown that the time-dependent stress variations in the polymer matrix are only of quantitative nature and depend both upon the strain characteristics of the components and the fiber volume content. Fig. 6.6 illustrates the dependence of the maximum stress concentration factor $\bar{\sigma}_{Ar}$ upon the polymer matrix creep.

It follows from Equation Systems (6.22) and (6.23) that under transversal loading of the reinforced plastic there is stress redistribution throughout the whole material. The changes in the ratio of the stresses to the mean stress $\langle \sigma_\perp \rangle$ are shown in Fig. 6.7.

The results discussed pertain to the appearance of a constant stress state in a unidirectionally reinforced plastic. If this stress state varies with time in a unidirectionally reinforced plastic layer ($\langle \sigma_\perp(t) \rangle$,

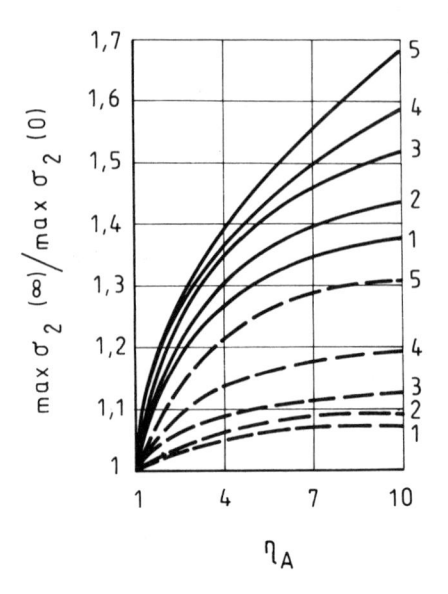

**FIGURE 6.6.** The dependence of maximum stress concentration in reinforced plastics containing anisotropic (——) and isotropic (- - -) fibers under long-term transversal loading upon the degree of creep of the polymer matrix: $E_{Br}/E_A = 2$ (——), $E_{Br}/E_A = 20$ (- - -), $\Psi = 0.4$ (1); 0.5 (2); 0.6 (3); 0.65 (4); 0.7 (5).

$\langle \tau_{\|\perp}(t) \rangle \neq$ const) it will be convenient to apply a phenomenological form of representation for the relationship $\sigma_{Ai}(t)$

$$\sigma_{Ai}(t) = J_i^* \{ \langle \sigma_\perp(t) \rangle, \langle \tau_{\|\perp}(t) \rangle \} \tag{6.25}$$

$$i = r, \theta, z, rz$$

where $J_i^*$ are the functionals governed by viscoelastic properties of the ply components and their volumetric ratio. The form of the functionals $J_i^*$ will be defined as Stieltjes integral series of growing multiplicity

$$\sigma_{Ai}(t) = \int_0^t q_{Aij}(t - \theta) d\langle \sigma_j \theta_1 \rangle$$
$$+ \int_0^t \int_0^t q_{Aijk}(t - \theta_1, t - \theta_2) d$$
$$\times \langle \sigma_j(\theta_1) \rangle d\langle \sigma_k(\theta_2) \rangle + \ldots \tag{6.26}$$

$$i = r, \theta, z, rz; \qquad j, k = \perp, \|\perp$$

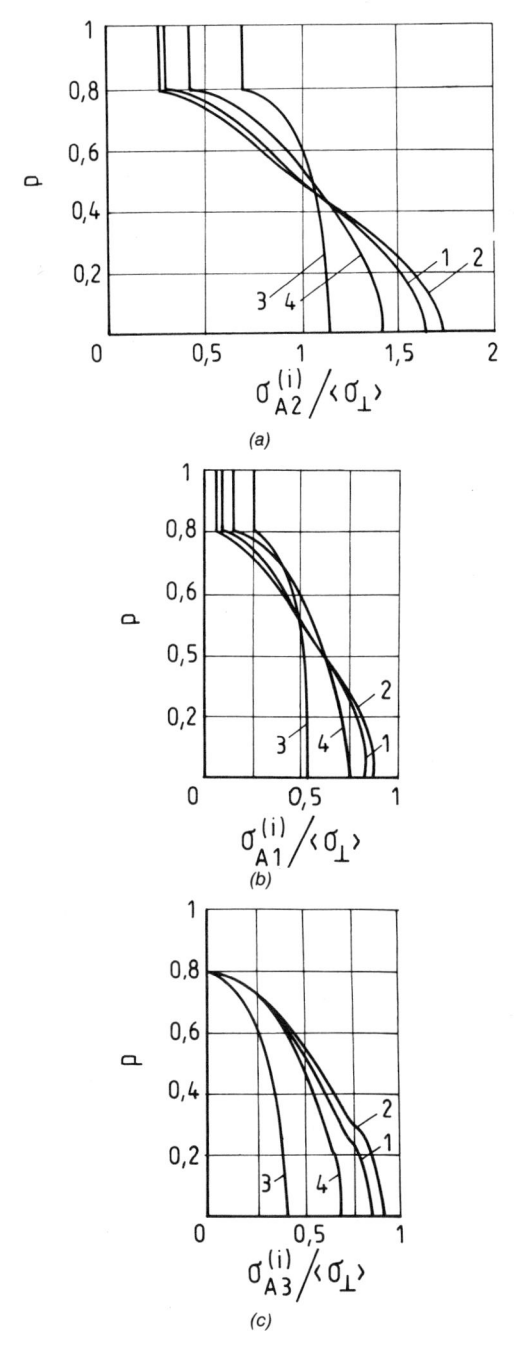

**FIGURE 6.7.** Time-dependent variations in the distribution of stresses $\sigma_{A2}^{(i)}$ (a), $\sigma_{A1}^{(i)}$ (b), $\sigma_{A3}^{(i)}$ (c) in the polymer matrix for glass- (1, 2) and carbon (3, 4) fiber reinforced plastics under transverse loading: $t = 0$ (1, 3), $t \to \infty$ (2, 4).

Using only the first term of Equation (6.26) and taking into account as stated in Skudra & Buluas [57] that under transversal loading $\tau_{Arz} = 0$ but under longitudinal shear the stress components $\sigma_{Ar}$ and $\sigma_{A\theta}$ may be neglected in the point of the highest load, we obtain

$$\sigma_{Ar}(t) = \int_0^t q_{r\perp}(t-\theta)d\langle\sigma_\perp(\theta)\rangle$$

$$\sigma_{A\theta}(t) = \int_0^t q_{\theta\perp}(t-\theta)d\langle\sigma_\perp(\theta)\rangle$$

$$\sigma_{Az}(t) = \int_0^t q_{z\perp}(t-\theta)d\langle\sigma_\perp(\theta)\rangle \tag{6.27}$$

$$\tau_{Arz}(t) = \int_0^t q_{A\|\perp}(t-\theta)d\langle\tau_{\|\perp}(\theta)\rangle$$

Integrating Equation System (6.27) we obtain the dependence of $\sigma_{Ai}(t)$ upon the loading time, expressed as

$$\sigma_{Ar}(t) = \bar{\sigma}_{Ar}(0)\langle\sigma_\perp(t)\rangle + \int_0^t K_{A\|}(t-\theta)\langle\sigma_\perp(\theta)\rangle d\theta$$

$$\sigma_{A\theta}(t) = \bar{\sigma}_{A\theta}(0)\langle\sigma_\perp(t)\rangle + \int_0^t K_{A\perp}(t-\theta)\langle\sigma_\perp(\theta)\rangle d\theta$$

$$\sigma_{Az}(t) = \bar{\sigma}_{Az}(0)\langle\sigma_\perp(t)\rangle + \int_0^t K_{A\|}(t-\theta)\langle\sigma_\perp(\theta)\rangle d\theta \tag{6.28}$$

$$\tau_{Arz}(t) = \bar{\tau}_{Arz}(0)\langle\tau_{\|\perp}(t)\rangle + \int_0^t K_{A\|\perp}(t-\theta)\langle\tau_{\|\perp}(\theta)\rangle d\theta$$

where

$$K_{Ai}(t) = \frac{dq_i(t)}{dt}$$

The functions $K_{A\|}$, $K_{A\perp}$, $K_\|$ and $K_{A\|\perp}$ are determined from the relationships of the layer stress state under constant loading ($\langle\sigma_\perp(t)\rangle$ = const, $\langle\tau_{\|\perp}(t)\rangle$ = const). According to (6.28) we write

$$K_{Ai}(t) = \frac{d\bar{\sigma}_{Ai}(t)}{dt}$$

Instantaneous structural parameters under variable loading $\sigma^*_{Ai}(t)$ are given by

$$\bar{\sigma}^*_{Ai}(t) = \sigma_{Ai}(t)/\langle \sigma_j(0) \rangle$$

$$j = \perp, \parallel, \perp$$

Consider stress relaxation in reinforced plastics components under transversal loading by constant mean strain when $\langle \epsilon_\perp \rangle$ = const. The Laplace transformation $\langle \bar{\epsilon}_\perp \rangle = \frac{1}{p} \langle \epsilon_\perp \rangle$ and the analytical relationships of the stress components with a prescribed creep kernel of the polymer matrix are determined from Equation System (6.21) by the reverse Laplace transformation. Thus, for the case when the creep kernel of the polymer matrix is an exponential function, Equation System (6.21) enables us to write for stresses in the loading direction

$$\sigma_2(t) = \frac{F_3}{D_3} \left[ 1 + \frac{C_A \left( \frac{F_1}{F_3} - \frac{D_1}{D_3} \right) + \alpha_A \left( \frac{F_2}{F_3} - \frac{D_2}{D_3} \right)}{C_A D_1 + \alpha_A D_2 + \alpha_A^2 \frac{D_3}{C_A}} D_3 \right.$$

$$\left. + \sum_{n=1}^{2} \frac{P_1(p_n)}{p_n P_2(p_n)} \exp(p_n t) \right] \langle \epsilon_\perp \rangle \quad (6.29)$$

The symbols are those used in Equation (6.22).

By averaging the stresses acting in the loading direction of the reinforced plastic we arrive at the expression

$$\langle \sigma_\perp(t) \rangle = \frac{r_B}{l} \int_0^{\pi/2} \sigma_2^{(i)}(t) \cos \varphi \, d\varphi + \left( 1 - \frac{r_B}{l} \right) \frac{S_{A11} - \nu_{\parallel\perp} S_{A12}}{S_{A11}^2 - S_{A12}^2}$$

$$\times (\alpha_A + C_A \exp[-(\alpha_A + C_A)t]) \frac{\langle \epsilon_\perp \rangle}{\alpha_A + C_A} \quad (6.30)$$

Equation (6.30) establishes the law of mean external stress relaxation for a unidirectionally reinforced ply under transversal loading. It should

## 6.4 STRESS STATE OF REINFORCED PLASTIC LAMINATES UNDER LONG-TERM LOADING

Section 5.3 describes general methods for determining the stress-strain state of reinforced plastic laminates under long-term loading. According to this methodology, strains in a laminated system are closely linked with the stress distribution law in individual plies.

The governing equations in the method are Equations (5.46) through (5.48). After applying the Laplace transformation, Equations (5.46) through (5.48) are transformed into a system of easily solved algebraic equations. The problem is reduced to a linear-elastic problem with respect to strain and stress modes of Laplace. The principal difficulty lies in performing the reverse Laplace transformations for Equations (5.51) through (5.53). The problem is only insignificantly simplified by the assumption that ply strains in the reinforcement direction are of elastic character and consequently $K_{\|}(t - \theta) = K_{\|\bot}(t - \theta) = 0$. It is possible to obtain explicit solutions only for the simplest reinforcement designs and exponential creep kernels, therefore a numerical implementation of the problem solution should be considered for determining stress state in the individual plies of a laminate under long-term loading.

For every ply of the laminate, the equation

$$[\langle \epsilon_i \rangle_k] = [s_{ij}^{(k)}(t)][\langle \sigma_j(t) \rangle_k]$$

$$i = 1, 2, 6$$

(6.31)

is true where the compliance matrix $[s_{ij}^{(k)}(t)]$ is continuously varying according to the law

$$[s_{ij}^{(k)}(t)] = \left[ s_{ij}^{(k)}(0)\left[ 1 + \frac{1}{\langle \sigma_j(t) \rangle_k} \times \int_0^t \langle \sigma_j(\theta) \rangle_k K_{ij}^{(k)}(t - \theta) d\theta \right] \right]$$

(6.32)

Here $s_{ij}^{(k)}(0)$ is the short-term value of the $k$-th compliance matrix components. The matrix $[s_{ij}^{(k)}(t)]$ will be called pseudo-elastic in difference from the quasi-elastic matrix where the components are determined regardless of time-dependent stress variations

$$s_{ij}^{(k)}(t) = s_{ij}^{(k)}(0)\left[1 + \int_0^t K_{ij}^{(k)}(t-\theta)d\theta\right] \quad (6.33)$$

Let us use the step-by-step calculation method. Its essence lies in splitting the range of loading time into small intervals $\Delta t_n$ ($1 \leq m \leq N$). Within these intervals ($t_{m-1} \leq t \leq t_m$) stress variations may be neglected and stresses may be regarded as constant. Then Equation (6.32) enables us to write

$$\begin{aligned}s_{ij}^{(k)}(t) \approx s_{ij}^{(k)}(0)&\bigg\{1 + \frac{1}{\langle\sigma_j(t)\rangle_k} \\ &\times\left[\sum_{m=1}^N \langle\sigma_j(t_m)\rangle_k \int_{t_{m-1}}^{t_m} K_{ij}^{(k)}(t-\theta)d\theta\right]\bigg\}\end{aligned} \quad (6.34)$$

It should be noted that with $N \to \infty$ we obtain an accurate solution, i.e., by varying the splitting frequency of $N$ or the interval $\Delta t_m$ it is possible to achieve any degree of calculation accuracy. At every computational stage ($1 \leq m \leq N$) the pseudo-elastic matrix $[s_{ij}^{(k)}(t_m)]$ is calculated, and then the elastic problem is solved where the matrix $[s_{ij}^{(k)}(t_m)]$ is the instantaneous elastic compliance matrix of the $k$-th ply. Finally we determine the values of $\langle\sigma_j(t_m)\rangle_k$, $j = \|, \perp, \|\perp$, $k = 1, \ldots, n$. In computing the new values of $[s_{ij}^{(k)}] = [s_{ij}^{(k)}(t_{m+1})]$ we iterate the process until $m = N$. The result is a numerical dependence of stress state in the plies and of its strain upon the loading time. The interval $\Delta t_m$ should not be made equal throughout the whole range of $1 \leq m \leq N$. In the initial stage of loading the variation in the stress state of the plies is most pronounced, and it tends to become asymptotic with time. In order to save computation time and without any sacrifice in the accuracy of the result, it is recommended that the interval $\Delta t_m$ be increased with loading time.

There is significant saving in computer memory capacity and the operations required if the creep kernels of the plies are selected as

$$K_{ij}^{(k)}(t-\theta) = f_{ij}^{(k)}(t)w_{ij}^{(k)}(\theta) \quad (6.35)$$

For the above case, with $t = t_m$, Equation (6.35) has the form

$$s_{ij}^{(k)}(t_m) = s_{ij}^{(k)}(0)\left[1 + \frac{f_{ij}^{(k)}(t_m)}{\langle \sigma_j(t_{m-1})\rangle_k} \Phi_{im}^{(k)}\right] \quad (6.36)$$

where

$$\Phi_{im}^{(k)} = \sum_{l=1}^{m} \langle \sigma_j(t_{l-1})\rangle_k \int_{t_{l-1}}^{t_l} w_{ij}^{(k)}(\theta) d\theta \quad (6.37)$$

Then

$$s_{ij}^{(k)}(t_{m+1}) = s_{ij}^{(k)}(0)\left[1 + \frac{f_{ij}^{(k)}(t_{m+1})}{\langle \sigma_j(t_m)\rangle_k} \right.$$
$$\left. \times \left(\Phi_{im}^{(k)} + \langle \sigma_j(t_m)\rangle_k \int_{t_m}^{t_{m+1}} w_{ij}^{(k)}(\theta) d\theta\right)\right] \quad (6.38)$$

According to Equation (6.38) we have

$$\Phi_{im}^{(k)} = \left[\frac{s_{ij}^{(k)}(t_m)}{s_{ij}^{(k)}(0)} - 1\right] \frac{\langle \sigma_j(t_{m-1})\rangle_k}{f_{ij}^{(k)}(t_m)} \quad (6.39)$$

Inserting Equation (6.39) into Equation (6.38) we obtain a recurrent formula for computing $s_{ij}^{(k)}(t_{m+1})$:

$$s_{ij}^{(k)}(t_{m+1}) = s_{ij}(0)\left\{1 + \frac{f_{ij}^{(k)}(t_{m+1})\langle \sigma_j(t_{m-1})\rangle_k}{f_{ij}^{(k)}(t_m)\langle \sigma_j(t_m)\rangle_k}\right.$$
$$\left. \times \left[\frac{s_{ij}^{(k)}(t_m)}{s_{ij}^{(k)}(0)} - 1\right] + f_{ij}^{(k)}(t_{m+1}) \int_{t_m}^{t_{m+1}} w_{ij}^{(k)}(\theta) d\theta\right\} \quad (6.40)$$

So, for example, if the creep kernels are represented as exponential functions,

$$f_{ij}^{(k)}(t) = C_{ij}^{(k)} \exp(-\alpha_{ij}^{(k)} t)$$
$$w_{ij}^{(k)}(\theta) = \exp(-\alpha_{ij}^{(k)} \theta) \quad (6.41)$$

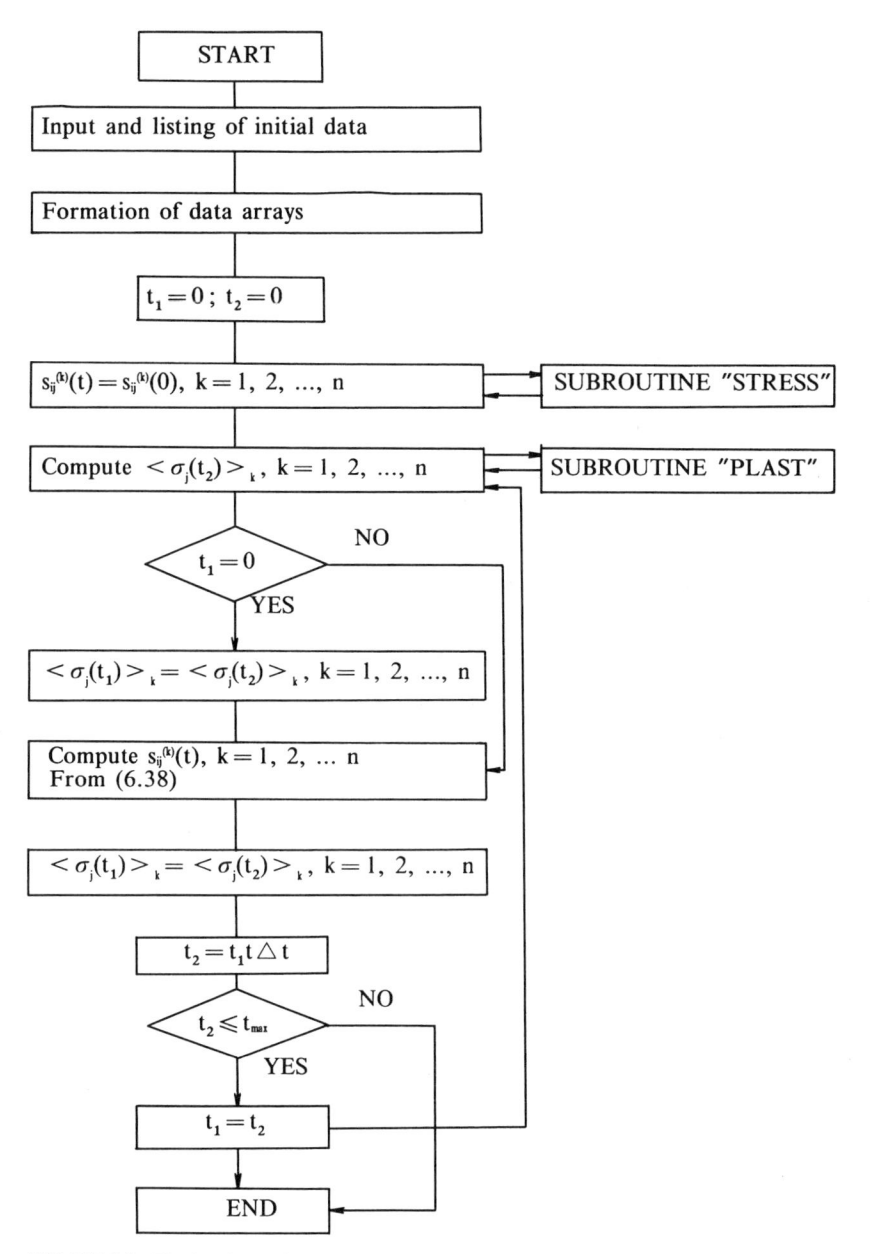

**FIGURE 6.8.** Block-scheme for computing stress-strain state of reinforced plastic laminates under long-term loading.

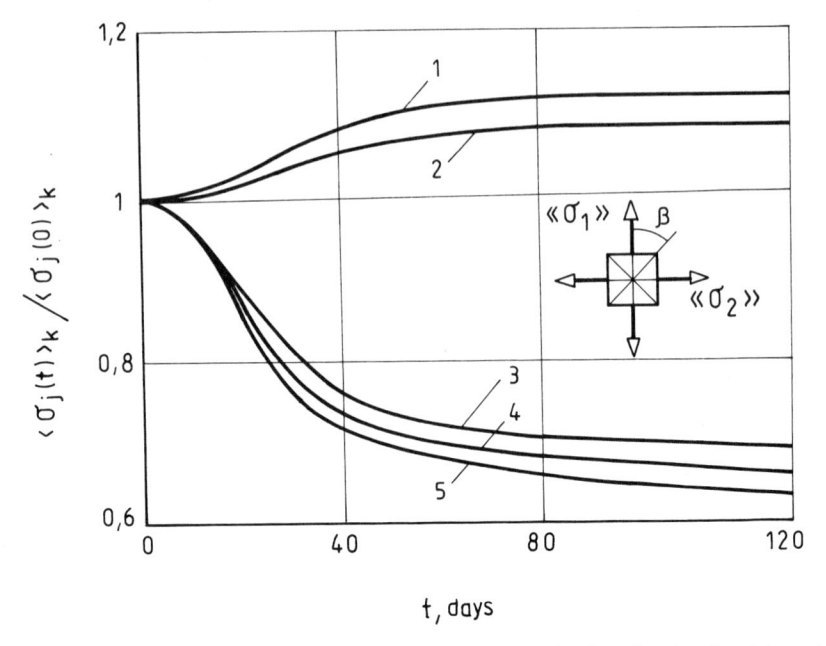

**FIGURE 6.9.** Time-dependent stress variations in the plies of a four-directionally reinforced glass-epoxy AG-4S under biaxial loading.

then according to Equation (6.40) we obtain

$$s_{ij}^{(k)}(t_{m+1}) = s_{ij}^{(k)}(0)\left\{1 + \frac{C_{ij}^{(k)}}{\alpha_{ij}^{(k)}}[1 - \exp[-\alpha_{ij}^{(k)}(t_{m+1} - t_m)]]\right.$$

$$+ \exp[-\alpha_{ij}^{(k)}(t_{m+1} - t_m)] \quad (6.42)$$

$$\left.\times \frac{\langle \sigma_i(t_{m-1})\rangle_k}{\langle \sigma_i(t_m)\rangle_k}\left(\frac{s_{ij}^{(k)}(t_m)}{s_{ij}^{(k)}(0)} - 1\right)\right\}$$

The recurrent Equation (6.40) essentially reduces the amount of computation work and, what is most important, saves the need to use large computer memory capacity because all the intermediate results need not be stored, the initial and the previous results being sufficient.

The given algorithm for the numerical calculation of stress-strain state of reinforced plastic laminates was implemented by means of an appropriate program compiled in FORTRAN-IV algorithm language and debugged on a series EC computer. The block-scheme of the program is given in Fig. 6.8. After the input and listing of initial data characterizing the structure of the reinforced plastic and the prescribed loading, and after forming operational arrays, the pseudo-elastic compliance matrix is assigned the initial values. Then the elastic calculation is made by means of program PLAST which makes it possible to compute instantaneous values of stresses and strains in the plies. The subprogram STRESS serves the determination of elastic and viscoelastic characteristics of the plies. The subprogram PLAST was compiled using the standard subprograms of matrix algebra.

After the printout of the results, the values of the pseudo-elastic compliance matrix components are recalculated according to the loading time $t_2 = t_1 + \Delta t$. The computation is iterated up to $t_1 > t_{max}$.

To illustrate the application of the proposed method, a calculation was made for the long-term stress-strain state of a four-directionally reinforced glass-epoxy AG-4S under biaxial stress state ($\langle\langle\sigma_2\rangle\rangle = 0.5\langle\langle\sigma_1\rangle\rangle; \langle\langle\sigma_{12}\rangle\rangle = 0$) with the following design structure: $m_{\beta=0°} = 0.2$; $m_{\beta=30°} = m_{\beta=-30°} = 0.2$; $m_{\beta=90°} = 0.4$ (Fig. 6.9). An analytical solution of such a problem would require a considerable effort. We should remark that in the first approximation some types of reinforced plastic laminates can be satisfactorily calculated by the quasi-elastic method where the components of the ply compliance matrix are determined from Equation System (6.33).

CHAPTER 7

# Structural Theory of Long-Term Strength

## 7.1 CRITERIA OF LONG-TERM STRENGTH OF REINFORCED PLASTICS COMPONENTS

The investigation of the long-term strength of reinforced plastics is based on data characterizing the stress-strain state of the structural elements and includes sequential analysis of the long-term strength of the material components (the fibers, the polymer matrix, and their bond), of unidirectionally reinforced elementary plies, and of laminates.

The components of any ply are in a complicated time-variant stress state even if they are subjected to uniaxial sustained load. This happens due to the internal structural geometry of a reinforced plastic and the rheonomic properties of its components. Therefore the strength criteria of reinforced plastics components under long-term loading should be such that they include time-dependent variations in their stress-strain state.

### 7.1.1 Long-Term Strength Criterion for the Polymer Matrix

The equation determining the long-term strength surface of the polymer matrix under combined loading is in the general case formulated on the basis of an experimental study of the material's behavior under long-term uniaxial tension, shear, compression, and their combinations. For a more convenient application of the long-term strength relationships some simplifying hypotheses may be introduced associated with the cause (or causes) of the material failure. Thus, the following hypothesis was adopted and proved experimentally for viscoelastic polymer matrices under uniaxial loading in Korf & Skudra [33] and Skudra et al. [56]: time-dependent failure of materials becomes possible when the viscoelastic work of stresses reaches some critical value. An investigation of uniaxial tension was made for the polyester resin PN-1 within the loading time of 90 days [33]. Time-variant loading was analyzed in uniformly growing uniaxial tension. Within the rate limit of up to 2.2 MPa/min and the loading time of up to 10 hrs the above hypothesis was proved to be applicable for the polyester resin PN-1 [56]. It has been established in Auzukalns [4] that for the epoxy resin ED-6 under uniaxial compression within a loading period of up to 83 hours the specific viscoelastic work of the compressive stress in the first approximation is also a strength characteristic of the material.

It has been proved by the analysis of polymer matrix failure that matrices fail in the areas subjected to the principal tensile stresses [44, 65]. Therefore the criterion of Gurvuch [19] will suitably describe the ultimate state of the polymer matrix, indicating the initiation of failure when the specific viscoelastic work of the principal time-dependent tensile stresses $W^+(t)$ reaches its ultimate value $W_R^+$.

In this way the strength surface area is given by

$$W^+(t) \leq W_R^+ \qquad (7.1)$$

where $W_R^+$ is the strength characteristic of the polymer matrix under tension to be determined by a uniaxial loading test.

Consider the case when the stresses $\sigma_r(t)$, $\sigma_\theta(t)$, $\sigma_z(t)$ and $\tau_{rz}(t)$ act upon the polymer matrix. This is the stress state in which the polymer matrix of the reinforced plastics finds itself when it is subjected to loading in the reinforcement plane. Then the expression $W^+(t)$ is written according to the sign of the principal stresses.

Let $\sigma_r(t)$, $\sigma_\theta(t)$ and $\sigma_z(t) > 0$. Then the principal stress may be positive or negative depending on the stress $\tau_{rz}(t)$:

$$\sigma_1(t) = \tfrac{1}{2}[\sigma_r(t) + \sigma_z(t) + \sqrt{[\sigma_r(t) - \sigma_z(t)]^2 + 4\tau_{rz}^2(t)}\,]$$

$$\sigma_2(t) = \tfrac{1}{2}[\sigma_r(t) + \sigma_z(t) - \sqrt{[\sigma_r(t) - \sigma_z(t)]^2 + 4\tau_{rz}^2(t)}\,] \quad (7.2)$$

$$\sigma_3(t) = \sigma_\theta(t)$$

If $\tau_{rz}(t) < \sqrt{\sigma_r(t)\sigma_\theta(t)}$ (Fig. 7.1 (b)) the principal stress is positive.

The strength surface of the polymer matrix according to Equation (7.1) is written as

$$W[\sigma_1(t), \sigma_2(t), \sigma_3(t)] = \int_0^{\epsilon_1(t_*)} \sigma_1 d\epsilon_1 + \int_0^{\epsilon_2(t_*)} \sigma_2 d\epsilon_2 \quad (7.3)$$

$$+ \int_0^{\epsilon_3(t_*)} \sigma_3 d\epsilon_3 = W_R^+$$

where $\epsilon_1(t_*)$, $\epsilon_2(t_*)$, $\epsilon_3(t_*)$ are strains of the polymer matrix; $t_*$ is the loading time until failure. The values of $W_R^+$ are easily determined by short-term uniaxial tension test

$$W_R^+ = \frac{(R_A^+)^2}{2E_A} \quad (7.4)$$

where $R_A^+$ and $E_A$ are tensile strength and elasticity modulus of the polymer matrix respectively. Then for polymer matrices deforming practically in a linear way up to failure we have

$$\varphi^*\{\sigma_1^2(t) + \sigma_2^2(t) + \sigma_3^2(t) - 2\nu_A[\sigma_1(t)\sigma_2(t)$$
$$+ \sigma_1(t)\sigma_3(t) + \sigma_2(t)\sigma_3(t)]\} = (R_A^+)^2 \quad (7.5)$$

Here $\varphi^*$ is an operator of rule

$$\varphi^*[F(t)] = F(t) + 2\int_0^t F(\tau)K_A(t-\tau)d\tau \quad (7.6)$$

where $K_A(t-\tau)$ and $\nu_A$ are the creep kernel and Poisson's ratio of the polymer matrix. If $\tau_{rz}(t) > \sqrt{\sigma_r(t)\sigma_z(t)}$, the principal stress $\sigma_2(t)$ is

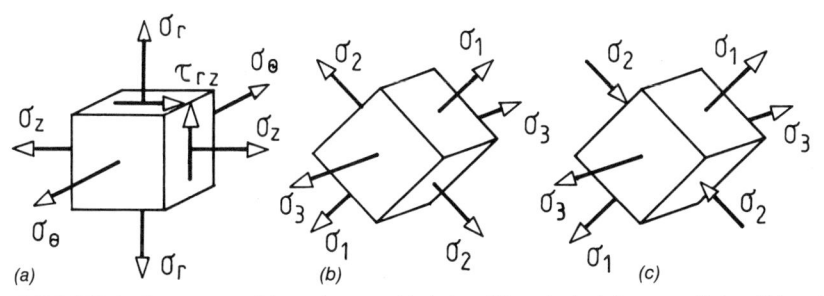

FIGURE 7.1. Stress state of the polymer matrix (a) and the principal stresses with low (b) and high (c) values of $\tau_{rz}$.

**146** STRUCTURAL MECHANICS OF A MATERIAL

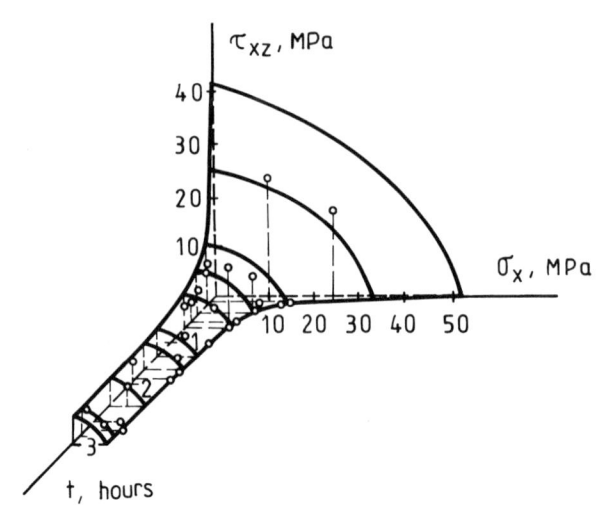

**FIGURE 7.2.** Long-term strength surface under combined loading (tension and shear) for the polyester resin PN-1.

negative. According to Equation (7.1) we have (see Fig. 7.1 (c))

$$W[\sigma_1(t), \sigma_3(t)] = \int_0^{\epsilon_1(t_*)} \sigma_1 d\epsilon_1 + \int_0^{\epsilon_3(t_*)} \sigma_3 d\epsilon_3 = W_R^+ \quad (7.7)$$

or, considering Equation (7.6), we obtain

$$\varphi^* \{\sigma_1^2(t) + \sigma_3^2(t) \\ - 2\nu_A[2\sigma_1(t)\sigma_3(t) + \sigma_1(t)\sigma_2(t) + \sigma_2(t)\sigma_3(t)]\} = (R_A^+)^2 \quad (7.8)$$

Let us consider the particular case when the stress state of the polymer matrix is determined by one normal and one tangential stress. Taking into account Equation (7.8), Equation (7.2) has the form

$$\varphi^* \{\sigma_r^2(t) + 2(1 + \nu_A)\tau_{rz}^2(t) \\ + \sigma_r(t)\sqrt{\sigma_r^2(t) + 4\tau_{rz}^2(t)}\} = 2(R_A^+)^2 \quad (7.9)$$

Fig. 7.2 shows the long-term strength surface of the polyester resin PN-1 in a constant combined stress state; the diagram was plotted according to Equation (7.9) and the test data of Skudra et al. [56] obtained from long-term constant loading were included. In the case of pure shear Equation (7.9) yields

$$\varphi^*[\tau_{rz}^2(t)] = (R_A^+)^2(1 + \nu_A) \qquad (7.10)$$

### 7.1.2 Long-Term Strength Criterion for the Bond and the Fibers

When continuous load is applied to a reinforced plastic, the contact surface is in a condition of complex (Fig. 7.3) continuously time-variant stress. Therefore, in order to describe the ultimate state of the bond in the general case, the strength is determined by the beginning of the loading time $t = 0$ up to the instant of failure $t = t_*$. Here the form of the functional is determined by strength properties of the bond and by the laws of time variations of the stress state. Any continuous functional above the continuous function field may be represented in a Stieltjes integral series of increasing multiplicity. Assuming that the beginning of the loading time coincides with the starting point of time reading, we write the equation of long-term strength surface of the bond in a general form

$$\int_0^{t_*} p_i[t_* - \tau_1]d\sigma_i(\tau_1) + \int_0^{t_*}\int_0^{t_*} p_{ij}[t_* - \tau_1, t_* - \tau_2]$$

$$\times d\sigma_i(\tau_1)d\sigma_j(\tau_2) + \ldots = 1 \qquad (7.11)$$

$$i, j = r, rz, r\theta$$

where $p_i, p_{ij} \ldots$ are functions of time-dependent changes in the bond strength properties.

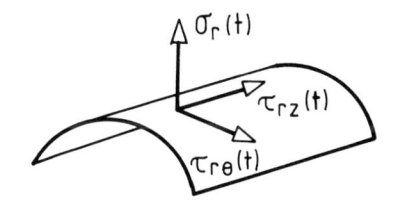

FIGURE 7.3. Stress state on the contact surface.

The determination of long-term strength surface parameters is made on the basis of tests conducted with simple (constant) loads. Limiting ourselves to the first two terms of Equation (7.11) and by analogy with short-term criterion of bond strength [31, 55, 57] we shall proceed to analyze five independent types of loading that result in the following stress states

- $\sigma_r(t) = \text{const} > 0$
- $\tau_{rz}(t) = \text{const} \neq 0$
- $\tau_{r\theta}(t) = \text{const} \neq 0$
- $\sigma_r(t) = \text{const} \neq 0; \tau_{rz} = \text{const} \neq 0; \tau_{r\theta}(t) = \text{const} \neq 0$
- $\sigma_r(t) = \text{const} < 0$

The axis $r$ here is normal, and axes $z$ and $\theta$ are tangential to the contact surface.

The other stress state components (except $\sigma_r(t)$, $\tau_{rz}(t)$, $\tau_{r\theta}(t)$) in all the loading processes are equal to zero. Then, proceeding from the symmetry properties of the bond strength (with respect to the normal to the contact surface—axis $r$) and analysing its failure under the given types of loading, we arrive at Equation (7.11) in the form

$$\int_0^{t_*} p_r[t_* - \tau_1] d\sigma_r(\tau_1)$$

$$+ \int_0^{t_*} \int_0^{t_*} p_{rz,rz}[t_* - \tau_1, t_* - \tau_2] d\tau_{rz}(\tau_1) d\tau_{rz}(\tau_2) \qquad (7.12)$$

$$+ \int_0^{t_*} \int_0^{t_*} p_{r\theta,r\theta}[t_* - \tau_1, t_* - \tau_2] d\tau_{r\theta}(\tau_1) d\tau_{r\theta}(\tau_2) = 1$$

Using the assumption that

$$p_{ii}[t_* - \tau_1, t_* - \tau_2] = p_i[t_* - \tau_1] p_i[t_* - \tau_2] \qquad (7.13)$$

and taking into account that

$$\int_0^{t_*} \int_0^{t_*} p_i(t_* - \tau_1) p_i(t_* - \tau_2) d\sigma_i(\tau_1) d\sigma_i(\tau_2)$$

$$= \left[ \int_0^{t_*} p_i(t_* - \tau) d\sigma_i(\tau) \right]^2 \qquad (7.14)$$

we obtain

$$\int_0^{t_*} p_r[t_* - \tau]d\sigma_r(\tau) + \left\{\int_0^{t_*} p_{rz}[t_* - \tau]d\tau_{rz}(\tau)\right\}^2 \qquad (7.15)$$
$$+ \left\{\int_0^{t_*} p_{r\theta}[t_* - \tau]d\tau_{r\theta}(\tau)\right\}^2 = 1$$

Integrating the expression of type $\int_0^{t_*} F(t_* - \tau)d\sigma(\tau)$ we have

$$\int_0^{t_*} F(t_* - \tau)d\sigma(\tau) = F(0)\sigma(t_*) + \int_0^{t_*} \sigma(\tau)F'(t_* - \tau)d\tau \qquad (7.16)$$

Hence for the case of constant loading ($\sigma(t) = \sigma = \text{const}$)

$$\int_0^{t_*} F(t_* - \tau)d\sigma(\tau) = F(t_*)\sigma \qquad (7.17)$$

Consequently if $R_b(t)$, $T_b(t)$ are laws of bond strength variation under rupture and shear depending upon the duration of constant loading, Equation (7.15) finally has the form [20]:

$$\int_0^{t_*} \frac{d\sigma_r(\tau)}{R_b[t_* - \tau]} + \left\{\int_0^{t_*} \frac{d\tau_{rz}(\tau)}{T_b[t_* - \tau]}\right\}^2$$
$$+ \left\{\int_0^{t_*} \frac{d\tau_{r\theta}(\tau)}{T_b[t_* - \tau]}\right\}^2 = 1 \qquad (7.18)$$

Correspondingly the long-term rupture strength criterion is written as

$$\int_0^{t_*} \frac{d\sigma_r(\tau)}{R_b[t_* - \tau]} = 1 \qquad (7.19)$$

whereas the long-term shear strength criterion is

$$\int_0^{t_*} \frac{d\tau_{r\theta}(\tau)}{T_b[t_* - \tau]} = 1 \qquad (7.20)$$

In the case of constant loading ($\sigma_r = $ const, $\tau_{rz} = $ const, $\tau_{r\theta} = $ const) Equation (7.18) yields

$$\frac{\sigma_r}{R_b(t_*)} + \left[\frac{\tau_{rz}}{T_b(t_*)}\right]^2 + \left[\frac{\tau_{r\theta}}{T_b(t_*)}\right]^2 = 1 \qquad (7.21)$$

Under short-term loading ($t_* = 0$) and with $t_* \to \infty$ the bond strength criterion is

$$\frac{\sigma_r(0, \infty)}{R_b(0, \infty)} + \left[\frac{\tau_{rz}(0, \infty)}{T_b(0, \infty)}\right]^2 + \left[\frac{\tau_{r\theta}(0, \infty)}{T_b(0, \infty)}\right]^2 = 1 \qquad (7.22)$$

The approximation of kernels $R_b^{-1}(t_* - \tau)$, $T_b^{-1}(t_* - \tau)$ is conveniently made as

$$R(t) = \frac{R(0)}{1 + \sum_{i=1}^{N} b_i [1 - \exp(-\alpha_i t)]} \qquad (7.23)$$

where $R = \{R_b, T_b\}$; $R(0)$ is short-term bond strength; $b_i$, $a_i$ are some constants of the material characterizing time-dependent strength variations.

The numerical determination of the functions $R_b$, $T_b$ may be achieved in either a direct or an indirect way. The direct method requires the direct investigation of the adhesive strength of the polymer matrix-fiber system under constant long-term tensile (shearing) load by means of various test methods that are discussed in, for example, Kirulis [31]. The indirect method is implemented on the basis of testing unidirectionally reinforced plastics. The values of the constants $a_i$, $b_i$, $i = 1$, ..., $N$, $R(0)$, which were found for some specific loading mode, permit us by using Equation (7.18) to calculate the bond strength with another law of $\sigma_i(t)$ changes. Thus, for a uniform loading law $\tau_{r\theta}(t) = vt$ with $N = 1$ integration of Equation (7.20) with $\tau_{r\theta}(t = 0) = 0$ taken into account yields

$$T_b(t_*) = T_b(0)t_* \left\{t_* + \frac{b_1}{a_1}[t_* a_1 - 1 + \exp(-a_1 t_*)]\right\}^{-1} \qquad (7.24)$$

Note that the limit of long-term bond strength $T_b(t_* \to \infty)$, $R_b(t_* \to \infty)$ is not dependent upon the loading law and is determined from

Equation (7.24) with $t_* \to \infty$. For the bond between epoxy (EDT-10) and phenol (BF-4) matrices with glass fibers a comparison was made between test relationships [18] and theoretical relationships $T_b(t)$ with a uniform loading law (Fig. 7.4 (a)). The theoretical curves were constructed according to Equation (7.24) with the following initial data: for EDT-10 $T_b(0) = 36.9$ MPa; $a_1 = 0.186$ sec$^{-1}$; $b_1 = 0.455$; for BF-4 $T_b(0) = 38.3$ MPa; $a_1 = 10.165$ sec$^{-1}$; $b_1 = 0.567$. Figure 7.4 represents long-term strength $T_b(t)$ curves for constant stress state.

The fibers are the principal components sustaining the load applied to reinforced plastics structural members. Therefore the determination of fiber strength properties both under short-term and long-term loading is of utmost importance for the structural strength theory of reinforced plastics.

It has been shown by numerous authors [1, 55, 56] that the ultimate fiber strain does not practically depend upon the loading time. Considering also the fact that creep strains are practically absent in glass-, carbon- and boron fibers [56, 71], the long-term fiber strength will be viewed from the theory of injury accumulation.

Confining ourselves to the first two terms of Equation (7.11) and considering the case of uniaxial tension (compression) ($\sigma_z \neq 0$), we arrive at the following long-term fiber strength criterion

$$\int_0^{t_*} p_z[t_* - \tau] d\sigma_z(\tau) = 1 \qquad (7.25)$$

where $p_z$ is the function of variations in time-dependent strength prop-

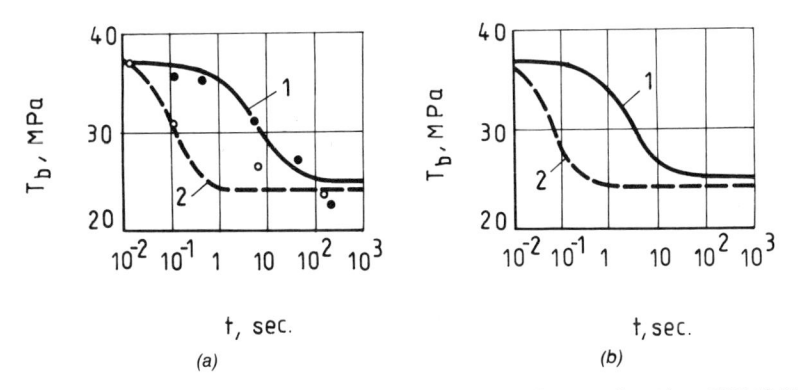

FIGURE 7.4. Dependence of bond shear strength of glass fibers and matrices EDT-10 (1) and BF-4 (2) upon loading duration: a — $\tau = vt$; B — $\tau =$ const; × — EDT-10; + — BF-4.

erties. If $R_B(t)$ is the long-term tensile strength variation law for fibers under constant stress, then Equation (7.25) has the following form

$$\int_0^{t_*} \frac{d\sigma_z(\tau)}{R_B(t_* - \tau)} = 1 \qquad (7.26)$$

Here $z$ denotes the reinforcement direction, and it is assumed that the compressive strength of the fibers is realistic only when the fibers are regarded as reinforced plastic components. In the ultimate case with $t_* = 0$, $t_* \to \infty$, Equation (7.26) is written as an inequality

$$\sigma_z(0, \infty) \leq R_B(0, \infty) \qquad (7.27)$$

In order to determine kernels $R_B^{-1}(t_* - \tau)$, Equation (7.24) will be used. Determination of long-term tensile strength properties of fibers is possible in a direct or an indirect way (by testing a unidirectionally reinforced plastic).

Fig. 7.5 gives test results on long-term strength of glass- (1) and boron (2) fibers under constant loading [56]. The theoretical curves

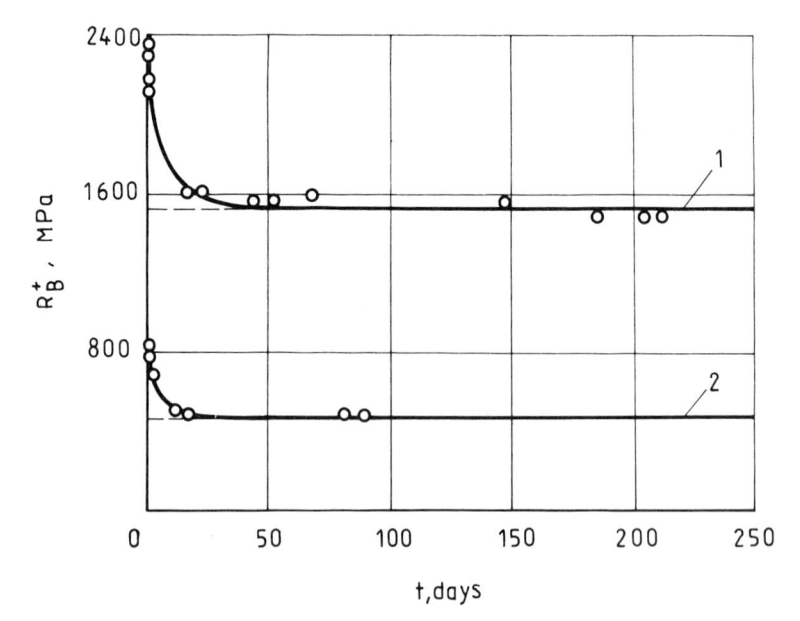

**FIGURE 7.5.** Long-term tensile strength curves of the boron fiber (1) and the glass filament (2).

have been plotted according to Equation (7.26): 1-with $R_B(0) = 0.80$ GPa; $N = 1$; $a_1 = 0.129$ days$^{-1}$; $b_1 = 0.689$; 2-with $R_B(0) = 2.17$ GPa; $N = 1$; $a_1 = 0.098$ days$^{-1}$; $b_1 = 0.417$.

## 7.2 LONG-TERM STRENGTH OF A UNIDIRECTIONALLY REINFORCED PLY WITH THE POLYMER MATRIX FAILURE

The strength of a unidirectionally reinforced ply that fails due to the polymer matrix failure under a plane stress state is determined by the value of and the ratio between mean stresses in the ply reinforcement plane $\langle \sigma_\perp \rangle$ and $\langle \tau_{\|\perp} \rangle$ because the stresses acting in the reinforcement direction are practically sustained only by high-strength fibers [11]. In works [55, 57] it has been shown that in the case of short-term loading the polymer matrix failure in the unidirectionally reinforced ply under transversal tension starts with the failure of the most highly loaded matrix regions and proceeds in an avalanche way. Developing the results for the case of long-term combined loading, we assume that the time-dependent failure of the ply occurs when the regions of utmost loading fail. The instant of the first failures initiates an avalanche-type failure in the polymer matrix and at the same time it signals the failure of the whole ply.

Under longitudinal shear the components of a unidirectionally reinforced plastic are also in non-uniform stress-state conditions. The reaching of the ultimate state in the most loaded regions, however, does not always initiate avalanche-type failure of the whole material [55, 57]. Further loading leads to the redistribution of the stress field owing to the conditional flow of the polymer matrix, and as shown in [55, 70], the shear strength of a unidirectionally reinforced ply is practically equal to the shear strength of the polymer matrix.

It follows from the above that at the instant of ply failure the stress state in the most highly loaded ply areas is described by the relationships

$$\sigma_r(t) = \bar{\sigma}_r(t)\langle \sigma_\perp \rangle$$

$$\sigma_\theta(t) = \bar{\sigma}_\theta(t)\langle \sigma_\perp \rangle$$

$$\sigma_z(t) = \bar{\sigma}_z(t)\langle \sigma_\perp \rangle$$

$$\tau_{rz}(t) = \langle \tau_{\perp\|} \rangle$$

(7.28)

The structural approach to the determination of the dimensionless functions of stress concentrations $\bar{\sigma}_r(t)$, $\bar{\sigma}_\theta(t)$, $\bar{\sigma}_z(t)$ is discussed in Chapter 6.

Under transversal loading the strains in the ply are insignificant in the reinforcement direction and for practical purposes can be neglected. The validity of such a supposition increases with the higher longitudinal elasticity modulus of the fibers. In this case the relationship

$$\sigma_z(t) = \nu_A[\sigma_r(t) + \sigma_\theta(t)] \tag{7.29}$$

holds true.

Taking into account the stress state of Equation (7.28) and the relationships in Equation (7.29), the principal stresses in the initial failure regions of the polymer matrix are given by the expressions

$$\sigma_1(t) = \frac{\langle \sigma_\perp \rangle}{2}\{(1 + \nu_A)\bar{\sigma}_r(t) + \nu_A \bar{\sigma}_\theta(t)$$

$$+ \sqrt{[(1 - \nu_A)\bar{\sigma}_r(t) - \nu_A \bar{\sigma}_\theta(t)]^2 + 4k^2}\}$$

$$\sigma_2(t) = \frac{\langle \sigma_\perp \rangle}{2}\{(1 + \nu_A)\bar{\sigma}_r(t) + \nu_A \bar{\sigma}_\theta(t) \tag{7.30}$$

$$- \sqrt{[(1 - \nu_A)\bar{\sigma}_r(t) - \nu_A \bar{\sigma}_\theta(t)]^2 + 4k^2}\}$$

$$\sigma_3(t) = \langle \sigma_\perp \rangle \bar{\sigma}_\theta(t)$$

where

$$k = \frac{\langle \tau_{\|\perp} \rangle}{\langle \sigma_\perp \rangle}$$

It follows from the analysis of the expressions in Equation System (7.30) that the sign of the principal stresses depends not only upon the sign of the stresses $\langle \sigma_\perp \rangle$ but also upon the relative magnitude of the tangential stresses $\langle \tau_{\|\perp} \rangle$ (ratio $k$). Consequently, the long-term strength condition in the polymer matrix in a stress state as shown in Equation System (7.30) according to Equation (7.1) will be determined by several equations, depending upon the loading conditions of the ply. Let us discuss in a greater detail some kinds of combined loading with transversal tension plus shear. Under transversal tension, the principal stress

$\sigma_2(t)$ has a positive or a negative value depending upon the ratio of $\langle \tau_{\|\perp} \rangle$ to $\langle \sigma_\perp \rangle$.

The first case corresponds to the ratios of stresses $\langle \sigma_\perp \rangle$ and $\langle \tau_{\|\perp} \rangle$ for which the following condition holds true

$$0 \leq k \leq \sqrt{\nu_A \bar{\sigma}_r(t)[\bar{\sigma}_r(t) + \bar{\sigma}_\theta(t)]} \tag{7.31}$$

Here all the principal stresses are positive (see Fig. 7.1 (b)). Using Equation (7.1) as a long-term strength criterion of the polymer matrix, we obtain an equation for the long-term strength surface of a ply in the coordinates of $\langle \sigma_\perp \rangle$, $\langle \tau_{\|\perp} \rangle$ and $t$

$$\langle \sigma_\perp \rangle^2 \varphi^* \{ [\bar{\sigma}_r^2(t) + \bar{\sigma}_\theta^2(t)](1 - \nu_A) \\ - 2[\nu_A \bar{\sigma}_r(t)\bar{\sigma}_\theta(t) - k^2] \} = \frac{(R_A^+)^2}{1 + \nu_A} \tag{7.32}$$

where $\varphi^*$ is the linear operator in the form of Equation (7.6).

For the case described by the inequality

$$k > \sqrt{\nu_A \bar{\sigma}_r(t)[\bar{\sigma}_r(t) + \bar{\sigma}_\theta(t)]}$$

the principal stress $\sigma_2(t)$ is a compressive stress (see Fig. 7.1 (c)). Thus the long-term strength surface of the ply, according to Equation (7.7), is described by the equation

$$\frac{\langle \sigma_\perp \rangle}{2} \varphi^* \{ [\bar{\sigma}_r^2(t) + 2\bar{\sigma}_\theta^2(t)](1 - \nu_A) - 3\nu_A \bar{\sigma}_r(t)\bar{\sigma}_\theta(t) \\ + \bar{\sigma}_r(t)\sqrt{[\bar{\sigma}_r(t)(1 - \nu_A) - \nu_A \bar{\sigma}_\theta(t)]^2 + 4k^2} + 2k^2 \} = \frac{(R_A^+)^2}{1 + \nu_A} \tag{7.33}$$

As shown above, the long-term strength surface of a unidirectionally reinforced ply under combined longitudinal shear and transversal tensile loading consists of two sections described by Equations (7.32) and (7.33). Taking into account that for realistic cases with $|k| > \sqrt{\nu_A \bar{\sigma}_r(t)[\bar{\sigma}_r(t) + \bar{\sigma}_\theta(t)]}$

$$\sqrt{[\bar{\sigma}_r(t)(1 - \nu_A) - \nu_A \bar{\sigma}_\theta(t)]^2 + 4k^2} \approx 2k \tag{7.34}$$

the set of Equations (7.32) and (7.33) describing the long-term strength surface of a ply in transversal tension combined with shear is written as $0 \leq k \leq a_1(t)$

$$\varphi^* \{a_2(t)\langle\sigma_\perp\rangle^2 + a_3\langle\tau_{\|\perp}\rangle^2\} = (R_A^+)^2 \tag{7.35}$$

or with $k > a_1(t)$

$$\varphi^* \{a_4(t)\langle\sigma_\perp\rangle^2 + a_5(t)\langle\sigma_\perp\rangle\langle\tau_{\|\perp}\rangle \\ + a_6(t)\langle\tau_{\|\perp}\rangle^2\} = (R_A^+)^2 \tag{7.36}$$

Here $a_i(t)$, $i = 1, \ldots, 6$, are dimensionless time functions

$$\begin{aligned} a_1(t) &= \sqrt{\nu_A \bar{\sigma}_r(t)[\bar{\sigma}_r(t) + \bar{\sigma}_\theta(t)]} \\ a_2(t) &= [\bar{\sigma}_r^2(t) + \bar{\sigma}_\theta^2(t)](1 - \nu_A^2) \\ &\quad - 2\nu_A(1 + \nu_A)\bar{\sigma}_r(t)\bar{\sigma}_\theta(t) \\ a_3(t) &= 2(1 + \nu_A) = \text{const} \\ a_4(t) &= 0.5[\bar{\sigma}_r(t) + 2\bar{\sigma}_\theta^2(t)](1 - \nu_A^2) \\ &\quad - 1.5\nu_A(1 + \nu_A)\bar{\sigma}_r(t)\bar{\sigma}_\theta(t) \\ a_5(t) &= \bar{\sigma}_r(t)(1 + \nu_A) \\ a_6(t) &= (1 + \nu_A) = \text{const} \end{aligned} \tag{7.37}$$

Fig. 7.6 represents test data [84] obtained in testing unidirectionally reinforced wound tubular samples made of glass-epoxy under combined torsion and tension with various longitudinal shear and transversal tensile stress ratios. The theoretical curves were plotted according to Equations (7.35) and (7.36) with the following initial data: $R_A^+ = 65$ MPa; $\nu_A = 0.35$; $\nu_B = 0.23$; $E_B/E_A = 20$; $\psi = 0.65$. The polymer matrix creep kernel was selected as

$$K_A(t - \tau) = C_A \exp[-\alpha_A(t - \tau)]$$

$$(C_A = 0.00675 \text{ hrs}^{-1}; \alpha_A = 0.00435 \text{ hrs}^{-1})$$

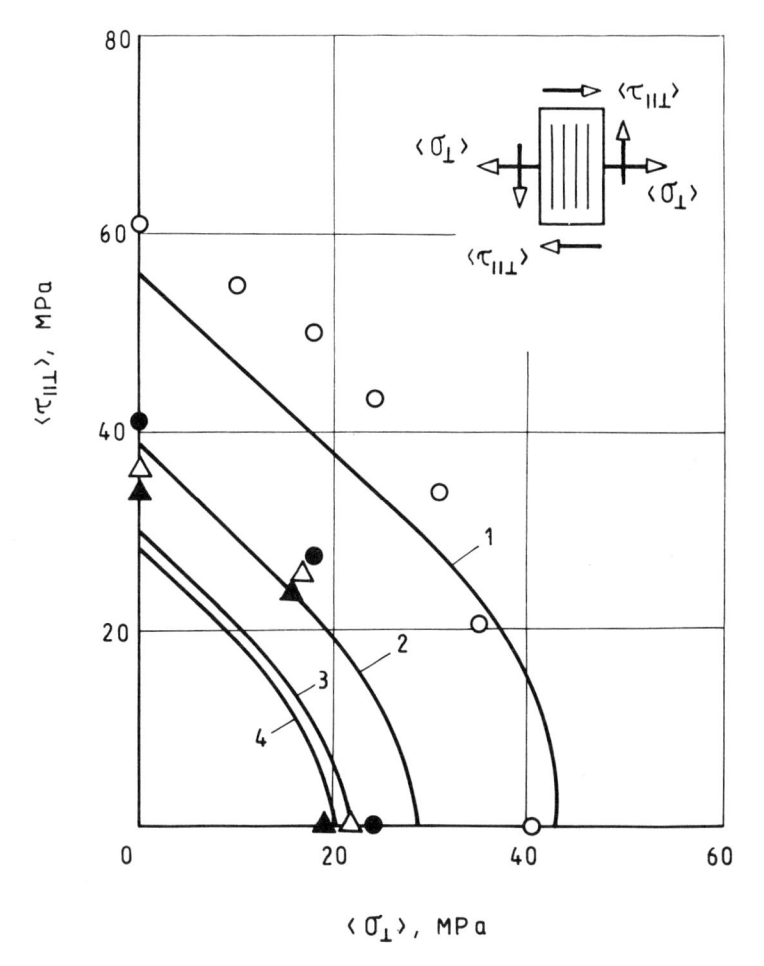

**FIGURE 7.6.** Long-term strength curves for a unidirectionally reinforced glass-epoxy with fixed time values of combined loading: $t = 0$ (1), 100 hrs (2), 500 hrs (3), 1000 hrs (4) (test data): $t$ — 0 (○), 100 hrs (●), 500 hrs (△), 1000 hrs (▲).

The ultimate short-term and long-term strengths of a unidirectionally reinforced ply with $t_* = 0$ and $t_* \to \infty$ are determined by Equations (7.35) and (7.36) substituting $a_i(0)$ or $a_i(\infty)$ for $a_i(t)$ respectively and using the following expression for $\varphi^*$ as given by

$$\varphi^*[F(t)]_{t_*=0, t_* \to \infty} \to F(0),$$

$$F(\infty)\left[1 + 2\int_0^\infty K_A(t_* - \tau)d\tau\right] \tag{7.38}$$

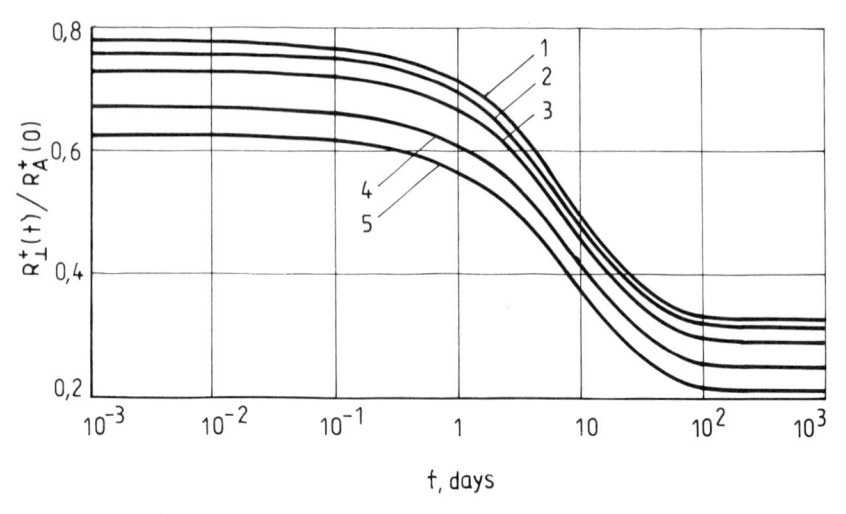

**FIGURE 7.7.** Time-dependent transversal tensile strength variation of a glass-epoxy with $R_A^+(0) = 1$ for various fibre contents: $\Psi = 0.4$ (1), 0.5 (2), 0.6 (3), 0.7 (4), 0.75 (5), $E_B/E_A = 20$, $\nu_A = 0.35$, $\nu_B = 0.23$, $C_A = 0.0872$ days$^{-1}$, $\alpha_A = 0.0436$ days$^{-1}$.

With respect to boron fiber reinforced plastics for which time-dependent changes in structural parameters $\bar{\sigma}_r(t)$, $\bar{\sigma}_\theta(t)$ may be neglected, the relationships $a_i(t) = a_i(0) = $ const, $i = 1, \ldots, 6$ hold true. Then Equation (7.6) has the form

$$\varphi^*[c] = c\left[1 + 2\int_0^{t_*} K_A(t_* - \tau)d\tau\right] \quad (7.39)$$

$$c = \text{const.}$$

Let us consider two particular cases of loading: transversal tension and longitudinal shear. In the first case $\langle \tau_{\|\perp} \rangle = 0$ and according to Equation (7.35) the following relationship is obtained for the time-dependent transversal tensile strength variation for a ply

$$R_\perp^+(t_*) = \frac{R_A^+}{\sqrt{\varphi^*\{a_2(t)\}}} \quad (7.40)$$

or in a developed form

$$R_\perp^+(t_*) = \frac{R_A^+}{\sqrt{\varphi^*\{[\bar{\sigma}_r^2(t) + \bar{\sigma}_\theta^2(t)](1 - \nu_A^2) - 2\nu_A(1 + \nu_A)\bar{\sigma}_r(t)\bar{\sigma}_\theta(t)\}}} \quad (7.41)$$

*Structural Theory of Long-Term Strength* 159

According to Equation (7.41) the analysis of long-term strength variations in a unidirectionally reinforced ply shows that the more the transversal strength decreases with time, the higher is the degree of creep in the polymer matrix. The relative fiber volume content $\psi$ for composites with high modulus ratio $E_B/E_A$ in the transversal direction (glass- and boron fiber reinforced plastics) only slightly influences the law of transversal strength variation $R_\perp^+(t_*)$ (Fig. 7.7).

For carbon fiber reinforced plastics ($E_{Br}/E_A \approx 2$) the changes in $R_\perp^+(t_*)$ become more pronounced with a high reinforcing factor $\psi$. Fig. 7.8 illustrates the relative long-term strength for a unidirectionally reinforced boron epoxy under transversal tension; the curve has been plotted according to Equation (7.41) with $E_B/E_A = 100$; $\nu_A = 0.35$; $\nu_B = 0.23$; $\psi = 0.6$; $C_A = 0.0033$ min$^{-1}$; $\alpha_A = 0.0289$ min$^{-1}$. The diagram also includes the test results from reference [76].

In the case of transversal shear ($\langle \sigma_\perp \rangle = 0$) according to Equation (7.36) the long-term strength of a ply is given as

$$T_{\parallel\perp}(t_*) = \frac{R_A^+}{\sqrt{1+\nu_A}} \left[ 1 + 2\int_0^{t_*} K_A(t_* - \tau)d\tau \right]^{-1/2} \quad (7.42)$$

or, taking into account Equation (7.39),

$$T_{\parallel\perp}(t_*) = T_A\{\varphi^*[1]\}^{-1/2} = T_A(t_*) \quad (7.43)$$

The above long-term strength expressions pertain to loading specifically in the elastic symmetry axis of the ply.

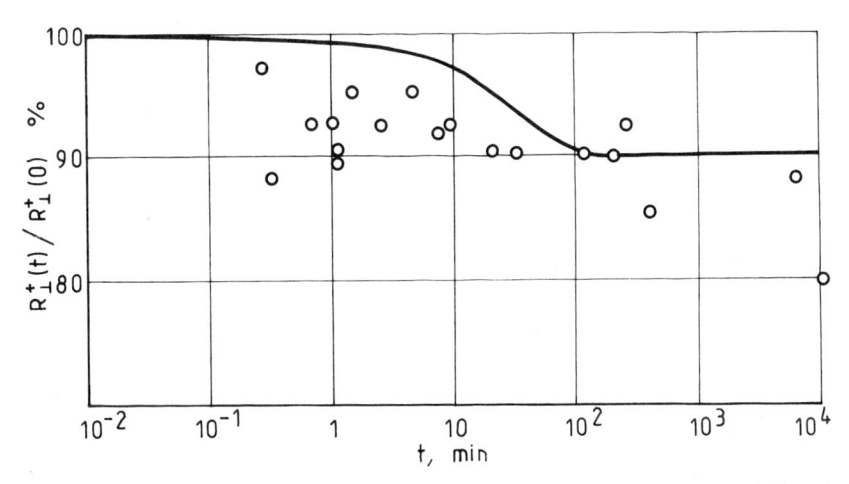

FIGURE 7.8. Relative long-term transversal tensile strength curve of a unidirectionally reinforced boron-epoxy.

Now consider the case of a uniaxial long-term tension of the unidirectionally reinforced ply loaded at an angle $\beta$ to the reinforcement direction. In the axes coinciding with the reinforcement direction, stresses are determined by

$$\langle \sigma_{\|} \rangle = \cos^2 \beta \langle \sigma_\beta \rangle$$

$$\langle \sigma_\perp \rangle = \sin^2 \beta \langle \sigma_\beta \rangle \quad (7.44)$$

$$\langle \tau_{\|\perp} \rangle = \cos \beta \sin \beta \langle \sigma_\beta \rangle$$

At the instant of failure, $\langle \sigma_\beta \rangle = R_\beta^+$ and the dependence of long-term tensile strength of the ply upon its lay-up structure and the angle $\beta$ as seen from Equations (7.35) and (7.36) taking into account Equation System (7.44) has the following form

with ctg $\beta \leqslant a_1(t)$

$$R_\beta^+(t_*) = \frac{R_A^+}{\sin \beta \{\varphi^*[a_2(t) \sin^2 \beta + a_3(t) \cos^2 \beta]\}^{1/2}} \quad (7.45)$$

with ctg $\beta > a_1(t)$

$$R_\beta^+(t_*) = \frac{R_A^+}{\sin \beta \{\varphi^*[a_4(t) \sin^2 \beta + a_5(t) \cos \beta \sin \beta + a_6(t) \cos^2 \beta]\}^{1/2}} \quad (7.46)$$

Fig. 7.9 represents the dependence of a unidirectionally reinforced glass epoxy tensile strength upon the time of constant loading and the angle $\beta$; the diagram has been plotted from Equations (7.45) and (7.46) with $E_B/E_A = 20$; $\nu_A = 0.35$; $\nu_B = 0.23$; $\psi = 0.51$; $R_A^+ = 63$ MPa; $C_A = 0.0143$ min$^{-1}$; $\alpha_A = 0.0315$ min$^{-1}$. The test data have been borrowed from Wu & Ruhman [99].

If parameter $\bar{\sigma}_\theta(t)$ is ignored compared with $\bar{\sigma}_r(t)$ as proposed for the case of short-term loading by Skudra & Bulvas [57], the following practicable expressions for evaluating long-term strength of a unidirectionally reinforced plastic are obtained

with $\langle \tau_{\|\perp} \rangle / \langle \sigma_\perp \rangle < \bar{\sigma}_r(t)\sqrt{\nu_A}$

$$\varphi^*\{\langle \sigma_\perp \rangle^2[\bar{\sigma}_r^2(t)(1 - \nu_A^2)] + \langle \tau_{\|\perp} \rangle^2[2(1 + \nu_A)]\} = (R_A^+)^2 \quad (7.47)$$

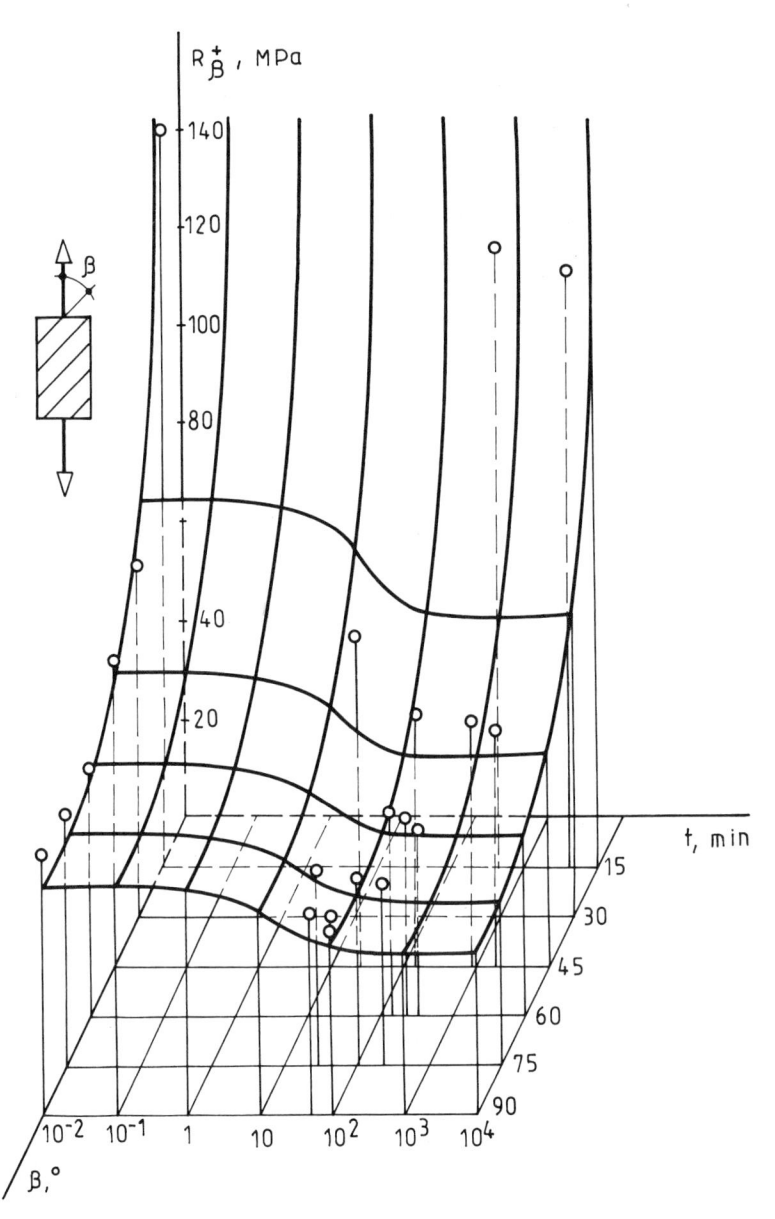

**FIGURE 7.9.** The dependence of tensile long-term strength of a unidirectionally reinforced glass-epoxy upon the orientation angle of the reinforcement in the case of the polymer matrix failure.

with $\langle \tau_{\|\perp} \rangle / \langle \sigma_\perp \rangle > \bar{\sigma}_r(t)\sqrt{v_A}$

$$\varphi^*\{\langle\sigma_\perp\rangle^2[0.5\bar{\sigma}_r^2(t)(1-v_A^2)] + \langle\sigma_\perp\rangle\langle\tau_{\|\perp}\rangle \\ \times [\bar{\sigma}_r(t)(1+v_A)] + \langle\tau_{\|\perp}\rangle^2[1+v_A]\} = (R_A^+)^2 \quad (7.48)$$

In addition, the long-term strength surface of the ply in the axes of $\langle\sigma_\perp\rangle, \langle\tau_{\|\perp}\rangle$ and $t$ is restricted to two sections.

## 7.3 DEPENDENCE OF LONG-TERM STRENGTH OF A UNIDIRECTIONALLY REINFORCED PLY UPON THE FIBER STRENGTH PROPERTIES AND THE INTER-COMPONENT BOND

Long-term strength of unidirectionally reinforced plies in the case of the bond failure is determined by the magnitude of stresses $\langle\sigma_\perp\rangle$ and $\langle\tau_{\|\perp}\rangle$ and the ratio between them. For a short-term ply strength problem, the results of the numerical simulation of the failure process have shown that the bond failure in the most highly loaded points of the ply leads to the ply failure [31]. For long-term loading it is assumed that the bond failure in the most highly loaded points also initiates the avalanche type of failure in the ply. Equation (7.18) provides the basis for obtaining the long-term strength condition for a unidirectionally reinforced ply under a plane stress state in the case of the bond failure [20]

$$\langle\sigma_\perp\rangle\int_0^{t_*}\frac{d\bar{\sigma}_r(\tau)}{R_b(t_*-\tau)} + \left\{\langle\tau_{\|\perp}\rangle\int_0^{t_*}\frac{d\bar{\tau}_{rz}(\tau)}{T_b(t_*-\tau)}\right\}^2 = 1 \quad (7.49)$$

For reinforced plastics in which time-dependent variations in the stress concentration factor may be neglected (boron fiber reinforced plastics), Equation (7.49) is simplified to

$$\frac{\langle\sigma_\perp\rangle\bar{\sigma}_r}{R_b(t_*)} + \left\{\frac{\langle\tau_{\|\perp}\rangle\bar{\tau}_{rz}}{T_b(t_*)}\right\}^2 = 1 \quad (7.50)$$

Specific cases of the strength criterion described in Equation (7.49) are those when $t_* = 0$ and $t_* \to \infty$.

$$\langle\sigma_\perp\rangle\frac{\bar{\sigma}_r(0,\infty)}{R_b(0,\infty)} + \left\{\langle\tau_{\|\perp}\rangle\frac{\bar{\tau}_{rz}(0,\infty)}{T_b(0,\infty)}\right\}^2 = 1 \quad (7.51)$$

For cases of long-term transversal tension or longitudinal shear, Equation (7.49) yields

$$R_\perp^+(t_*) = \left[ \int_0^{t_*} \frac{d\bar{\sigma}_r(\tau)}{R_b(t_* - \tau)} \right]^{-1} \tag{7.52}$$

$$T_{\|\perp}(t_*) = \left[ \int_0^{t_*} \frac{d\bar{\tau}_{rz}(\tau)}{T_b(t_* - \tau)} \right]^{-1} \tag{7.53}$$

If a unidirectionally reinforced ply is loaded at an angle $\beta$ to the reinforcement direction, then the stress state in the elastic symmetry axes is given by Equation (7.44), and the strength criterion of Equation (7.49) has the form

$$\langle \sigma_\beta \rangle A(t_*) + \langle \sigma_\beta \rangle^2 B(t_*) = 1 \tag{7.54}$$

$$A(t_*) = \sin^2 \beta \int_0^{t_*} \frac{d\bar{\sigma}_r(\tau)}{R_b(t_* - \tau)} \tag{7.55}$$

$$B(t_*) = \sin^2 \beta \cos^2 \beta \left\{ \int_0^{t_*} \frac{d\bar{\sigma}_{rz}(\tau)}{T_b(t_* - \tau)} \right\}^2 \tag{7.56}$$

In the ultimate state $\langle \sigma_\beta \rangle = R_\beta^+$ and the solution of Equation (7.54) with respect to $R_\beta^+$ yields

$$R_\beta^+(t_*) = \frac{\sqrt{A^2(t_*) + 4B(t_*)} - A(t_*)}{2B(t_*)} \tag{7.57}$$

Testing of uniaxially reinforced plastics for longitudinal tension at an angle $\beta$ to the reinforcement permits indirect determination of long-term strength characteristics of the bond. The unknown variables of Equation (7.23) are found by the least-squares method, and examination of the long-term curves for the minimum value of $S$

$$S = \sqrt{\frac{1}{M} \sum_{j=1}^{M} \left\{ \frac{R_{\beta,\text{test}}^{+(j)} - R_{\beta,\text{theor}}^{+(j)}}{R_{\beta,\text{test}}^{+(j)}} \right\}^2} \to \min \tag{7.58}$$

where $R_{\beta,\text{test}}^{+(j)}$ is the test value of the ply strength; $R_{\beta,\text{theor}}^{+(j)}$ is the theoretical value of the ply strength calculated from Equation (7.57) with the corresponding test variables $(t_*, \beta, \ldots)$; $M$ is the number of test values.

Table 7.1. Rupture Strength Characteristics of the Bond.

| T, °C | $a_1$, hrs$^{-1}$ | $b_1$ | $R_b(o)$ MPa | M | S, % |
|---|---|---|---|---|---|
| 114.7 | 0.9630 | 0.1887 | 60.389 | 5 | 1.36 |
| 128.0 | 0.6205 | 0.2158 | 52.509 | 6 | 2.96 |
| 141.3 | 0.7016 | 0.2558 | 41.190 | 5 | 3.79 |
| 154.6 | 0.3828 | 0.0812 | 20.485 | 6 | 2.85 |

In order to determine long-term bond strength parameters ($R_b(0)$, $T_b(0)$, $a_1$, $b_1$, ...) in the algorithm language FORTRAN-IV an adequate program has been developed and debugged on a series EC-computer. The network approach was used for the numerical implementation of the search for a maximum of value $S$; the choice of the method was governed by the multiextremality of Equation (7.58) with approximation of $R_b(t)$ and $T_b(t)$ in Equation (7.24). Angle $\beta = 90°$ leads to the characteristics of the function $R_b(t)$ whereas with angle $0° < \beta < 90°$ we obtain the function $T_b(t)$. The utilization of the compiled program is only possible with the assurance that the reinforced plastic fails as a result of the bond failure. The review of various methods of analyzing the causes of reinforced plastics failure is given by Kirulis [31].

Brinson et al. [78] present test data on the long-term strength of unidirectionally reinforced heat-resistant epoxy-carbon in the temperature range of 110–160°C. The authors' attempt to describe the temperature-time dependence of the strength of the plastic under discussion as a homogeneous material at various loading angles $\beta$ by using the Larson-Miller parameter has not been very successful. The bond strength characteristics have been approximated in the form of Equation (7.23) with $N = 1$ and calculated by means of the program compiled for the test results [78] (Tables 7.1, 7.2).

Table 7.2. Shear Strength Characteristics of the Bond.

| T, °C | $a_1$, hrs$^{-1}$ | $b_1$ | $T_b(o)$ MPa | M | S, % |
|---|---|---|---|---|---|
| 114.7 | 1.0100 | 0.3546 | 112.516 | 7 | 2.65 |
| 128.0 | 1.0190 | 1.0055 | 93.002 | 9 | 2.85 |
| 141.3 | 1.9221 | 1.0555 | 82.495 | 11 | 7.64 |
| 154.6 | 0.9298 | 0.5444 | 37.055 | 9 | 4.65 |

Figs. 7.10 through 7.13 illustrate the long-term tensile strength dependence of a unidirectionally reinforced carbon-epoxy ($E_{Br}/E_A = 2$; $E_{Bz}/E_{Br} = 20$; $\psi = 0.6$; $\nu_A = 0.35$; $\nu_{B\theta r} = 0.40$; $\nu_{Bzr} = 0.25$) upon the angle of its orientation in the case of bond failure at various temperatures. The theoretical curves have been plotted from Equation (7.57) with the values of the parameters from Tables 7.1 and 7.2 whereas the test data have been borrowed from Brinson et al. [78].

The analysis of data shown in Figs. 7.10 through 7.13 indicate that the long-term tensile strength of a ply depends essentially upon the loading angle. The share of stresses taken by the fibers increases with the smaller angle $\beta$. The ultimate long-term strength $R_\beta^+(t_* \to \infty)$ constitutes 75–80 percent of the short-time values in the range $45° \leq \beta \leq 90°$.

The long-term strength of a unidirectionally reinforced ply in the case of fiber failure is determined by the magnitude and the sign of $\langle \sigma_\| \rangle$ since the stresses $\langle \sigma_\perp \rangle$ and $\langle \tau_{\|\perp} \rangle$ have an insignificant influence upon the fiber stress state in glass-, boron- and carbon fiber reinforced plastics [55]. In Bulvas & Radinsh [10] it has been shown that constant loading in the ply reinforcement direction causes time-dependent stress $\sigma_{Bz}(t)$ redistribution in the fibers that does not exceed 2–3 percent of $\sigma_{Bz}(0)$. An analogous test result was mentioned in reference [87] where it was shown by analysing a unidirectional glass-epoxy that strength reduction under tension in the reinforcement direction is practically equal to creep and relaxation. Consequently, the stress state of fibers under constant loading may be considered to be time-invariant. Thus,

$$\sigma_{Bz}(t) = \sigma_{Bz}(0) = \frac{[(1-\psi)E_A + \psi E_B]}{E_B} \langle \sigma_\| \rangle \qquad (7.59)$$

but for high-strength glass-, carbon- and boron fiber reinforced plastics

$$\sigma_{Bz} \approx \psi \langle \sigma_\| \rangle \qquad (7.60)$$

Applying Equation (7.26) we obtain the long-term strength condition for a unidirectionally reinforced ply in the case of fiber failure under tension

$$\frac{\langle \sigma_\| \rangle}{\psi R_B(t)} = 1 \qquad (7.61)$$

Fig. 7.14 gives a comparison between test data [56] on the long-term tensile strength of a boron-epoxy ($\psi = 0.67$) and the theoretical dependence plotted according to Equation (7.61) with $R_B = 2.17$ GPa; $N = 1$; $a_1 = 0.098$ days$^{-1}$; $b_1 = 0.417$; $R_B(\infty) = 1.53$ GPa.

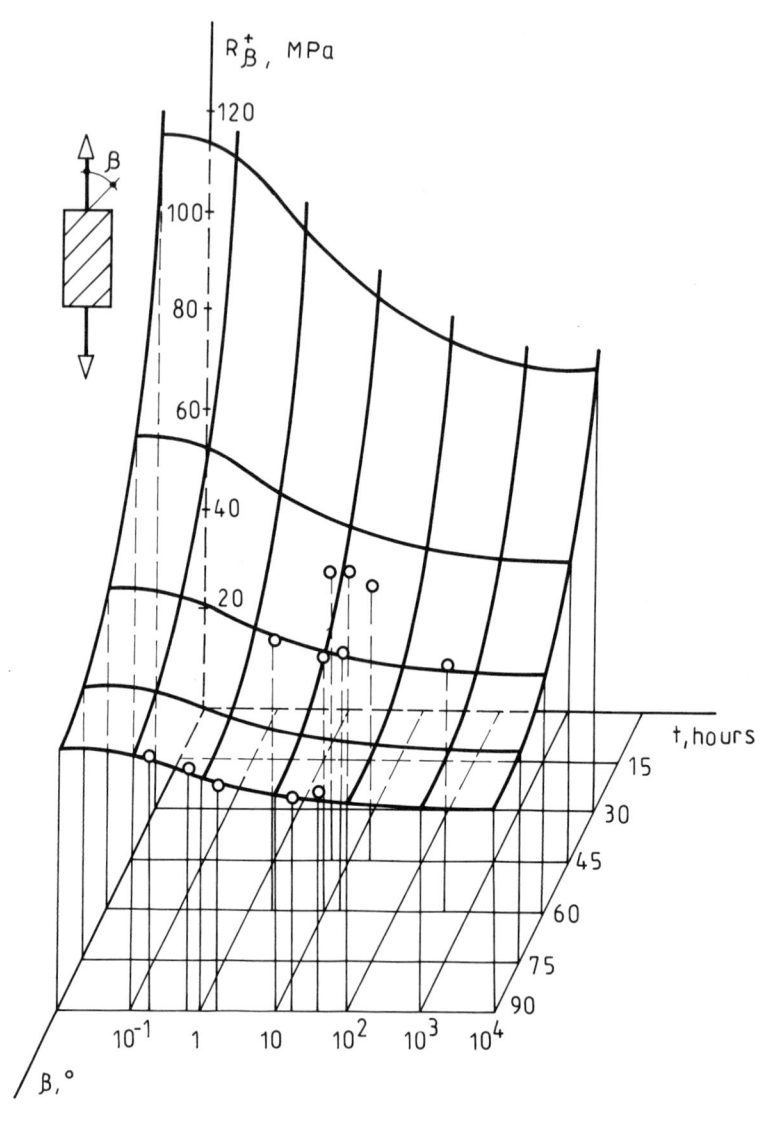

**FIGURE 7.10.** Dependence of long-term tensile strength of a unidirectionally reinforced carbon fiber epoxy resin upon its orientation angle to the reinforcement direction in the case of bond failure with $T = 114.7°C$.

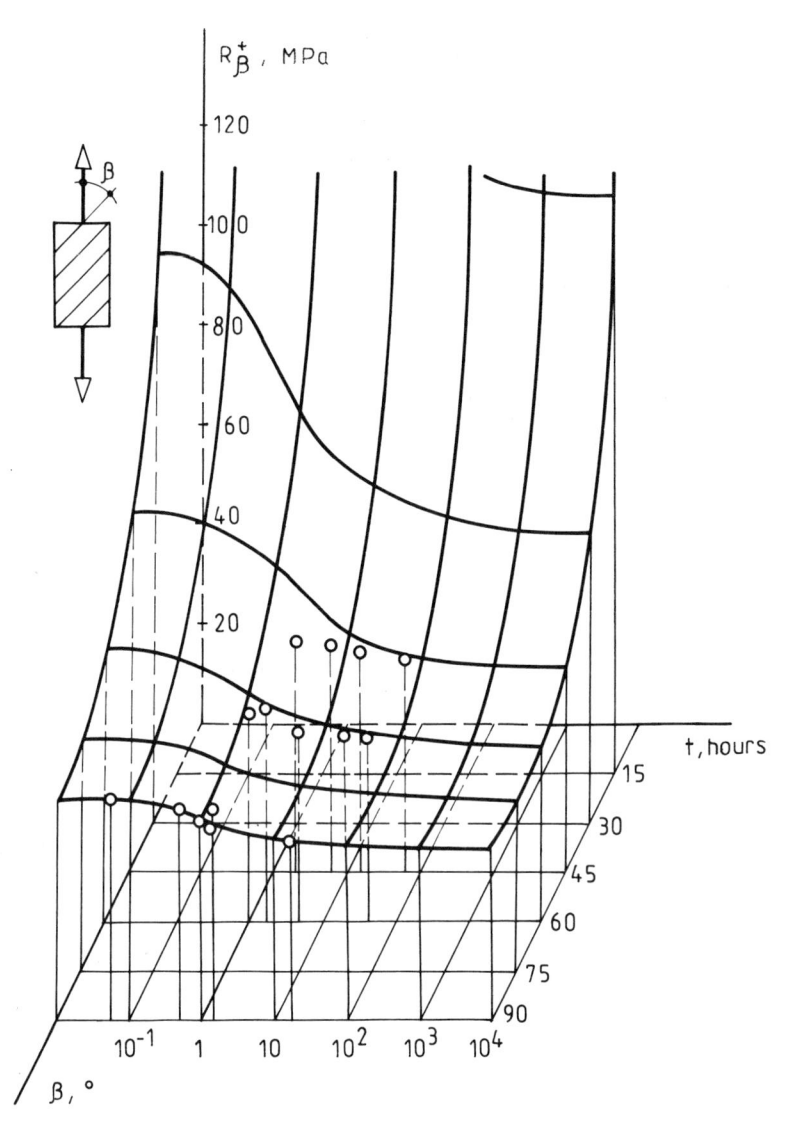

**FIGURE 7.11.** Dependence of long-term tensile strength of a unidirectionally reinforced carbon-epoxy upon its angle of orientation to the reinforcement direction in the case of bond failure with $T = 128.0°C$.

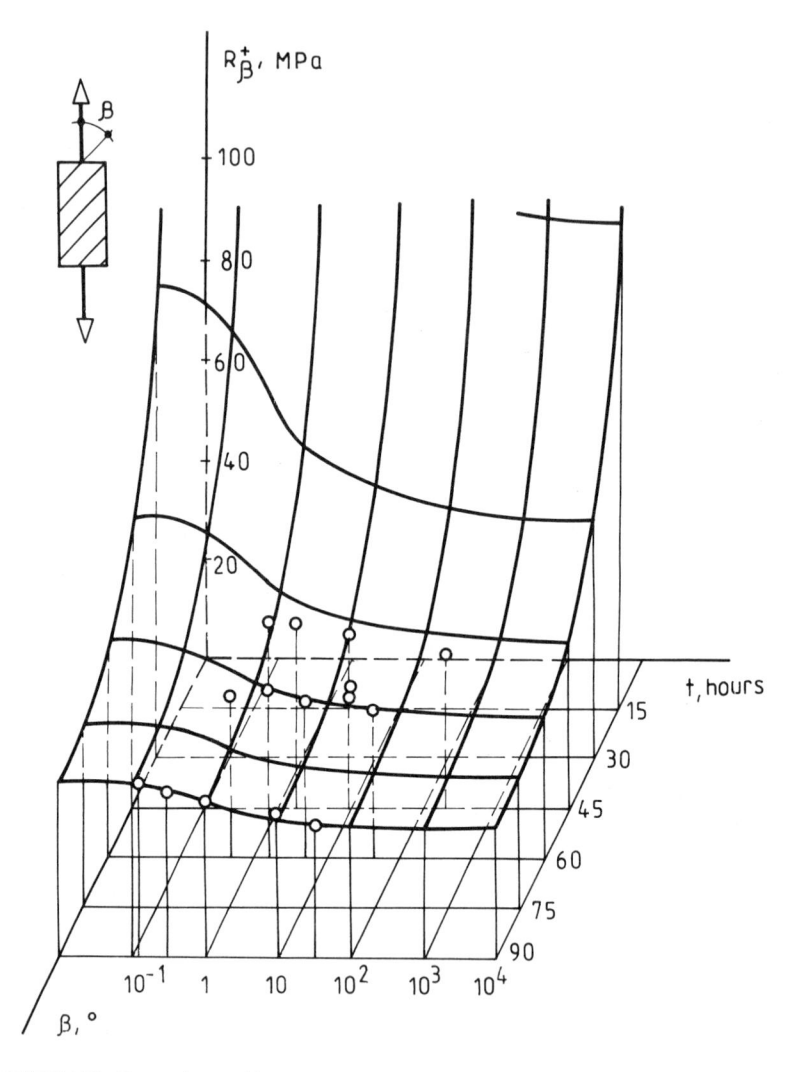

**FIGURE 7.12.** Dependence of long-term tensile strength of a unidirectionally reinforced carbon-epoxy upon the angle of its orientation to the reinforcement direction in the case of bond failure with $T = 141.3°C$.

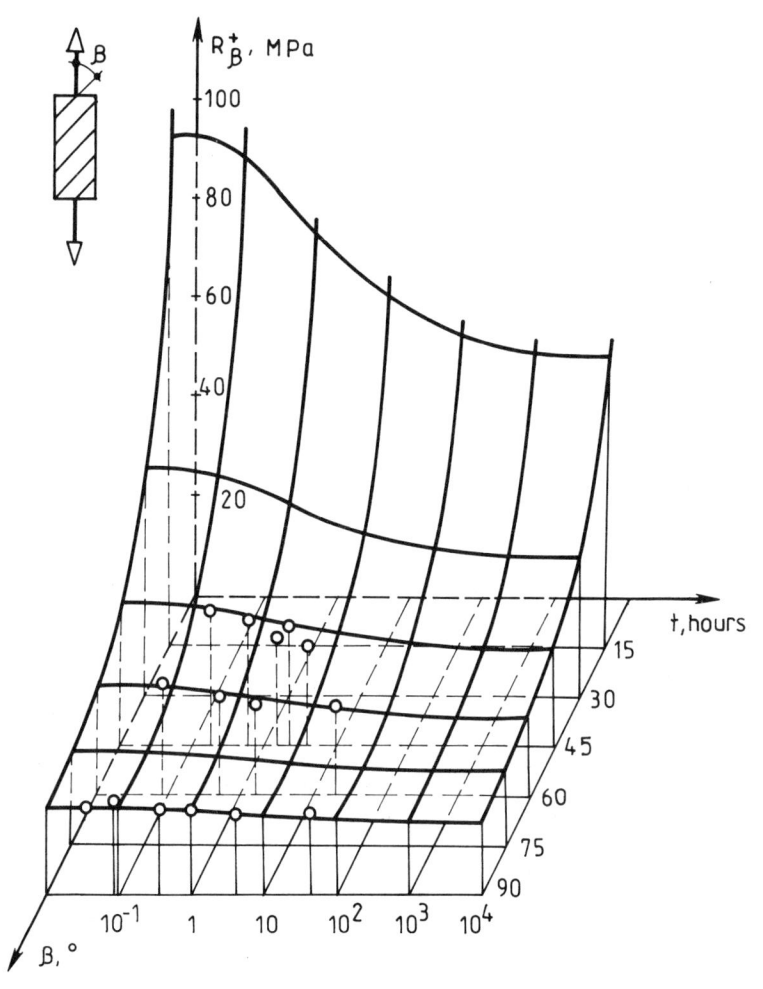

**FIGURE 7.13.** Dependence of long-term tensile strength of a unidirectionally reinforced carbon-epoxy upon its angle of orientation to the reinforcement direction in the case of bond failure with $T = 154.6°C$.

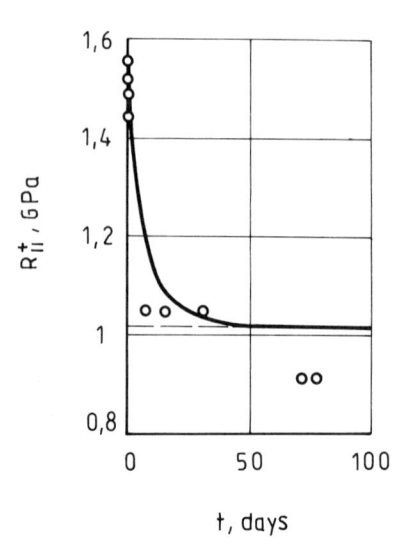

**FIGURE 7.14.** The long-term tensile strength curve in the reinforcement direction of a boron-epoxy.

## 7.4 CONTINUITY LOSS OF LAMINATED REINFORCED PLASTICS UNDER LONG-TERM PLANE STRESS STATE

Continuity loss in reinforced plastics laminates is caused by the failure of some of the most highly loaded plies. Continuity loss in multidirectionally ($n \geq 3$) reinforced plastics does not necessarily lead to their complete failure (the reaching of ultimate state). Cracking of plastics, however, changes their deformation properties, causes damage to the structural members and dramatically reduces their resistance to the environmental attack. Therefore the continuity loss in certain field conditions may be regarded as a strength criterion. In addition, the continuity loss of angle-ply laminates under arbitrary loads or cross-ply laminates in the presence of shear stresses in the reinforcement axes is a joint condition for their reaching the ultimate state. Therefore the construction of the continuity loss surface is a significant problem in reinforced plastic laminate mechanics.

Let us consider a multidirectionally reinforced plastic of prescribed structural design (see Fig. 4.1). In accordance with the methodology discussed in Chapter 6, from the known viscoelastic properties of the plies, determinable either by test or analytically from the component properties, it is possible to calculate the long-term stress state of a ply in the following way

$$\langle \sigma_\|(t)_i \rangle = \langle \sigma_\|(0)_i \rangle f_{\|i}(t)$$

$$\langle \sigma_\perp(t)_i \rangle = \langle \sigma_\perp(0)_i \rangle f_{\perp i}(t) \quad (7.62)$$

$$\langle \tau_{\|\perp}(t)_i \rangle = \langle \tau_{\|\perp}(0)_i \rangle f_{\|\perp i}(t)$$

where $f_{ji}(t), j = \|, \perp, \|\perp$ are dimensionless time functions.

When machine-computing the strength numerically, it is effective to use the set of discrete values $\langle \sigma_j(t)_i \rangle, j = \|, \perp, \|\perp, i = 1, \ldots, n$, with step $\Delta t$ determinable by the methodology described in Gurvich [24] instead of Equation (7.62). Then the long-term strength surface of a reinforced plastic is determined as the intercrossing of ply long-term strength surfaces in the axes $N_x$, $N_y$, $N_{xy}$, $t$. Likewise the ply failure may be caused by the failure of the fibers, the polymer matrix, or the bond between them. Therefore the long-term strength criterion of a reinforced plastic laminate in the case of continuity loss has one of the following forms depending upon the cause of failure [5]:

(1) Failure of the $i$-th ply fibers ($i = 1, \ldots, n$)

$$\int_0^{t_*} \frac{d\langle \sigma_\|(\tau)_i \rangle}{\psi R_B(t_* - \tau)} = \langle \sigma_\|(0)_i \rangle \int_0^{t_*} \frac{df_{\|i}(\tau)}{R_B(t_* - \tau)\psi} \leq 1 \quad (7.63)$$

(2) Failure of the polymer matrix of the $i$-th ply ($i = 1, \ldots, n$)

$$F^*\{\langle \sigma_\perp(t)_i \rangle, \langle \tau_{\|\perp}(t)_i \rangle\} \leq 1 \quad (7.64)$$

or in the developed form in accordance with Equations (7.35) and (7.36), provided

$$\langle \tau_{\|\perp}(t)_i \rangle / \langle \sigma_\perp(t)_i \rangle \leq a_1^*(t)$$

we have

$$\varphi^*\{\langle \sigma_\perp(0)_i \rangle^2 a_2^*(t) + \langle \tau_{\|\perp}(0)_i \rangle^2 a_3^*(t)\} \leq (R_A^+)^2 \quad (7.65)$$

but in the case of

$$\langle \tau_{\|\perp}(t)_i \rangle / \langle \sigma_{\perp}(t)_i \rangle a_1^*(t)$$

we have

$$\varphi^* \{ \langle \sigma_{\perp}(0)_i \rangle^2 a_4^*(t) + \langle \sigma_{\perp}(0)_i \rangle \langle \tau_{\|\perp}(0)_i \rangle a_5^*(t) \\ + \langle \tau_{\|\perp}(0)_i \rangle^2 a_6^*(t) \} \leq (R_A^+)^2 \quad (7.66)$$

Here the expressions $a_k^*(t)$, $k = 1, \ldots, 6$, are determined from the corresponding Equation (7.37) with a substitution according to the law

$$\bar{\sigma}_r(t) \to \bar{\sigma}_r^*(t)$$
$$\bar{\sigma}_\theta(t) \to \bar{\sigma}_\theta^*(t) \quad (7.67)$$

The expressions $\bar{\sigma}_r^*(t)$, $\bar{\sigma}_\theta^*(t)$ are obtained in accordance with the methodology discussed in Section 6.2:

$$\sigma_r^*(t)_i = f_{\perp i}(t)\bar{\sigma}_r(0) + \int_0^t f_{\perp i}(\tau) \bar{K}_r(t - \tau) d\tau$$
$$\sigma_\theta^*(t)_i = f_{\perp i}(t)\bar{\sigma}_\theta(0) + \int_0^t f_{\perp i}(\tau) \bar{K}_\theta(t - \tau) d\tau \quad (7.68)$$

(3) Bond failure in the $i$-th ply ($i = 1, \ldots, n$)

$$\int_0^{t_*} \frac{d[\bar{\sigma}_r^*(\tau)_i \langle \sigma_{\perp}(0)_i \rangle]}{R_b(t_* - \tau)} \\ + \left\{ \int_0^{t_*} \frac{d[\tau_{rz}^*(\tau)_i \langle \tau_{\|\perp}(0)_i \rangle]}{T_b(t_* - \tau)} \right\}^2 \leq 1 \quad (7.69)$$

and hence

$$\langle \sigma_{\perp}(0)_i \rangle \int_0^{t_*} \frac{d[\bar{\sigma}_r^*(\tau)_i]}{R_b(t_* - \tau)} \\ + \left\{ \langle \tau_{\|\perp}(0)_i \rangle \int_0^{t_*} \frac{d[\bar{\tau}_{rz}^*(\tau)_i]}{T_b(t_* - \tau)} \right\}^2 \leq 1 \quad (7.70)$$

where the methodology of Section 6.1 and 6.2 enables us to write

$$\tau_{rz}^*(t)_i = f_{\|\perp i}(t)\tau_{rz}(0) + \int_0^t f_{\|\perp i}(\tau) K_{rz}(t-\tau) d\tau \quad (7.71)$$

Let us consider long-term ultimate state of an angle-ply laminate under constant uniaxial loading in the elastic symmetry axis ($N_2 = N_{12} = 0$). The stress states in the plies according to Equation System (5.58) are determined by

$$\langle \sigma_\|(t) \rangle = N_1[\cos^2\beta - 2\sin\beta\cos\beta \bar{v}_{12}(t)]/h$$

$$\langle \sigma_\perp(t) \rangle = N_1[\sin^2\beta + 2\sin\beta\cos\beta \bar{v}_{12}(t)]/h \quad (7.72)$$

$$\langle \tau_{\|\perp}(t) \rangle = N_1[\sin\beta\cos\beta + (\cos^2\beta - \sin^2\beta)\bar{v}_{12}(t)]/h$$

where $h$ is the laminate thickness

$$\bar{v}_{12}(t) = \langle \tau_{12}(t) \rangle h/N_1 \quad (7.73)$$

Taking into account that at the instant of failure $R_1 = N_1/h$, according to Equation (7.63) the strength in the case of fiber failure is given by

$$R_1^+(t_*) = \left\{ \int_0^{t_*} \frac{d[\cos^2\beta - 2\sin\beta\cos\beta \bar{v}_{12}(\tau)]}{\psi R_B(t_* - \tau)} \right\}^{-1} \quad (7.74)$$

In the case of polymer matrix failure with $\langle \sigma_\perp(t) \rangle > 0$ (which corresponds to $\bar{v}_{12}(t) > 0{,}5 tg\beta$) Equations (7.65) and (7.66) lead to the following expression for the angle-ply laminate strength

$$R_1^+(t_*) = \frac{R_A^+}{\varphi^* \{g_1 + g_2\bar{v}_{12}(t) + g_3\bar{v}_{12}^2(t)\}^{1/2}} \quad (7.75)$$

with

$$\bar{v}_{12}(t) \leq \frac{a_1^*(t)\sin^2\beta - \sin\beta\cos\beta}{\cos^2\beta - \sin^2\beta - 2a_1^*(t)\sin\beta\cos\beta}$$

or

$$R_1^+(t_*) = \frac{R_A^+}{\varphi^* \{g_4 + g_5\bar{v}_{12}(t) + g_6\bar{v}_{12}^2(t)\}^{1/2}} \quad (7.76)$$

## 174  STRUCTURAL MECHANICS OF A MATERIAL

with

$$\bar{v}_{12}(t) > \frac{a_1^*(t) \sin^2 \beta - \sin \beta \cos \beta}{\cos^2 \beta - \sin^2 \beta - 2a_1^*(t) \sin \beta \cos \beta}$$

The designations used here are

$$g_1 = a_2^*(t) \sin^4 \beta + a_3^*(t) \sin^2 \beta \cos^2 \beta$$

$$g_2 = a_2^*(t) 4 \sin^3 \beta \cos \beta + a_3^*(t) \cdot 2 \sin \beta \cos \beta (\cos^2 \beta - \sin^2 \beta)$$

$$g_3 = a_2^*(t) \cdot 4 \sin^2 \beta \cos^2 \beta + a_3^*(t)(\cos^2 \beta - \sin^2 \beta)^2$$

$$g_4 = a_4^*(t) \sin^4 \beta + a_5^*(t) \sin^3 \beta \cos \beta + a_6^*(t) \sin^2 \beta \cos^2 \beta$$

$$g_5 = a_4^*(t) \cdot 4 \sin^3 \beta \cos \beta + a_5^*(t)(3 \sin^2 \beta \cos^2 \beta - \sin^4 \beta)$$
$$+ a_6^*(t) \cdot 2 \sin \beta \cos \beta (\cos^2 \beta - \sin^2 \beta)$$

$$g_6 = a_4^*(t) 4 \sin^2 \beta \cos^2 \beta + a_5^*(t) \cdot 2 \sin \beta \cos \beta (\cos^3 \beta - \sin^2 \beta)$$
$$+ a_6^*(t)(\cos^2 \beta - \sin^2 \beta)^2$$

For the bond failure between the fibers and the matrix, Equation (7.70) yields

$$R_1^+(t_*) = \frac{\sqrt{g_7^2 + 4g_8} - g_2}{2g_8} \qquad (7.77)$$

where

$$g_7 = \int_0^{t_*} \frac{d\{[\cos^2 \beta - 2 \sin \beta \cos \beta \bar{v}_{12}(\tau)]\bar{\sigma}_r^*(\tau)\}}{R_b(t_* - \tau)}$$

$$g_8 = \int_0^{t_*} \frac{d\{[\sin \beta \cos \beta + (\cos^2 \beta - \sin^2 \beta)\bar{v}_{12}(\bar{\tau})]\tau_z^*(\tau)\}}{T_b(t_* - \tau)}$$

Fig. 7.15 gives a comparison between test results [76] for the long-term strength of boron fiber reinforced angle-ply laminate with an angle

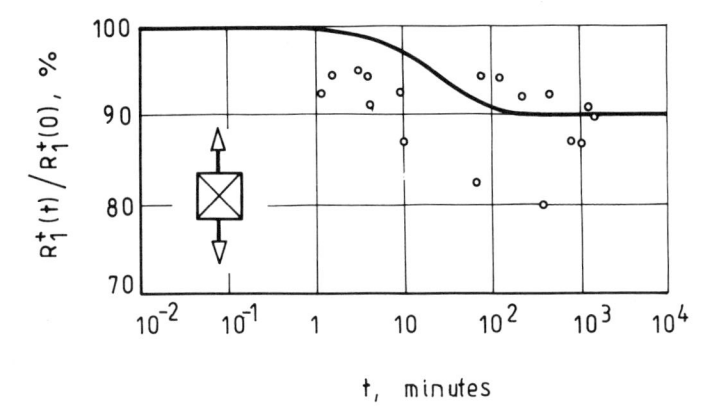

FIGURE 7.15. Time-dependent changes in the relative long-term tensile strength of a boron-epoxy angle-ply laminate.

of fiber orientation $\beta = 45°$ under uniaxial loading and a theoretical dependence plotted from Equations (7.75) and (7.76) with $E_B/E_A = 100$; $\psi = 0.6$; $\nu_A = 0.35$; $\nu_B = 0.23$; $K_A(t - \tau) = C_A \exp[-\alpha_A(t - \tau)]$; $C_A = 0.0033$ min$^{-1}$; $\alpha_A = 0.0289$ min$^{-1}$. Here, we apply a condition that the bond strength between boron fibers and the epoxy matrix is higher than the strength of this matrix. Note that the fiber failure in an angle-ply laminate takes place only at low values of angle $\beta$. Under biaxial loading ($N_{12} = 0$) the ply stress state in angle-ply laminates is given by

$$\langle \sigma_\parallel(t) \rangle = \{N_1[\cos^2\beta - 2\sin\beta\cos\beta\bar{v}_{12}(t)]$$
$$+ N_2[\sin^2\beta - 2\sin\beta\cos\beta\bar{v}'_{12}(t)]\}\frac{1}{h}$$

$$\langle \sigma_\perp(t) \rangle = \{N_1[\sin^2\beta + 2\sin\beta\cos\beta\bar{v}_{12}(t)]$$
$$+ N_2[\cos^2\beta + 2\sin\beta\cos\beta\bar{v}'_{12}(t)]\}\frac{1}{h}$$
(7.78)

$$\langle \tau_{\parallel\perp}(t) \rangle = \{N_1[\sin\beta\cos\beta + (\cos^2\beta - \sin^2\beta)\bar{v}_{12}(t)]$$
$$+ N_2[\sin\beta\cos\beta - (\cos^2\beta - \sin^2\beta)\bar{v}'_{12}(t)]\}\frac{1}{h}$$

**176** STRUCTURAL MECHANICS OF A MATERIAL

where

$$\bar{v}'_{12}(t, \beta) = \bar{v}_{12}(t, 90° - \beta)$$

By inserting Equation System (7.78) into the strength criteria of Equations (7.63) through (7.70) an equation is obtained for the strength surface of an angle-ply laminate under a biaxial stress state. As an example, consider a tube made of reinforced plastic laminate with the structure design [±β] loaded by internal pressure $p$. Then $N_1/h = p$; $N_2/h = 0.5p$. Using Equations (7.63) through (7.70) the fracture value is written as

(1) for fiber failure

$$p_*(t_*) = \left\{ \int_0^{t_*} \frac{d[\cos^2 \beta - 2 \sin \beta \cos \beta \bar{v}_{12}(\tau)]}{\psi R_B(t_* - \tau)} \right.$$

$$\left. + \frac{1}{2} \int_0^{t_*} \frac{d[\sin^2 \beta - 2 \sin \beta \cos \beta \bar{v}'_{12}(\tau)]}{\psi R_B(t_* - \tau)} \right\}^{-1} \quad (7.79)$$

(2) for the polymer matrix failure

$$p_*(t_*) = R_A^+ / \{ \varphi^* \{ g_1 + g_2 \bar{v}_{12}(t) + g_3 \bar{v}_{12}^2(t)$$

$$+ 0.25[g'_1 + g'_2 \bar{v}'_{12}(t) + g'_3(\bar{v}'_{12}(t))^2] + g_9 \quad (7.80)$$

$$+ g_{10} \bar{v}_{12}(t) + g_{11} \bar{v}'_{12}(t) + g_{12} \bar{v}_{12}(t) \bar{v}'_{12}(t) \} \}^{1/2}$$

if

$$\bar{v}_{12}(t)[\cos 2\beta - 4a_1^*(t) \sin 2\beta] - \bar{v}'_{12}(t)[0.5 \cos^2 \beta$$

$$+ 2a_1^*(t) \sin 2\beta] \leq 3a_1^*(t) - a_1^*(t) \cos 2\beta - 3 \sin 2\beta$$

otherwise

$$p_*(t_*) = R_A^+ / \{ \varphi^* \{ g_4 + g_5 \bar{v}_{12}(t) + g_6 \bar{v}_{12}^2(t)$$

$$+ 0.25[g'_4 + g'_5 \bar{v}'_{12}(t) + g'_6(\bar{v}'_{12}(t))^2] + g_{13} \quad (7.81)$$

$$+ g_{14} \bar{v}_{12}(t) + g_{15} \bar{v}'_{12}(t) + g_{16} \bar{v}_{12}(t) \bar{v}'_{12}(t) \} \}^{1/2}$$

Here

$$g_9 = \sin^2 \beta \cos^2 \beta [a_2^*(t) + a_3^*(t)]$$

$$g_{10} = \sin \beta \cos \beta [2 \cos^2 \beta a_2^*(t) + \cos 2\beta a_3^*(t)]$$

$$g_{11} = \sin \beta \cos \beta [2 \sin^2 \beta a_2^*(t) - \cos 2\beta a_3^*(t)]$$

$$g_{12} = 4 \sin^2 \beta \cos^2 \beta a_2^*(t) - \cos^2 2\beta a_3^*(t)$$

$$g_{13} = \sin \beta \cos \beta \{\sin \beta \cos \beta [a_4^*(t) + a_6^*(t)] + 0.5 a_5^*(t)\}$$

$$g_{14} = 2 \sin \beta \cos^3 \beta a_4^*(t) + \sin \beta \cos \beta \cos^2 \beta a_6^*(t)$$
$$+ 0.5 \cos^2 \beta a_5^*(t)$$

$$g_{15} = 2 \sin^3 \beta \cos \beta a_4^*(t) - \sin \beta \cos \beta \cos 2\beta a_6^*(t)$$
$$+ 0.5 \sin^2 \beta a_5^*(t)$$

$$g_{16} = 4 \sin^2 \beta \cos^2 \beta a_4^*(t) - \cos^2 2\beta a_6^*(t)$$

$$g_j'(\beta, t) = g_j(90° - \beta, t)$$

$$j = 1, \ldots, 16$$

(3) for the bond failure

$$p_*(t_*) = \frac{\sqrt{[g_7 + 0.5 g_7']^2 + 4 g_8 + 2 g_8'} - g_7 - 0.5 g_7'}{2 g_8 + g_8'} \quad (7.82)$$

Fig. 7.16 shows the calculated results of the dependence of $p_*$ upon loading time at angle $\beta = 50°$ for glass-epoxy ($E_B/E_A = 20$; $\psi = 0.6$; $C_A = 0.00245$ hrs$^{-1}$; $\alpha_A = 0.00418$ hrs$^{-1}$) and phenol formaldehyde based plastics ($E_B/E_A = 20$; $\psi = 0.6$; $C_A = 0.00293$ hrs$^{-1}$; $\alpha_A = 0.00763$ hrs$^{-1}$). The theoretical curves are based on Equations (7.80) and (7.81); the test data are from Bax [77]. Analyzing the data presented in Figs. 7.15 and 7.16 it may be concluded that the relative time-dependent change in the reinforced plastic laminate strength (specifically in that

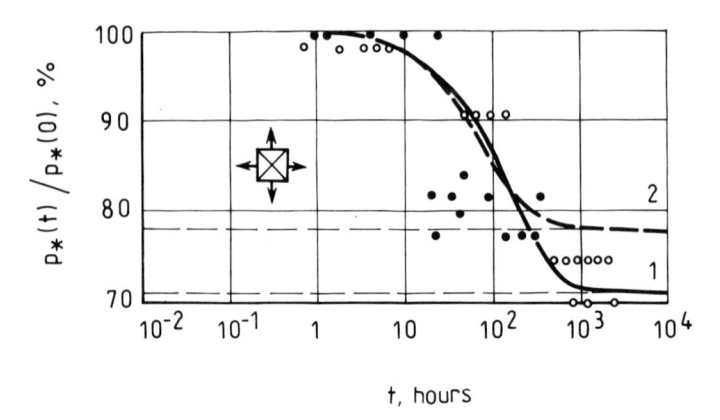

**FIGURE 7.16.** Relative dependence of the internal destructive pressure for unsymmetrically reinforced glass plastic tubes containing epoxy (1) and phenol formaldehyde (2) matrices upon the duration of loading: ○—epoxy matrix, ●—phenol formaldehyde matrix.

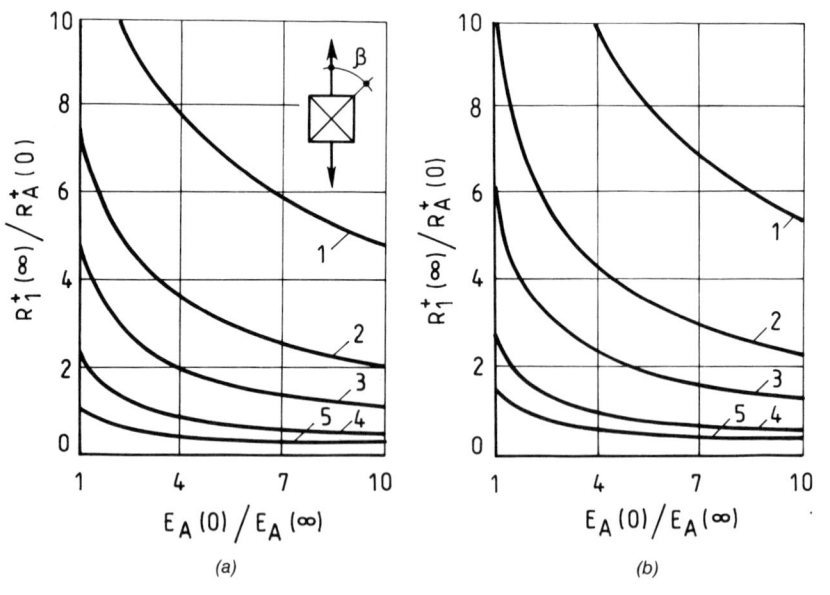

**FIGURE 7.17.** Ultimate strength dependence of angle-ply glass- (a) and carbon (b) fiber reinforced laminates with $R_A^+(0) = 1$ upon the degree of creep of the polymer matrix in the case of uniaxial tension: $B = 10°$ (1), 15 (2), 20 (3), 30 (4), 45° (5).

of an angle-ply laminate) under long-term loading is lower than the change in the unidirectionally reinforced material strength (the mechanical properties of the components being equal). This is explained by the redistribution of ply stress states during the viscoelastic deformation of a laminate. It was noted in Chapter 6 that under constant loading, stresses $\langle \sigma_\parallel(t)_i \rangle$ usually increase in the reinforcement direction and stresses $\langle \sigma_\perp(t)_i \rangle$ and $\langle \tau_{\parallel\perp}(t)_i \rangle$ decrease due to the increased components of the compliance tensor of the plies $S_{22}^{(i)}(t)$, $S_{66}^{(i)}(t)$. Therefore the reduction in the long-term strength is partly balanced by the laminated structure geometry. This is shown by an example of the ultimate tensile strength dependence ($t_* \to \infty$) of an angle-ply glass- ($E_B/E_A = 20$; $\psi = 0.6$) and carbon ($E_{Br}/E_A = 2$; $E_{Bz}/E_{Br} = 20$; $\psi = 0.6$) fiber reinforced laminates upon the degree of creep in the polymer matrix (supposing that the matrix failure is the cause of the whole failure) (Fig. 7.17). It should be remarked that time-dependent changes in the long-term strength surface of reinforced plastic laminates are close to affine transformation. Besides, the more essential the rheonomic properties of the plastics components, the higher the anisotropy of relative long-term strength properties ($R_{ij}(t)/R_{ij}(0)$).

## 7.5 ULTIMATE STATE OF REINFORCED PLASTIC LAMINATES UNDER LONG-TERM PLANE STRESS STATE

The continuity loss in reinforced plastic laminates, such as angle-ply laminates, in the general case of plane stress state, is the cause of complete failure of the whole material. The failure process of three- or more directionally reinforced materials or orthogonally reinforced materials under biaxial loading in the reinforcement direction is of a multistage character. After a continuity loss due to bond or matrix failure in some mostly loaded plies, the material is able to sustain an external load because the injured plies continue to take up the load in the reinforcement direction. With a further increase in the load or in the duration of loading, cracking takes place (bond failure or matrix failure) in the other plies, and finally the fiber failure initiates the ultimate state of the reinforced plastic. Every failure stage is marked by an abrupt redistribution of stresses, and it is only after the mostly loaded fibers have failed that an avalanche-type failure of the whole material starts.

The determination of the stress state in the plies at the instant before the initiation of its ultimate state should generally be made by analyzing the failure of the individual plies. In the first approximation, however, all the plies may be considered incapable of sustaining shear and transversal stresses, and the plies work only in the reinforcement direction.

It has been shown by various researchers [52, 55, 70] that for realistic multidirectionally glass-, boron- and carbon fiber reinforced plastics, the loads capable of causing the ultimate state and failure of the first ply differ by approximately one order. Therefore, in determining the ultimate state under long-term static loading, the strength properties of a ply in the transversal direction and under shear will be neglected, i.e., it is assumed that components of ply stiffness matrix $Q_{22}$, $Q_{12}$, $Q_{66}$ equal zero.

Let us consider the ultimate state in a long-term arbitrary plane stress state of a four-directionally reinforced laminate comprising a combination of angle-ply and orthogonal reinforcement design with the following structure: $m_{\beta=0°} = m_1$; $m_\beta = m_{-\beta} = (1 - m_1 - m_2)/2$; $m_{\beta=90°} = m_2$. In accordance with the assumptions made, the stiffness matrices of plies $[\bar{Q}_\beta]$ are written as

$$[\bar{Q}_\beta] = \begin{bmatrix} \cos^4 \beta & \sin^2 \beta \cos^2 \beta & \sin \beta \cos^3 \beta \\ \sin^2 \beta \cos^2 \beta & \sin^4 \beta & \sin^3 \beta \cos \beta \\ \sin \beta \cos^3 \beta & \sin^3 \beta \cos \beta & \sin^2 \beta \cos^2 \beta \end{bmatrix} E_\|(t) \quad (7.83)$$

where $E_\|(t)$ is the elasticity modulus of the ply in the reinforcement direction neglecting ply creep in the reinforcement direction as was shown above, i.e., $E_\|(t) = E_\|$. According to Equation (8.4) the elasticity matrix of a four-directionally reinforced laminate has the form

$$[Q] = E_\| \begin{bmatrix} m_1 + 4m \cos^4 \beta & m \sin^2 2\beta & 0 \\ m \sin^2 2\beta & m_2 + 4m \sin^4 \beta & 0 \\ 0 & 0 & m \sin^2 2\beta \end{bmatrix} \quad (7.84)$$

where $m = \frac{1}{4}(1 - m_1 - m_2)$. In such a case the strains of the laminate are determined by the relationships

$$\langle\langle \epsilon_1 \rangle\rangle = \frac{N_1}{hE_\|} \frac{[m_2 + (1 - m_1 - m_2) \sin^4 \beta]}{\Delta}$$

$$- \frac{N_2}{hE_\|} \frac{[1 - m_1 - m_2] \sin^2 \beta \cos^2 \beta}{\Delta}$$

$$\langle\langle \epsilon_2 \rangle\rangle = \frac{N_2}{hE_\|} \frac{[m_1 + (1 - m_1 - m_2) \cos^4 \beta]}{\Delta} \quad (7.85)$$

$$- \frac{N_1}{hE_\|} \frac{[1 - m_1 - m_2] \sin^2 \beta \cos^2 \beta}{\Delta}$$

$$\langle\langle \gamma_{12} \rangle\rangle = \frac{N_{12}}{hE_\|} \frac{1}{(1 - m_1 - m_2) \sin^2 \beta \cos^2 \beta}$$

where

$$\Delta = m_1 m_2 + (1 - m_1 - m_2)(m_1 \sin^4 \beta + m_2 \cos^4 \beta)$$

whereas the stresses in the reinforcement direction of the plies are given [6] by Equation System

$$\langle \sigma_\| \rangle_{\beta=0°} = [N_1 b_{11} + N_2 b_{12}] \frac{1}{h}$$

$$\langle \sigma_\| \rangle_{\beta=90°} = [N_1 b_{12} + N_2 b_{22}] \frac{1}{h}$$

$$\langle \sigma_\| \rangle_{\beta} = [N_1 b_{31} + N_2 b_{32} + N_{12} b_{33}] \frac{1}{h}$$

$$\langle \sigma_\| \rangle_{-\beta} = [N_1 b_{31} + N_2 b_{32} - N_{12} b_{33}] \frac{1}{h}$$

(7.86)

Here

$$b_{11} = [m_2 + (1 - m_1 - m_2) \sin^4 \beta]/\Delta$$

$$b_{22} = [m_1 + (1 - m_1 - m_2) \cos^4 \beta]/\Delta$$

$$b_{12} = -[(1 - m_1 - m_2) \sin^2 \beta \cos^2 \beta]/\Delta$$

$$b_{31} = m_2 \cos^2 \beta / \Delta$$

$$b_{32} = m_1 \sin^2 \beta / \Delta$$

$$b_{33} = [(1 - m_1 - m_2) \sin \beta \cos \beta]^{-1}$$

The ultimate state of a laminate is determined by the intersection of planes

$$\langle \sigma_\| \rangle_i \leq R_\|^+ = \psi R_B^+$$

$$i = 1, \ldots, 4,$$

(7.87)

as the stresses $\langle \sigma_\| \rangle_i$ are constant under constant loading. Finally we arrive at the long-term strength surface of a four-directionally reinforced laminate calculated according to the ultimate state in the form of the following system of equations

$$N_1 b_{11} + N_2 b_{12} \leq R_{\|}^{+}(t_*)h$$
$$N_1 b_{12} + N_2 b_{22} \leq R_{\|}^{+}(t_*)h \quad (7.88)$$
$$N_1 b_{31} + N_2 b_{32} + N_{12} b_{33} \leq R_{\|}^{+}(t_*)h$$

Fig. 7.18 illustrates long-term strength surfaces of a four-directionally reinforced laminate for $N_1, N_2 > 0$ with the following structure design: $m_1 = 0.4$, $m_2 = 0.2$; $\beta = 45°$. The dashed lines are plotted from Equation System (7.88) and correspond to the failure surfaces of each lay-up direction. The dashed area of the diagram shows the strength surface of the whole laminate which is affinely transformed in time.

In the particular case of three-directional structures ($m_2 = 0$) Equation System (7.86) enables us to write

$$\langle \sigma_{\|} \rangle_{\beta=0°} = \frac{N_1}{m_1 h} - \frac{N_2 \cos^2 \beta}{m_1 h \sin^2 \beta}$$

$$\langle \sigma_{\|} \rangle_{\pm \beta} = \frac{N_2}{(1-m_1)h \sin^2 \beta} \mp \frac{N}{(1-m_1)h \sin \beta \cos \beta} \quad (7.89)$$

(analogous expressions for three-directional structures have been shown in paper [48]). The long-term strength condition for three-directionally

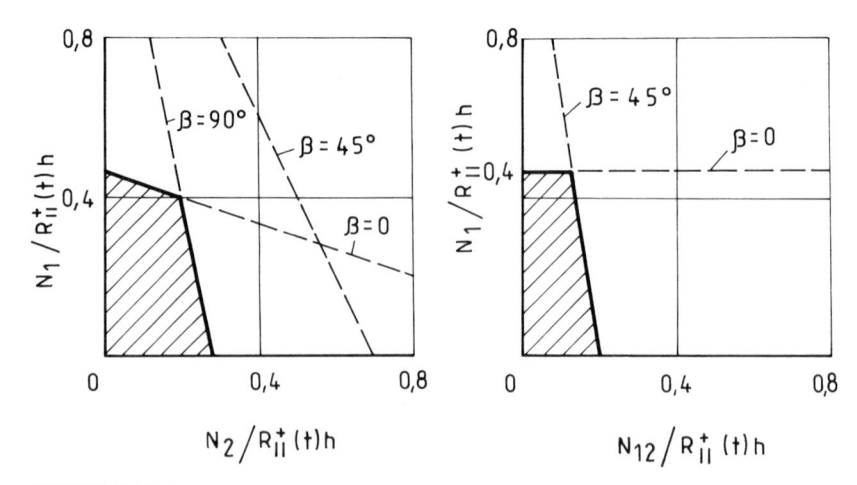

FIGURE 7.18. Long-term strength surfaces of four-directionally reinforced laminates plotted according to the ultimate state condition with $N_{12} = 0$ (a) and $N_2 = 0$ (b).

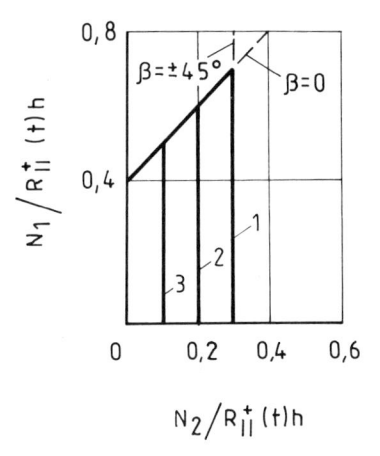

**FIGURE 7.19.** Long-term strength surface of a three-directionally reinforced laminate plotted according to the ultimate state condition with $N_{12} = 0$ (1); $0.1 \, R_{11}^+(t)h$ (2); $0.2 \, R_{11}^+(t)h$ (3).

reinforced plastics expressed through their ultimate state and taking into account Equation Systems (7.87) and (7.88) takes the form

$$\frac{N_1}{m_1} - \frac{N_2 \cos^2 \beta}{m_1 \sin^2 \beta} \leq R_{\|}^+(t_*)h$$

$$\frac{N_2}{(1 - m_1) \sin^2 \beta} \mp \frac{N_{12}}{(1 - m_1) \sin \beta \cos \beta} \leq R_{\|}^+(t_*)h \quad (7.90)$$

As an example, Fig. 7.19 illustrates the long-term strength surface of a three-directionally reinforced plastic with the following structure design: $m_1 = 0.4$; $\beta = 45°$. It should be noted that the expressions in Equation System (7.89) may also be obtained from equilibrium equations because a three-directionally reinforced plastic calculated by means of its ultimate state is a statically determinate system.

When the calculation scheme for reinforced plastic laminates is based on the initiation of their ultimate state, it is convenient to apply reinforcement in at least three directions, otherwise the continuity loss of the material will at the same time cause the ultimate state. An exception to the case is orthogonally reinforced plastics loaded biaxially in the reinforcement direction.

# Part II

# MECHANICS OF ELASTIC COMPOSITE BEAM

CHAPTER 8

# Transverse Bending of Beam

## 8.1 NORMAL STRESSES IN THE INDIVIDUAL PLIES

The geometric lay-up configuration of a beam is shown in Fig. 8.1. The stresses $\langle \sigma_z \rangle_k$ and $\langle \tau_{yz} \rangle_k$ are considered insignificant.
A generalized laminate deformation law is written as

$$\left\{ \frac{N}{M} \right\} = \left[ \begin{array}{c|c} A & B \\ \hline B & D \end{array} \right] \left\{ \frac{\langle E^0 \rangle}{k} \right\} \quad (8.1)$$

$$N = \int_{-h/2}^{h/2} \left[ \begin{array}{c} \langle \sigma_x \rangle \\ \langle \sigma_y \rangle \\ \langle \tau_{xy} \rangle \end{array} \right]_k dz \quad (8.2)$$

$$M = \int_{-h/2}^{h/2} \left[ \begin{array}{c} \langle \sigma_x \rangle \\ \langle \sigma_y \rangle \\ \langle \tau_{xy} \rangle \end{array} \right]_k z\, dz \quad (8.3)$$

$$A_{ij}; B_{ij}; D_{ij} = \sum_{k=1}^{n} (\bar{Q}_{ij})_k \left[ (h_k - h_{k-1}); \right.$$

$$\left. \frac{1}{2}(h_k^2 - h_{k-1}^2); \frac{1}{3}(h_k^3 - h_{k-1}^3) \right] \quad (8.4)$$

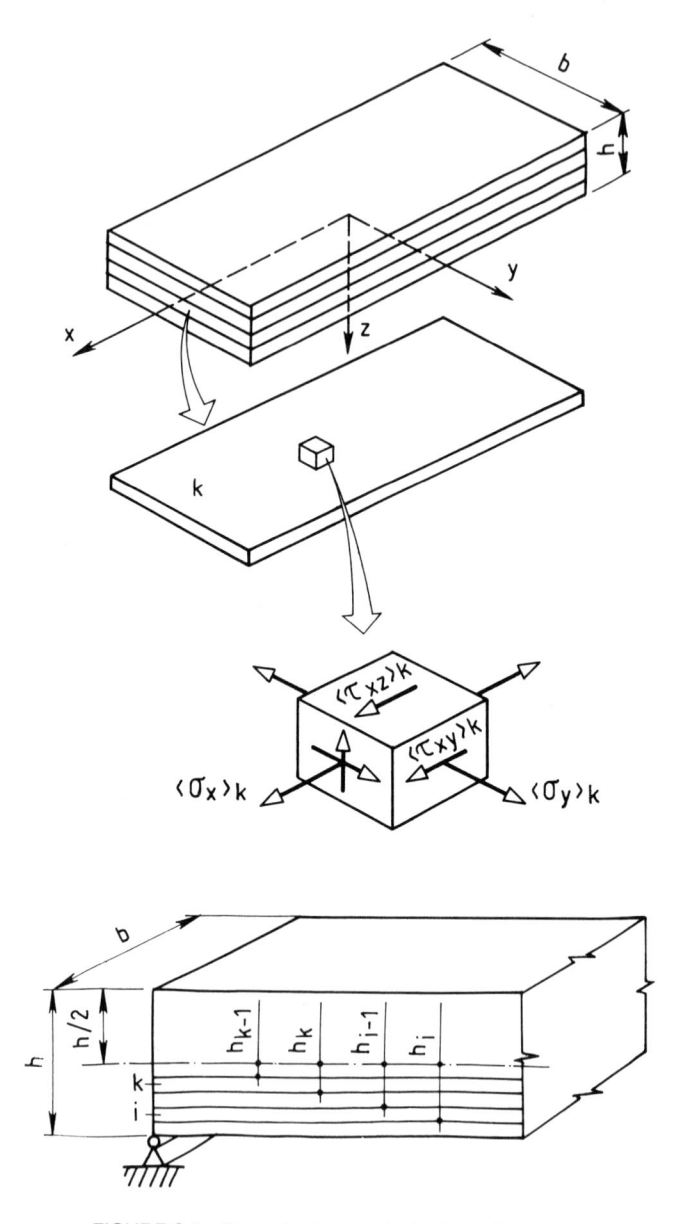

**FIGURE 8.1.** Geometric lay-up of a laminated beam.

The deformation law of Equation (8.1) can be used in an inverted form

$$\left\{ \frac{\langle E^0 \rangle}{k} \right\} = \left[ \begin{array}{c|c} A' & B' \\ \hline B' & D' \end{array} \right] \left\{ \frac{N}{M} \right\}$$

$$\begin{Bmatrix} \langle E_x^0 \rangle \\ \langle E_y^0 \rangle \\ \langle \gamma_{xy}^0 \rangle \\ k_x \\ k_y \\ k_{xy} \end{Bmatrix} = \begin{bmatrix} A'_{11} & A'_{12} & A'_{16} & B'_{11} & B'_{12} & B'_{16} \\ A'_{21} & A'_{22} & A'_{26} & B'_{21} & B'_{22} & B'_{26} \\ A'_{61} & A'_{62} & A'_{66} & B'_{61} & B'_{62} & B'_{66} \\ B'_{11} & B'_{21} & B'_{61} & D'_{11} & D'_{12} & D'_{16} \\ B'_{12} & B'_{22} & B'_{62} & D'_{21} & D'_{22} & D'_{26} \\ B'_{16} & B'_{26} & B'_{66} & D'_{61} & D'_{62} & D'_{66} \end{bmatrix} \begin{Bmatrix} N_x \\ N_y \\ N_{xy} \\ M_x \\ M_y \\ M_{xy} \end{Bmatrix} \quad (8.5)$$

where $A'_{ij}$, $B'_{ij}$ and $D'_{ij}$ are the elements of the inverted matrix of Equation (8.1).

It follows from Equation (8.4) that for angle-ply laminated plastics the stiffness elements $D_{16}$ and $D_{26}$ are not equal to zero, and hence the matrices $D_{ij}$ and $D'_{ij}$ will be fully populated.

When transverse bending is coupled with an axial force, $N_y = N_{xy} = M_y = M_{xy} = 0$ whereas $M_x \neq 0$ and $N_x \neq 0$. From the deformation law in Equation (8.5) the following relationships are obtained, for evaluating the midplane strains and curvature $k$

$$\langle \epsilon_x^0 \rangle = A'_{11} N_x + B'_{11} M_x$$

$$k_x = B'_{11} N_x + D'_{11} M_x$$

$$\langle \epsilon_y^0 \rangle = A'_{21} N_x + B'_{21} M_x$$

$$k_y = B'_{12} N_x + D'_{21} M_x \quad (8.6)$$

$$\langle \gamma_{xy}^0 \rangle = A'_{61} N_x + B'_{61} M_x$$

$$k_{xy} = B'_{16} N_x + D'_{61} M_x$$

According to the lamination theory, stress in the $k$-th layer is calculated from the relationship

$$\{\langle \sigma \rangle_k\} = [\bar{Q}]_k \{\langle \epsilon^0 \rangle + zk\} \quad (8.7)$$

In the fully developed form, Equation (8.7) for the stress $\langle \sigma_x \rangle_k$, for instance, is written as

$$\langle \sigma_x \rangle_k = (\bar{Q}_{11})_k \langle \epsilon_x^0 \rangle + (\bar{Q}_{12})_k \langle \epsilon_y^0 \rangle + (\bar{Q}_{16})_k \langle \gamma_{xy}^0 \rangle \\ + z[(\bar{Q}_{11})_k k_x + (\bar{Q}_{12})_k k_y + (\bar{Q}_{16})_k k_{xy}] \quad (8.8)$$

In order to calculate $\langle \sigma_x \rangle_k$ under combined transverse bending and axial load, the expression for $\langle \epsilon^0 \rangle$ and $k$ from Equation System (8.6) should be inserted into Equation (8.8). Considering that $M = bM_x$ and $N = bN_x$ we obtain

$$\langle \sigma_x \rangle_k = \frac{N}{b}(q_{1k} + z_k q_{3k}) + \frac{M}{b}(q_{2k} + z_k q_{4k}) = NW_1 + MW_2 \quad (8.9)$$

where

$$q_{1k} = (\bar{Q}_{11})_k A'_{11} + (\bar{Q}_{12})_k A'_{21} + (\bar{Q}_{16})_k A'_{61}$$

$$q_{2k} = (\bar{Q}_{11})_k B'_{11} + (\bar{Q}_{12})_k B'_{21} + (\bar{Q}_{16})_k B'_{61}$$

$$q_{3k} = (\bar{Q}_{11})_k B'_{11} + (\bar{Q}_{12})_k B'_{12} + (\bar{Q}_{16})_k B'_{16}$$

$$q_{4k} = (\bar{Q}_{11})_k D'_{11} + (\bar{Q}_{12})_k D'_{12} + (\bar{Q}_{16})_k D'_{61}$$

$$W_1 = \frac{q_{1k} + z_k q_{3k}}{b}$$

$$W_2 = \frac{q_{2k} + z_k q_{4k}}{b}$$

Equation (8.9) is the normal stress distribution law for the individual plies across the thickness of an arbitrarily configured laminated beam. It is significant that in applying Equation (8.9) it is not necessary to calculate the position of the neutral axis. This position may be found from Equation (8.8) when $\langle \sigma'_x \rangle_k = 0$.

It should be noted that in cases when $B'_{ij} = 0$ the maximum values of the normal stress $\langle \sigma_x \rangle_k$ will coincide with the maximum of $z_k[(\bar{Q}_{11})_k D'_{11} + (\bar{Q}_{12})_k D'_{21} + (\bar{Q}_{16})_k D'_{61}]$. This feature is characteristic of the stress state in the laminated beams, in contrast to homogeneous beams where the maximum normal stresses always take place in the fibers located farthest away from the neutral axis.

## Transverse Bending of Beam 191

If the beam is symmetrically laminated about its midplane and consists of a sufficiently large number of plies $n$, then $B_{ij} = 0$, $A'_{16} = 0$ and $D'_{61} = 0$. Equation (8.9) is thus simplified

$$\langle \sigma_x \rangle_k = NW_3 + MW_4 \tag{8.10}$$

where

$$W_3 = \frac{1}{b}[(\bar{Q}_{11})_k A'_{11} + (\bar{Q}_{12})_k A'_{21}]$$

$$W_4 = \frac{z}{b}[(\bar{Q}_{11})_k D'_{11} + (\bar{Q}_{12})_k D'_{21}]$$

In the above formulas $\langle \sigma_x \rangle_k$ should be used for $N = 0$ in the case of simple transverse bending.

If the ratio between the span $l$ and width $b$ of the beam's cross-section is reasonably large, the effect of $D'_{21}$ in bending may be neglected

$$\langle \sigma_x \rangle_k = \frac{M}{b} z_k (\bar{Q}_{11})_k D'_{11} \tag{8.11}$$

To evaluate the compliance element $D'_{11}$ we shall make use of the expressions for $N_x$ and $M_x$ given in Equations (8.2) and (8.3). Inserting $\langle \sigma_x \rangle_k$ into these relationships, where $\langle \sigma_x \rangle_k$ is calculated from the formula

$$\langle \sigma_x \rangle_k = (\bar{Q}_{11})_k (\langle \epsilon_x^0 \rangle + z_k k_x) \tag{8.12}$$

we obtain after integration

$$N_x = A_{11} \langle \epsilon_x^0 \rangle + B_{11} k_x \tag{8.13}$$

$$M_x = B_{11} \langle \epsilon_x^0 \rangle + D_{11} k_x \tag{8.14}$$

where $A_{11}$, $B_{11}$ and $D_{11}$ are calculated according to Equation (8.4). If no axial force is applied to the beam, i.e., $N_x = 0$, then, taking into account that $M = bM_x$, Equations (8.13) and (8.14) yield

$$M = \left[\left(D_{11} - \frac{B_{11}^2}{A_{11}}\right)b\right]k_x \tag{8.15}$$

The expression between the brackets is the effective bending stiffness $D_{11}^*$ of a beam

$$D_{11}^* = \frac{b}{D'_{11}} \approx \left(D_{11} - \frac{B_{11}^2}{A_{11}}\right)b = E^* J \qquad (8.16)$$

From this formula we obtain the following result

$$D'_{11} = \frac{A_{11}}{A_{11}D_{11} - B_{11}^2}$$

If the beam width $b = 1$, Equation (8.16) yields another formula for determining the effective elasticity modulus $E^*$ for a laminated beam in free bending

$$E^* = \frac{12}{h^3}\left(D_{11} - \frac{B_{11}^2}{A_{11}}\right) \qquad (8.17)$$

It should be remembered that the above relationships apply to the case when the length to width ratio of a beam is significantly large. In this case, Poisson's effect ($D_{12}$) and bending anisotropy ($D_{16}$) may be neglected.

If the beam is symmetrically laminated about its midplane, then $B_{11} = 0$ and

$$D_{11}^* = D_{11}b \qquad (8.18)$$

$$E^* = \frac{12}{h^3}D_{11} \qquad (8.19)$$

$$\langle \sigma_x \rangle_k = z_k \frac{(\bar{Q}_{11})_k}{bD_{11}} M \qquad (8.20)$$

Many types of reinforced plastics are characterized by two- or three-stage failure under tension and one-stage failure under compression.

Figs. 8.2 and 8.3 show characteristic deformation diagrams for cross-ply reinforced plastics and also fabric-reinforced plastics. According to the diagrams the elasticity moduli of the material are different in tension and compression after the first stage of failure, i.e. the material has become bimodular and the structure of the beam is no longer symmetric

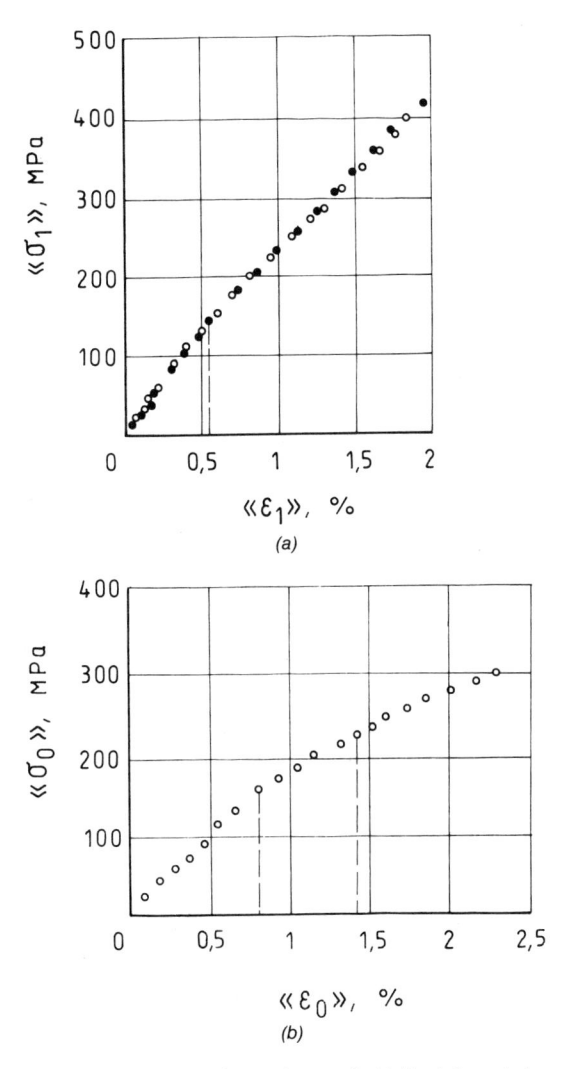

**FIGURE 8.2.** Deformation diagrams of an orthogonally (1:1) reinforced glass-epoxy and of a fabric reinforced glass-epoxy.

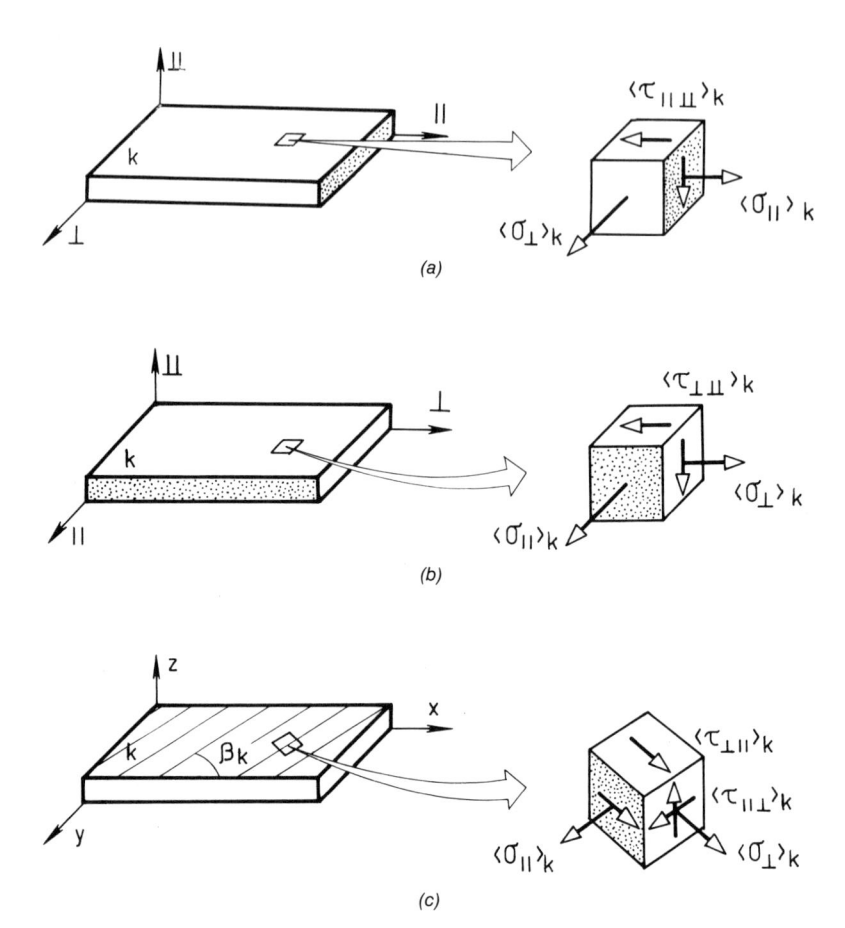

**FIGURE 8.3.** States of stress in individual plies: (a) axially reinforced ply; (b) ply reinforced perpendicular to the axis; (c) ply reinforced at an angle $B_x$ to the beam axis.

with respect to its midplane. For such a beam $B_{11} \neq 0$. Therefore Equation (8.18) is applicable only to the first stage of failure in the tensioned region of the beam. As the load continues to increase, the more general Equation (8.16) will be used, in which the effect of bimodularity has been included.

In addition to the stress $\langle \sigma_x \rangle_k$ Equation (8.7) yields formulas for calculating the distribution of the stresses $\langle \sigma_y \rangle_k$ and $\langle \tau_{xy} \rangle_k$ across the beam's thickness

$$\langle \sigma_y \rangle_k = \frac{N}{b}(q_{5k} + z_k q_{7k}) + \frac{M}{b}(q_{6k} + z_k q_{6k}) = NW_5 + MW_6 \quad (8.21)$$

where

$$W_5 = \frac{q_{5k} + z_k q_{7k}}{b}$$

$$W_6 = \frac{q_{6k} + z_k q_{8k}}{b}$$

$$q_{5k} = (\bar{Q}_{12})_k A'_{11} + (\bar{Q}_{22})_k A'_{21} + (\bar{Q}_{26})_k A'_{61}$$

$$q_{6k} = (\bar{Q}_{12})_k B'_{11} + (\bar{Q}_{22})_k B'_{21} + (\bar{Q}_{26})_k B'_{61}$$

$$q_{7k} = (\bar{Q}_{12})_k B'_{11} + (\bar{Q}_{22})_k B'_{12} + (\bar{Q}_{26})_k B'_{16}$$

$$q_{8k} = (\bar{Q}_{12})_k D'_{11} + (\bar{Q}_{22})_k D'_{21} + (\bar{Q}_{26})_k D'_{61}$$

$$\langle \tau_{xy} \rangle_k = \frac{N}{b}(q_{9k} + z_k q_{11k}) + \frac{M}{b}(q_{10k} + z_k q_{12k}) \quad (8.22)$$

$$= NW_7 + MW_8$$

where

$$W_7 = \frac{q_{9k} + z_k q_{11k}}{b}$$

$$W_8 = \frac{q_{10k} + z_k q_{12k}}{b}$$

$$q_{9k} = (\bar{Q}_{16})_k A'_{11} + (\bar{Q}_{26})_k A'_{21} + (\bar{Q}_{66})_k A'_{61}$$

$$q_{10k} = (\bar{Q}_{16})_k B'_{11} + (\bar{Q}_{26})_k B'_{21} + (\bar{Q}_{66})_k B'_{61}$$

$$q_{11k} = (\bar{Q}_{16})_k B'_{11} + (\bar{Q}_{26})_k B'_{12} + (\bar{Q}_{66})_k B'_{16}$$

$$q_{12k} = (\bar{Q}_{16})_k D'_{11} + (\bar{Q}_{26})_k D'_{21} + (\bar{Q}_{66})_k D'_{61}$$

## 8.2 SHEAR STRESSES IN THE INDIVIDUAL PLIES

To derive the distribution law for the shear stresses across the beam's thickness we shall use the equilibrium equation for an infinitely small volume of the $k$-th ply shown in Fig. 8.1. Assuming that the $k$-th ply is in a state of plane stress and that stresses do not depend on the coordinate $y$, the equilibrium equation is

$$\frac{\partial \langle \sigma_x \rangle_k}{\partial x} + \frac{\partial \langle \tau_{xz} \rangle_k}{\partial z} = 0$$

Integration of the equilibrium equation is performed to obtain an approximate expression for $\langle \tau_{xz} \rangle_k$

$$\langle \tau_{xz} \rangle_k = -\int_{-h/2}^{z} \frac{\partial \langle \sigma_x \rangle_k}{\partial x} dz$$

For a laminated beam of width $b$, this relationship is written as

$$\langle \tau_{xz} \rangle_k = -\frac{1}{b} \sum_{i=1}^{k} \int_{h_{i-1}}^{h_i} \frac{\partial \langle \sigma_x \rangle_i}{\partial x} dz \qquad (8.23)$$

The $\langle \sigma_x \rangle_i$ is obtained from a formula similar to Equation (8.8). Utilizing the familiar relationship between the bending moment $M$ and shear force $S$

$$S = \frac{\partial M}{\partial x}$$

we arrive at the following relationship for calculating the derivative of the stress $\langle \sigma_x \rangle_i$ with respect to $x$

$$\frac{\partial \langle \sigma_x \rangle_i}{\partial x} = S(q_{1i} + z_i q_{2i}) \qquad (8.24)$$

where

$$q_{1i} = (\bar{Q}_{11})_i B'_{11} + (\bar{Q}_{12})_i B'_{21} + (\bar{Q}_{16})_i B'_{61}$$

$$q_{2i} = (\bar{Q}_{11})_i D'_{11} + (\bar{Q}_{12})_i D'_{21} + (\bar{Q}_{16})_i D'_{61}$$

Taking into account Equation (8.24), Equation (8.23) becomes

$$\langle \tau_{xz} \rangle_k = \frac{1}{b} \sum_{i=1}^{k} \int_{h_{i-1}}^{h_i} (q_{1i} + zq_{2i}) S dz \qquad (8.25)$$

Integration across individual plies yields

$$\langle \tau_{xz} \rangle_k = SW_g \qquad (8.26)$$

where

$$W_9 = -\frac{1}{b} \sum_{i=1}^{k} \left[ q_{1i}(h_i - h_{i-1}) + \frac{1}{2} q_{2i}(h_i^2 - h_{i-1}^2) \right]$$

The shear stress $\langle \tau_{xz} \rangle_k$ will reach a maximum at a distance $h_k$ from the midplane of a beam, which is determined from the condition that the sum

$$\sum_{i=1}^{k} \left[ q_{1i}(h_i - h_{i-1}) + \frac{1}{2} q_{2i}(h_i^2 - h_{i-1}^2) \right]$$

attains its maximum value.

If the beam is symmetrically laminated with respect to its midplane and the number of plies is reasonably large, then

$$B'_{ij} = 0$$

$$D'_{61} = 0$$

$$q_{1i} = 0$$

$$q_{2i} = (\bar{Q}_{11})_i D'_{11} + (\bar{Q}_{12})_i D'_{21}$$

and Equation (8.26) is simplified as follows

$$\langle \tau_{xz} \rangle_k = -SW_{10} \qquad (8.27)$$

where

$$W_{10} = \frac{1}{2b} \sum_{i=1}^{k} [(\bar{Q}_{11})_i D'_{11} + (\bar{Q}_{12})_i D'_{21}](h_i^2 - h_{i-1}^2)$$

Further simplification of Equation (8.27) is obtained when $D'_{21} = 0$:

$$\langle \tau_{xz} \rangle_k = -\frac{S}{2b} \frac{D_{22}}{D_{11}D_{22} - D_{12}^2} \sum_{i=1}^{k} (\bar{Q}_{11})_i (h_i^2 - h_{i-1}^2) \quad (8.28)$$

Neglecting transverse effects, i.e., assuming that $D_{12} = 0$, we obtain

$$\langle \tau_{xz} \rangle_k = -\frac{S}{2bD_{11}} \sum_{i=1}^{k} (\bar{Q}_{11})_i (h_i^2 - h_{i-1}^2) \quad (8.29)$$

It follows from the above that in order to determine the shear stresses it is not necessary to also determine the position of the beam's neutral axis.

We shall note that for laminated structures the so-called edge effect induces a three-dimensional state of stress in the region adjacent to the free edge of the beam, which is approximately equal in length to the beam thickness. In order to reduce this edge effect, the width-to-depth ratio of the beam should be reasonably large.

## 8.3 STRENGTH CRITERIA OF A PLY

Assume that a laminated beam consisting of unidirectionally reinforced plies obeys general strength criteria as described by Skudra & Bulavs [55, 57, 58] and Sih & Skudra [80]. The possible various orientations of an arbitrary $k$-th ply are shown in Fig. 8.3. In general, the $k$-th ply will be subjected to the stresses $\langle \sigma_x \rangle_k$, $\langle \sigma_y \rangle_k$, $\langle \tau_{xy} \rangle_k$ and $\langle \tau_{xz} \rangle_k$.

When the reinforcement of the $k$-th ply coincides with the beam's longitudinal axis (Fig. 8.3(a)) there are two failure modes possible: (a) fiber failure under tension or compression and (b) matrix failure or failure of the bond between the fibers and the matrix.

In the case of fiber failure the effect of $\langle \tau_{\parallel\perp} \rangle_k$ may be disregarded, and subsequently the strength condition for the $k$-th ply can be expressed as

$$\langle \epsilon_x \rangle_k = \epsilon_{BR} \quad (8.30)$$

where $\langle \epsilon_x \rangle_k$ and $\epsilon_{BR}$ are the ultimate strains of the $k$-th ply and of the fibers, respectively.

The total strain of the $k$-th ply is expressed as

$$\langle \epsilon_x \rangle_k = \langle \epsilon_x^0 \rangle + z_k k_x \quad (8.31)$$

In a generalized way

$$\langle \epsilon_x^0 \rangle = A'_{11} N_x + B'_{11} M_x$$

$$k_x = B'_{11} N_x + D'_{11} M_x$$

By inserting the expressions for $\langle \epsilon_x^0 \rangle$ and $k_x$ into Equation (8.31) and taking into account that $N = bN_x$ and $M = bM_x$, Equation (8.30) becomes

$$A'_{11} N + B'_{11} M + z_k(B'_{11} N + D'_{11} M) = b\epsilon_{BR} \qquad (8.32)$$

In the case of matrix failure in the $k$-th ply due to the action of the stresses $\langle \sigma_{\|} \rangle_k$, $\langle \sigma_{\perp} \rangle_k$ and $\langle \tau_{\|\perp} \rangle_k$ (Fig. 8.3(a)), it is advisable to make the following assumptions:

- The stress $\langle \sigma_{\|} \rangle_k$ has a negligible effect on the stress state of the matrix.
- The interaction between the stresses $\langle \sigma_{\perp} \rangle_k$ and $\langle \tau_{\|\perp} \rangle_k$ has a negligible effect on the matrix strength.

With these assumptions, the $k$-th ply will fail due to longitudinal shear, and its strength can thus be written as

$$SW_9 \approx T_A \qquad (8.33)$$

where $T_A$ is the shear strength of the matrix.

Certain types of reinforced plastics, such as carbon/epoxy, are characterized by somewhat lower bond strength between the fibers and the matrix in comparison to matrix strength. Applying the criterion of the bond strength proposed [55, 57, 80], Equation (8.33) is written as

$$SW_9 = \frac{T_b}{\tau_{rz}} \qquad (8.34)$$

If the $k$-th ply is reinforced perpendicular to the beam's longitudinal axis (Fig. 8.3(b)), it will be subjected to the stresses $\langle \sigma_{\|} \rangle_k$, $\langle \sigma_{\perp} \rangle_k$ and $\langle \tau_{\perp\perp} \rangle_k$.

Assuming that the stress $\langle \sigma_{\|} \rangle_k$ has practically no effect on the stress state of the matrix, the strength criterion for the $k$-th ply is given by

$$\langle \sigma_{\perp}^2 \rangle_k + 2 \langle \tau_{\perp\perp}^2 \rangle_k (1 + \nu_{\perp\perp})$$
$$\pm \langle \sigma_{\perp} \rangle_k \sqrt{\langle \sigma_{\perp}^2 \rangle_k + 4 \langle \tau_{\perp\perp}^2 \rangle_k} = 2(R_{\perp}^+)_k^2 \qquad (8.35)$$

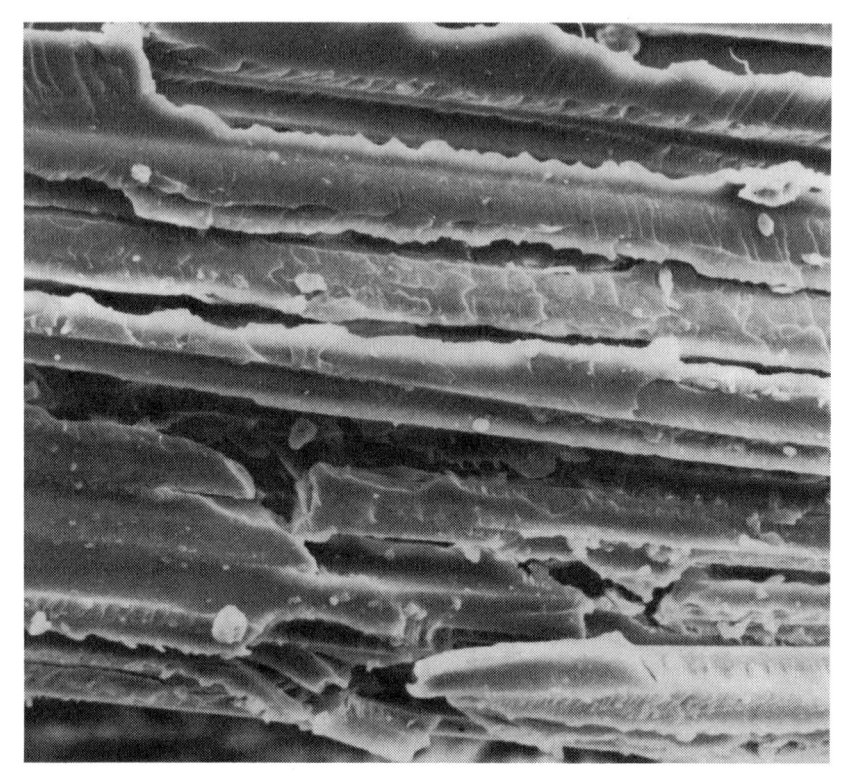

FIGURE 8.4. Failure mode of a unidirectionally reinforced glass-epoxy under transverse tension (enlarged ×500).

where $\nu_{\perp\parallel}$ and $(R_\perp^+)_k$ are the Poisson's ratio and the transverse tensile strength of a unidirectionally reinforced $k$-th ply, respectively. In this particular case $\beta = 90°$, and

$$\langle \sigma_\perp \rangle_k = NW_1 + MW_2$$
$$\langle \tau_{\perp\parallel} \rangle_k = SW_9 \tag{8.36}$$

Substitution of the expressions for the stresses $\pm\langle \sigma_\perp \rangle_k$ and $\langle \tau_{\perp\parallel} \rangle_k$ into Equation (8.35) yields

$$(NW_1 + MW_2)^2 + 2(1 + \nu_{\perp\parallel})S^2 W_9^2$$
$$\pm (NW_1 + MW_2)\sqrt{(NW_1 + MW_2)^2 + 4S^2 W_9^2} = 2(R_\perp^+)_k^2 \tag{8.37}$$

where [55, 57]

$$(R_\perp^+)_k = \frac{R_A^+}{\sqrt{\bar{\sigma}_r^2 + \bar{\sigma}_z^2 - 2\nu_A\bar{\sigma}_r\bar{\sigma}_z}} = \frac{R_A^2}{\bar{\sigma}_r\sqrt{1-\nu_A^2}} \qquad (8.38)$$

A characteristic mode of the matrix failure of a glass/epoxy under transverse tension is shown in Fig. 8.4.

If the bond strength is lower than the matrix strength, the strength is transverse tension included in Equation (8.38) is determined according to the formula [81]

$$(R_\perp^+)_k = \frac{R_b}{\bar{\sigma}_r} \qquad (8.39)$$

For the failure of the $k$-th ply in transverse compression, the strength criterion is given by

$$(NW_1 + MW_2)^2 + 2(1 + \nu_{\perp\parallel})S^2W_9^2 \\ + (NW_1 + MW_2)\sqrt{(NW_1 + MW_2)^2 + 4S^2W_9^2} = 2(R_\perp^-)_k^2 \qquad (8.40)$$

where the strength of the unidirectionally reinforced $k$-th ply in transverse compression $(R_\perp^-)_k$ is determined according to the formula [55, 57, 81]

$$(R_\perp^-)_k = 3{,}5\,\frac{(R_\perp^+)_k}{1 + \nu_{\perp\parallel}} \qquad (8.41)$$

If the $k$-th ply is reinforced at an arbitrary angle $\beta_k$ with respect to the beam's axis, then, as seen from Fig. 8.3(b), the stresses $\langle\sigma_y\rangle_k$, $\langle\tau_{xy}\rangle_k$, $\langle\sigma_x\rangle_k$ and $\langle\tau_{xz}\rangle_k$ will result in the following stresses in the material principal coordinate system of the ply

$$\langle\sigma_\parallel\rangle_k = \langle\sigma_x\rangle_k \cos^2\beta_k + \langle\sigma_y\rangle_k \sin^2\beta_x \\ + 2\langle\tau_{xy}\rangle_k \cos\beta_k \sin\beta_k \qquad (8.42)$$

$$\langle\sigma_\perp\rangle_k = \langle\sigma_x\rangle_k \sin^2\beta_k + \langle\sigma_y\rangle_k \sin^2\beta_k \\ + 2\langle\tau_{xy}\rangle_k \cos\beta_k \sin\beta_x \qquad (8.43)$$

$$\langle \tau_{\parallel\perp} \rangle_k = (\langle \sigma_y \rangle_k - \langle \sigma_x \rangle_k) \cos \beta_k \sin \beta_k \\ - \langle \tau_{xy} \rangle_k (\cos^2 \beta_k - \sin^2 \beta_k) \quad (8.44)$$

$$\langle \tau_{\parallel\perp\!\!\!\perp} \rangle_k \approx \langle \tau_{xz} \rangle_k \cos \beta_k \quad (8.45)$$

$$\langle \tau_{\perp\!\!\!\perp\perp} \rangle_k \approx \langle \tau_{xz} \rangle_k \sin \beta_k \quad (8.46)$$

In the relationships for $\langle \sigma_\parallel \rangle_k$, $\langle \sigma_\perp \rangle_k$ and $\langle \tau_{\parallel\perp} \rangle_k$ it is necessary to take into account the sign of the shear stress $\langle \tau_{xy} \rangle_k$ as shown schematically in Fig. 8.5.

Substituting the expressions for stresses $\langle \sigma_x \rangle_k$, $\langle \sigma_y \rangle_k$ and $\langle \tau_{xy} \rangle_k$ in Equations (8.42) through (8.46) we obtain

$$\langle \sigma_\perp \rangle_k = NF_1 + MF_2 \quad (8.47)$$

where

$$F_1 = W_1 \sin^2 \beta_k + W_5 \cos^2 \beta_k - 2W_7 \cos \beta_k \sin \beta_k$$

$$F_2 = W_2 \sin^2 \beta_k + W_6 \cos^2 \beta_k - 2W_8 \cos \beta_k \sin \beta_k$$

$$\langle \tau_{\parallel\perp} \rangle_k = NF_3 + MF_4 \quad (8.48)$$

where

$$F_3 = (W_5 - W_4) \cos \beta_k \sin \beta_k - W_7 \sin^2 \beta_k$$

$$F_4 = (W_6 - W_2) \cos \beta_k \sin \beta_k - W_8 \sin^2 \beta_k$$

$$\langle \sigma_\parallel \rangle_k = NF_5 + MF_6 \quad (8.49)$$

where

$$F_5 = W_1 \cos^2 \beta_k + W_5 \sin^2 \beta_k + 2W_7 \cos \beta_k \sin \beta_k$$

$$F_6 = W_2 \cos^2 \beta_k + W_6 \sin^2 \beta_k + 2W_8 \cos \beta_k \sin \beta_k$$

$$\langle \tau_{\parallel\perp\!\!\!\perp} \rangle_k \approx SW_9 \cos \beta_k \quad (8.50)$$

$$\langle \tau_{\perp\!\!\!\perp\perp} \rangle_k \approx SW_9 \sin \beta_k \quad (8.51)$$

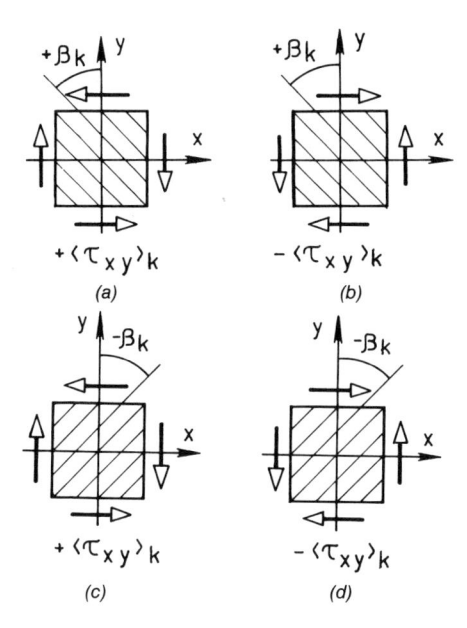

FIGURE 8.5. A scheme for determining the sign of the shear stress.

The contribution of the stress $\langle \sigma_\| \rangle_k$ to the matrix failure may be neglected in the case of polymeric matrices. Then the actual strength of the beam will be the lowest value determined from the following criteria:

(1) Failure due to the action of $\langle \tau_{\|\perp} \rangle_k$ and $\langle \sigma_\perp \rangle_k$
 • tensile failure of the matrix

$$(1 - \nu_A^2)\left(\frac{\bar{\sigma}_r}{R_A^+}\right)^2 (NF_1 + MF_2)^2 + \frac{1}{T_A^2}(NF_3 + MF_4)^2 = 1 \quad (8.52)$$

 • shear failure of the matrix

$$\begin{aligned}&(NF_1 + MF_2)^2(1 + \nu_A^2) + 2(1 + \nu_A)(NF_3 + MF_4)^2 \\&+ (NF_1 + MF_2)(1 + \nu_A) \\&\times \sqrt{(NF_1 + MF_2)^2(1 - \nu_A)^2 + 4(NF_3 + MF_4)^2} \\&= 2(R_A^+)^2\end{aligned} \quad (8.53)$$

- bond failure

$$(NF_1 + MF_2)T_b^2 \bar{\sigma}_r + (NF_3 + MF_4)^2 R_b \bar{\tau}_{rz}^2 = T_b^2 R_b \quad (8.54)$$

(2) Failure due to the action of $\langle \sigma_\perp \rangle_k$ and $\langle \tau_{||\perp} \rangle_k$:
- tensile failure of the matrix

$$(1 - \nu_A^2)\left(\frac{\bar{\sigma}_r}{R_A^+}\right)^2 (NF_1 + MF_2)^2 + \frac{1}{T_A^2} S^2 W_9^2 \sin^2 \beta_k = 1 \quad (8.55)$$

- shear failure of the matrix

$$\begin{aligned}
&(NF_1 + MF_2)^2(1 + \nu_A^2) + 2(1 + \nu_A)S^2 W_9^2 \sin^2 \beta_k \\
&+ (NF_1 + MF_2)(1 + \nu_A) \\
&\times \sqrt{(NF_1 + MF_2)^2(1 - \nu_A)^2 + 4S^2 W_9^2 \sin^2 \beta_k} \\
&= 2(R_A^+)^2
\end{aligned} \quad (8.56)$$

- bond failure

$$(F_1 N + MF_2)T_b^2 + \bar{\sigma}_r + S^2 W_9^2 \sin^2 \beta_k R_b \bar{\tau}_{rz}^2 = T_b^2 R_b \quad (8.57)$$

(3) Failure due to the action of $\langle \sigma_{||} \rangle_k$

$$\begin{aligned}
&(NW_1 + MW_2)\cos^2 \beta_k + (NW_5 + MW_6)\sin^2 \beta_k \\
&+ 2(NW_7 + NW_8)\sin \beta_k \cos \beta_k = E_{Bz} \epsilon_{BR}^+
\end{aligned} \quad (8.58)$$

The above relationships pertain to the case when the $k$-th ply lies in the tensile region of the laminated beam. In the compressed region the stresses $\langle \sigma_x \rangle_k$ and $\langle \sigma_\perp \rangle_k$ are compressive and thus the strength criteria for the $k$-th ply take the following form.

- For the failure of the $k$-th ply under transverse compression due to the joint action of $\langle \sigma_\perp \rangle_k$ and $\langle \tau_{||\perp} \rangle_k$

$$\begin{aligned}
&(NF_1 + MF_2)^2 + 2(1 + \nu_{\perp\perp})(NF_3 + MF_4)^2 \\
&+ (NF_1 + MF_2)\sqrt{(NF_1 + MF_2)^2 + 4(NF_3 + MF_4)^2} \\
&= 2(R_\perp^-)^2
\end{aligned} \quad (8.59)$$

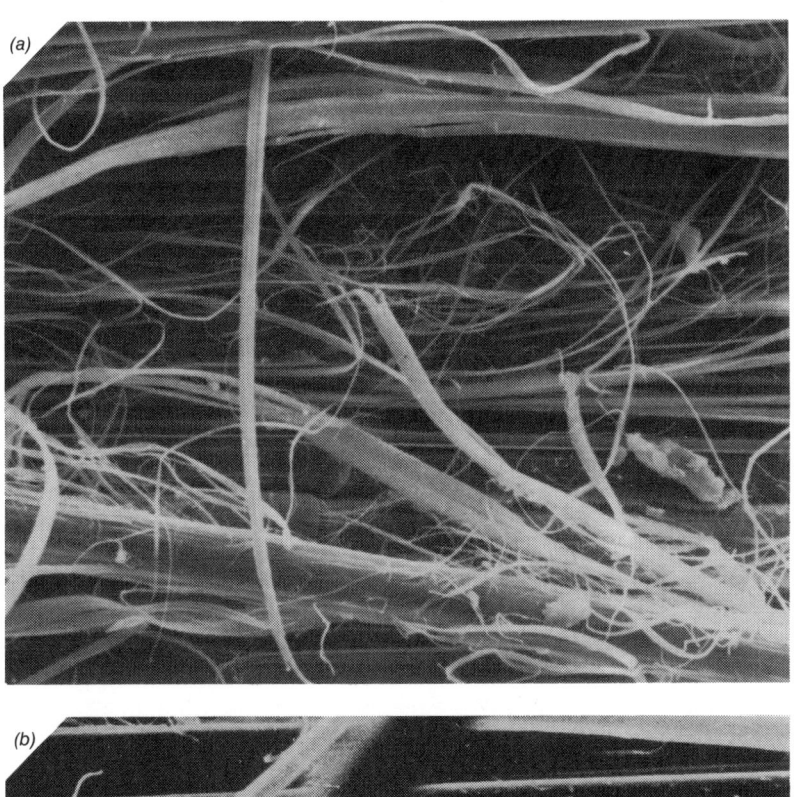

**FIGURE 8.6.** Failure modes of a unidirectionally reinforced organic plastic under (a) transverse tension and (b) compression (enlarged ×500).

- For the shear failure of the matrix due to the joint action of $\langle\sigma_\perp\rangle_k$ and $\langle\tau_{\|\perp}\rangle_k$

$$
\begin{aligned}
&(NF_1 + MF_2)^2(1 + \nu_A^2) + 2(1 + \nu_A)(NF_3 + MF_4)^2 \\
&- (NF_1 + MF_2)(1 + \nu_A) \\
&\times \sqrt{(NF_1 + MF_2)^2(1 - \nu_A)^2 + 4(NF_3 + MF_4)^2} \\
&= 2(R_A^+)^2
\end{aligned}
\quad (8.60)
$$

- Equation (8.30) for the axial compression failure of the fibers is

$$
\begin{aligned}
&(NW_1 + MW_2)\cos^2\beta_k + (NW_5 + MW_6)\sin^2\beta_k \\
&+ 2(NW_7 + NW_8)\sin\beta_k\cos\beta_k = E_{Bz}\bar{\epsilon}_{BR}
\end{aligned}
\quad (8.61)
$$

The above strength criteria are not applicable to plastics reinforced with organic fibers. This is explained by the fact that the transverse tensile and shear strengths of organic fibers are lower than the respective matrix strengths and therefore the failure of organic fiber reinforced plastics is always due to fiber failure. Characteristic types of failure of a unidirectionally reinforced organic plastic under tension and compression are shown in Fig. 8.6.

# CHAPTER 9

# Torsion of a Laminated Beam

## 9.1 THE COMPLEX NATURE OF INTER-PLY SHEAR IN REINFORCED PLASTIC LAMINATES

The structural peculiarities of reinforced plastic laminates (RPL) are responsible for higher demands on interply shear in designing RPL products. This type of consideration is presently based on the law for interply shear in the following form [86]:

$$\begin{Bmatrix} \langle\langle \gamma_{yz} \rangle\rangle \\ \langle\langle \gamma_{xz} \rangle\rangle \end{Bmatrix} = \begin{bmatrix} u_{44} & u_{45} \\ u_{54} & u_{55} \end{bmatrix} \begin{Bmatrix} S_{yz} \\ S_{xz} \end{Bmatrix}$$

or (9.1)

$$\{\langle\langle \gamma \rangle\rangle\} = [u]\{S\}$$

where $\langle\langle \gamma_{yz} \rangle\rangle$ and $\langle\langle \gamma_{xz} \rangle\rangle$ are the relative rotation angles of the normal and the tangent to the midplane surface of RPL in the planes $yz$ and $xz$ respectively (Fig. 9.1). Since the elastic symmetry axes of orthotropic composite layers usually lie at an angle to one another only within the lay-up plane, the interply shear of RPL is not interrelated with membrane-bending deformation of its midplane.

The values of compliance matrix components $[u]$ depend upon the elastic characteristics of a ply and also upon the laminated structure of the whole material, because it is this structure that governs the distribution of interply shear stresses $\langle \tau_{yz} \rangle$ and $\langle \tau_{xz} \rangle$ or $\{\langle \tau \rangle\}$ across the material. The distribution of stresses $\langle \tau_{yz} \rangle$ and $\langle \tau_{xz} \rangle$ may be determined

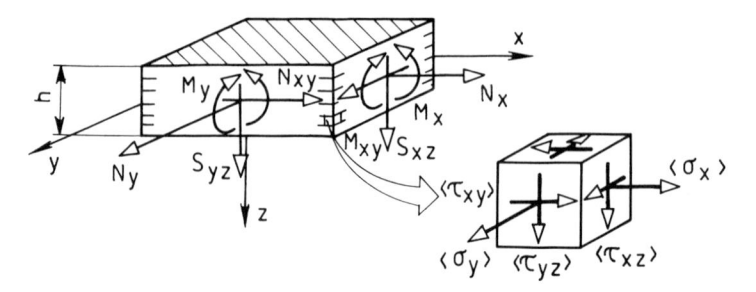

FIGURE 9.1. Calculation diagram of a reinforced plastic laminate.

by integrating differential equations of equilibrium across the thickness of the RPL (see Fig. 9.1).

$$\frac{\partial \langle \sigma_x \rangle}{\partial x} + \frac{\partial \langle \tau_{xy} \rangle}{\partial y} + \frac{\partial \langle \tau_{xz} \rangle}{\partial z} = 0$$

$$\frac{\partial \langle \sigma_y \rangle}{\partial y} + \frac{\partial \langle \tau_{xy} \rangle}{\partial x} + \frac{\partial \langle \tau_{yz} \rangle}{\partial z} = 0$$

(9.2)

Thus the following expression is obtained for $\langle \tau_{yz} \rangle$ and $\langle \tau_{xz} \rangle$

$$\left\{ \begin{array}{c} \langle \tau_{yz}(z) \rangle \\ \langle \tau_{xy}(z) \rangle \end{array} \right\} = -\int_{-h/2}^{z} \left\{ \begin{array}{c} \dfrac{\partial \langle \sigma_y \rangle}{\partial y} + \dfrac{\partial \langle \tau_{xy} \rangle}{\partial x} \\ \dfrac{\partial \langle \sigma_x \rangle}{\partial x} + \dfrac{\partial \langle \tau_{xy} \rangle}{\partial y} \end{array} \right\} dz \qquad (9.3)$$

where integration constants are absent due to the absence of shear stresses on the surface of the RPL. In order to express interply shear stresses through force factors and to include the structure of the RPL, the following relationships are inserted into Equation (9.3)

(1) The deformation law of component plies

$$\{\langle \sigma \rangle\} = [T]^{-1}[Q][T]^{-1T}\{\langle \epsilon \rangle\} = [\bar{Q}]\{\langle \epsilon \rangle\} \qquad (9.4)$$

where

$$\{\langle \sigma \rangle\}^T = \{\langle \sigma(z) \rangle\}^T = [\langle \sigma_x \rangle; \langle \sigma_y \rangle; \langle \tau_{xy} \rangle]$$

$$\{\langle \epsilon \rangle\}^T = \{\langle \epsilon(z) \rangle\}^T = [\langle \epsilon_x \rangle; \langle \epsilon_y \rangle; \langle \gamma_{xy} \rangle]$$

$$[T] = [T(z)]$$

$$[Q] = [Q(z)]$$

$[Q]$ is the stiffness matrix of a ply whereas matrix $[T]$ has been given in Chapter 1. Here $[T(z)]$ and $[Q(z)]$ refer to the ply containing coordinate $z$.

(2) Hypotheses of flat sections

$$\{\langle \epsilon(z) \rangle\} = \{\langle\langle \epsilon^0 \rangle\rangle\} + z\{k\} \tag{9.5}$$

(3) The law of membrane-bending deformation of a laminate

$$\left\{ \begin{array}{c} \{\langle\langle \epsilon^0 \rangle\rangle\} \\ \{k\} \end{array} \right\} = \left[ \begin{array}{cc} [A'] & [B'] \\ [B']^T & [D'] \end{array} \right] \left\{ \begin{array}{c} \{N\} \\ \{M\} \end{array} \right\} \tag{9.6}$$

where

$$\{N\}^T = [N_x; N_y; N_{xy}]$$

$$\{M\}^T = [M_x; M_y; M_{xy}]$$

By inserting Equations (9.4–9.6) into Equation (9.3) the interply shear stresses are expressed through all the derivatives of all the membrane-bending forces $\{\partial N/\partial x\}$, $\{\partial N/\partial y\}$, $\{\partial M/\partial x\}$ and $\{\partial M/\partial y\}$, i.e. any gradient of any internal force may in the general case cause interply shear stresses. It should be taken into account, however, that some part of these force derivatives are linked by the relationships

$$\frac{\partial N_x}{\partial x} + \frac{\partial N_{xy}}{\partial y} = 0; \quad \frac{\partial N_y}{\partial y} + \frac{\partial N_{xy}}{\partial x} = 0 \tag{9.7}$$

$$\frac{\partial M_x}{\partial x} + \frac{\partial M_{xy}}{\partial y} = S_{xzi}; \quad \frac{\partial M_y}{\partial y} + \frac{\partial M_{xy}}{\partial x} = S_{yz} \tag{9.8}$$

where these relationships are conditions of the plate element equilibrium. The equilibrium conditions in Equation (9.7) lead to the independence of interply shear stresses corresponding to the derivatives $\partial N_x/\partial x$ and $\partial N_y/\partial y$, on one hand, and to the derivatives $\partial N_{xy}/\partial y$ and $\partial N_{xy}/\partial x$ on the other hand. In a similar way the relationships in Equation (9.8) lead to the conclusion that only these interply shear stresses that correspond to the derivatives of moments $\partial M_x/\partial x$, $\partial M_y/\partial y$, $\partial M_{xy}/\partial y$ and $\partial M_{xy}/\partial x$ may yield effective shear forces, whereas the stresses corresponding to other derivatives are equalized across the RPL. The relations between derivatives of moments $\partial M_x/\partial x$, $\partial M_y/\partial y$, $\partial M_{xy}/\partial x$ and $\partial M_{xy}/\partial y$ may be of any kind, and this is why two types of shear forces are to be considered in an RPL. In order to explain this, we should refer to Equation (9.3), which illustrates the double nature of shear forces.

As is evident from Equation (9.3), the distribution of interply shear stresses in a RPL depends upon the distribution of normal and shear stresses acting in the planes of RPL plies. In a homogeneous or a laminated material consisting of isotropic layers the stress distribution curves across the thickness under bending and torsion are similar in form, but in a RPL that is anisotropic, these curves showing the distribution of normal and shear stresses acting in the layer planes differ substantially. This is due to the fact that, for instance, the plies having higher elasticity moduli may have low shear moduli in their planes. The different character of normal and shear stress distribution in the layer plane affects the distribution of interply shear stresses in a different way, and as a result affects the characteristics of the interply shear of the whole material, and we have to deal with two types of shear forces here. Firstly, these forces act in an equal manner and represent interply shear stresses that can be determined by integrating the distribution laws for normal and shear stresses acting in the ply planes $\langle \sigma_x \rangle$, $\langle \sigma_y \rangle$ and $\langle \tau_{xy} \rangle$ which appear under bending and torsion respectively, i.e.

$$\begin{Bmatrix} S_{yz}^b \\ S_{xz}^b \end{Bmatrix} = -\int_{-h/2}^{h/2} \int_{-h/2}^{z} \begin{Bmatrix} \dfrac{\partial \langle \sigma_y \rangle}{\partial y} \\ \dfrac{\partial \langle \sigma_x \rangle}{\partial x} \end{Bmatrix} dzdz$$

$$\begin{Bmatrix} S_{yz}^t \\ S_{xz}^t \end{Bmatrix} = -\int_{-h/2}^{h/2} \int_{-h/2}^{z} \begin{Bmatrix} \dfrac{\partial \langle \tau_{xy} \rangle}{\partial x} \\ \dfrac{\partial \langle \tau_{xy} \rangle}{\partial y} \end{Bmatrix} dzdz$$

(9.9)

Secondly, the shear forces of double character correspond to the derivatives of bending and torsional moments as indicated by superscripts "$b$" and "$t$," which means that the condition given in Equation (9.8) may be represented as

$$S_{yz}^b = \frac{\partial M_y}{\partial y} \; ; \quad S_{yz}^t = \frac{\partial M_{xy}}{\partial x}$$

$$S_{xz}^b = \frac{\partial M_x}{\partial x} \; ; \quad S_{xz}^t = \frac{\partial M_{xy}}{\partial y}$$

(9.10)

Taking into account the above, we obtain an expression for determining interply shear stresses from external force factors

$$-\{\langle \tau(z) \rangle\} = [V^b(z)]\{S^b\} + [V^t(z)]\{S^t\} + [V^\Phi(z)]\{\Phi\} \quad (9.11)$$

where

$$\{\Phi\}^T = \left[\frac{\partial N_x}{\partial x} = -\frac{\partial N_{xy}}{\partial y}; \frac{\partial N_y}{\partial y} = -\frac{\partial N_{xy}}{\partial x}; \frac{\partial N_x}{\partial y}; \frac{\partial N_y}{\partial x}; \frac{\partial M_x}{\partial y}; \frac{\partial M_y}{\partial x}\right]$$

$$[V^b(z)] = \begin{bmatrix} V_1 & V_2 \\ V_3 & V_4 \end{bmatrix}$$

$$[V^t(z)] = \begin{bmatrix} V_5 & V_6 \\ V_7 & V_8 \end{bmatrix}$$

$$[V^\Phi(z)] = \begin{bmatrix} V_9 & V_{10} & V_{11} & V_{12} & V_{13} & V_{14} \\ V_{15} & V_{16} & V_{17} & V_{18} & V_{19} & V_{20} \end{bmatrix}$$

$$V_1(z) = A_{2j}(z)B'_{j2} + B_{2j}(z)D'_{j2}$$

$$V_2(z) = A_{6j}(z)B'_{j1} - B_{6j}(z)D'_{j1}$$

$$V_3(z) = A_{6j}(z)B'_{j2} + B_{6j}(z)D'_{j2}$$

$$V_4(z) = A_{1j}(z)B'_{j1} - B_{1j}(z)D'_{j1}$$

$$V_5(z) = A_{6j}(z)B'_{j6} + B_{6j}(z)D'_{j6}$$

$$V_6(z) = A_{2j}(z)B'_{j6} + B_{2j}(z)D'_{j6}$$

$$V_7(z) = A_{1j}(z)B'_{j6} + B_{1j}(z)D'_{j6}$$

$$V_8(z) = A_{6j}(z)B'_{j6} + B_{6j}(z)D'_{j6}$$

$$V_9(z) = A_{6j}(z)A'_{j1} + B_{6j}(z)B'_{1j} - A_{2j}(z)A'_{j6} - B_{2j}(z)B'_{6j}$$

$$V_{10}(z) = -A_{6j}(z)A'_{j6} - B_{6j}(z)B'_{6j} + A_{2j}(z)A'_{j2} + B_{2j}(z)B'_{2j}$$

$$V_{11}(z) = A_{2j}(z)A'_{j1} + B_{2j}(z)B'_{1j}$$

$$V_{12}(z) = A_{6j}(z)A'_{j2} + B_{6j}(z)B'_{2j}$$

$$V_{13}(z) = A_{2j}(z)B'_{j1} + B_{2j}(z)D'_{j1}$$

$$V_{14}(z) = A_{6j}(z)B'_{j2} + B_{6j}(z)D'_{j2}$$

$$V_{15}(z) = A_{1j}(z)A'_{j1} + B_{1j}(z)B'_{1j} - A_{6j}(z)A'_{j6} - B_{6j}(z)B'_{6j}$$

$$V_{16}(z) = -A_{1j}(z)A'_{j6} - B_{1j}(z)B'_{6j} + A_{6j}(z)A'_{j2} + B_{6j}(z)B'_{2j}$$

$$V_{17}(z) = A_{6j}(z)A'_{j1} + B_{6j}(z)B'_{1j}$$

$$V_{18}(z) = A_{1j}(z)A'_{j2} + B_{1j}(z)B'_{2j}$$

$$V_{19}(z) = A_{6j}(z)B'_{j1} + B_{6j}(z)D'_{j1}$$

$$V_{20}(z) = A_{1j}(z)B'_{j2} + B_{1j}(z)D'_{j2}$$

$$A_{ij}(z) = \int_{-h/2}^{z} \bar{Q}_{ij}(z)dz$$

$$B_{ij}(z) = \int_{-h/2}^{z} \bar{Q}_{ij}(z)z\,dz$$

(summation is supposed to be made with respect to similar indices $i$, $j = 1, 2, 6$).

Fig. 9.2 shows distribution curves for interply shear stresses plotted according to Equation (9.11) for all of the ten factors separately.

The action of interply shear stresses causes interply shear strains $\langle \gamma_{yz} \rangle$ and $\langle \gamma_{xz} \rangle$ or $\{\langle \gamma \rangle\}$ in the plies, and may result in interply shear of the whole lay-up. The lay-up interply shear may be caused by shear forces as well as by force derivatives which result in interply shear stresses that balance themselves across the RPL surface (see the third right-hand term in Equation (9.11)).

First we shall determine the interply shear in a RPL due to the action of shear forces. With this aim, the equation expressing the action of external forces and the action of the internal stresses in the elementary area of RPL will be used. Here we have to distinguish between the two types of strains of the RPL interply shear: $\langle\langle \gamma^b_{yz} \rangle\rangle$, $\langle\langle \gamma^b_{xz} \rangle\rangle$ and $\langle\langle \gamma^t_{yz} \rangle\rangle$, $\langle\langle \gamma^t_{xz} \rangle\rangle$ or $\{\langle\langle \gamma^b \rangle\rangle\}$ and $\{\langle\langle \gamma^t \rangle\rangle\}$ originating from

∂Mₓ/∂x=1  ∂Mxy/∂y=1  ∂Mxy/∂x=1

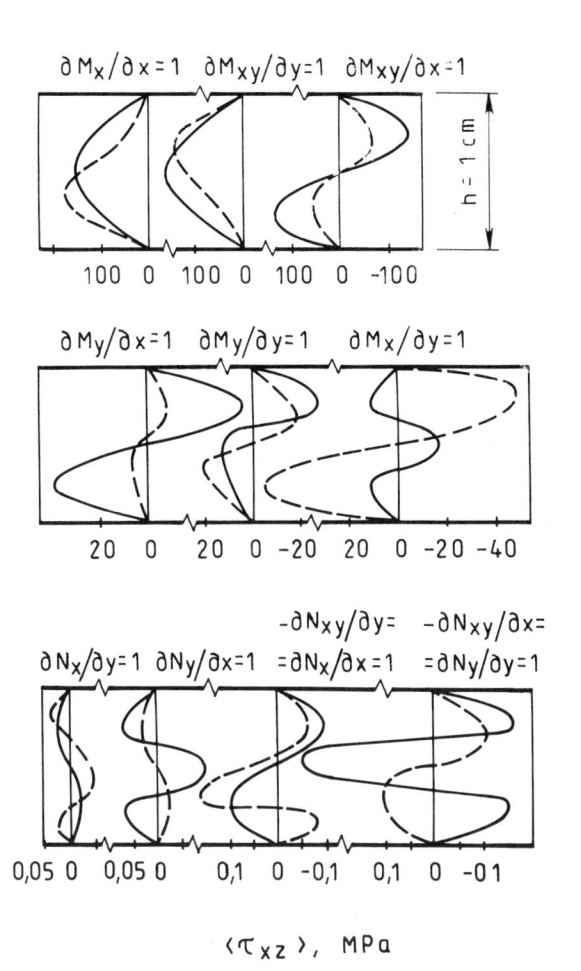

h = 1 cm

100  0  100  0  100  0  -100

∂My/∂x=1  ∂My/∂y=1  ∂Mₓ/∂y=1

20  0  20  0  -20  20  0  -20  -40

∂Nₓ/∂y=1  ∂Ny/∂x=1   -∂Nxy/∂y= =∂Nₓ/∂x=1   -∂Nxy/∂x= =∂Ny/∂y=1

0,05  0  0,05  0   0,1  0  -0,1   0,1  0  -0,1

⟨τ_xz⟩, MPa

**FIGURE 9.2.** Distribution of interply shear stresses $\langle \tau_{xz} \rangle$ across the RPL thickness having the structure (+25°/−25°) (1) and an RPL containing an aluminium ply and the structure (+45°/Al/0°) (2) under the action of unit forces. The curves have been built from formula (9.11) with the initial data [54]: $E_{A1} = 70$ EPa; $E_{A1} = 26$ EPa; $E_\parallel = 178$ EPa; $E_\perp = 7$ EPa; $\nu_{\parallel\perp} = 0.28$; $E_{\parallel\perp} = 3.8$ EPa; $E_{\perp\parallel} = 2.7$ EPa.

bending or torsion respectively. The equation expressing equality of work will be

$$\int \{\langle \tau \rangle\}^T \{\langle \gamma \rangle\} dz = \begin{Bmatrix} \{S^b\} \\ \{S^t\} \end{Bmatrix}^T \begin{Bmatrix} \{\langle\langle \gamma^b \rangle\rangle\} \\ \{\langle\langle \gamma^t \rangle\rangle\} \end{Bmatrix} \quad (9.12)$$

Further, the following relationships will be consequently inserted into Equation (9.12).

- The law of interply shear under the action of shear forces only and supposedly including a symmetric compliance matrix is to be determined as

$$\begin{Bmatrix} \{\langle\langle\gamma^b\rangle\rangle\} \\ \{\langle\langle\gamma^t\rangle\rangle\} \end{Bmatrix} = \begin{bmatrix} [u] & [g] \\ [g] & [p] \end{bmatrix} \begin{Bmatrix} \{S^b\} \\ \{S^t\} \end{Bmatrix} \qquad (9.13)$$

- The law of interply shear for a component ply is

$$\{\langle\gamma\rangle\} = [P]^T[S][P]\{\langle\tau\rangle\} = [\bar{S}]\{\langle\tau\rangle\} \qquad (9.14)$$

where

$$[S] = \begin{bmatrix} \dfrac{1}{G_{44}} & \\ & \dfrac{1}{G_{55}} \end{bmatrix}$$

whereas matrix $[P]$ may be found in Chapter 4.
- The expression for determining the interply shear stresses arising from shear forces (the first two right-hand terms of Equation (9.11)) will be

$$\{\langle\tau\rangle\} = [V^b]\{S^b\} + [V^t]\{S^t\} \qquad (9.15)$$

By substitution we obtain the submatrices of the sought compliance matrix of the law of RPL interply shear

$$[u] = \int_{-h/2}^{h/2} [V^b]^T[S][V^b]dz$$

$$[p] = \int_{-h/2}^{h/2} [V^t]^T[\bar{S}][V^t]dz \qquad (9.16)$$

$$[g] = \int_{-h/2}^{h/2} [V^b]^T[S][V^t]dz$$

In this way the symmetry of the complete compliance matrix in the interply shear law of Equation (9.13) is confirmed.

The practical determination of the submatrix components $[u]$, $[p]$ and $[g]$ from Equation System (9.16) is possible by using summation instead of integration. At first sight there seems to be an unexpected dependence of the result, namely, the elastic characteristics of RPL, upon the number of summands. Such dependence is explained by the non-linearity of the distribution of interply shear stresses across the thickness of RPL. Due to this non-linearity, the accuracy in describing the distribution of interply shear stresses and consequently the accuracy in determining the elastic characteristics of interply shear in a RPL is increased with the layer number of summands. Fig. 9.3 represents some components of submatrices $[u]$, $[p]$ and $[g]$ obtained from Equation System (9.16) for a RPL with various numbers of summands when summation was substituted for integration.

The following formula makes it possible to transform the compliance matrix of the interply shear of RPL

$$\begin{bmatrix} [\bar{u}] & [\bar{g}] \\ [\bar{g}]^T & [\bar{p}] \end{bmatrix} = \begin{bmatrix} [P] & \\ & [P] \end{bmatrix}^T \begin{bmatrix} [u] & [g] \\ [g]^T & [p] \end{bmatrix} \begin{bmatrix} [P] & \\ & [P] \end{bmatrix} \quad (9.17)$$

The analysis of the results thus obtained reveals quite an unexpected fact. It turns out that in one cross-section of the RPL the shear forces applied simultaneously in one direction, e.g. $S_{xz}^b$ and $S_{xz}^t$, may perform work on various displacements: $\langle\langle \gamma_{xz}^b \rangle\rangle dx$ and $\langle\langle \gamma_{xz}^t \rangle\rangle dx$ respectively. In addition, the angles $\langle\langle \gamma_{xz}^b \rangle\rangle$ and $\langle\langle \gamma_{xz}^t \rangle\rangle$ lying in the

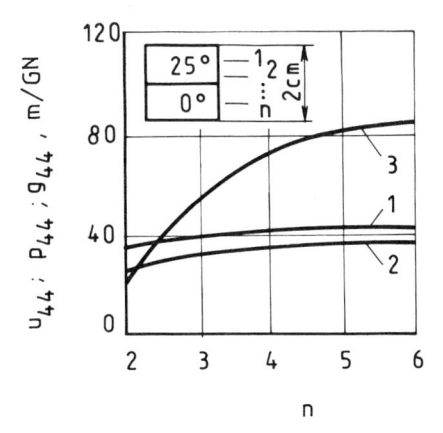

FIGURE 9.3. Dependence of calculated values of compliances $u_{44}$ (1), $p_{44}$ (2) and $g_{44}$ (3) of interply shear for RPL having the structure (25°/0°) upon the number of summation terms when integration was replaced by summation in (9.16). The initial data are those of Fig. 9.2.

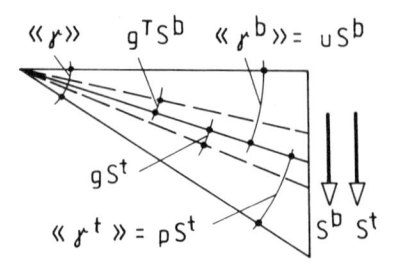

FIGURE 9.4. The structure of interply shear angles in RPL.

same plane cannot be summarized in order to obtain the virtually necessary angle of interply shear $\langle\langle \gamma_{xz} \rangle\rangle$ which is to be found along with the angle $\langle\langle \gamma_{yz} \rangle\rangle$ in the following way

$$\{\langle\langle \gamma \rangle\rangle\} = [u]\{S^b\} + [p]\{S^t\} \tag{9.18}$$

The above conclusions are based on the varying distributions of interply shear stresses across the thickness of the RPL, the stresses originating from bending and torsion. The structure of the actually observed shear angle is shown in Fig. 9.4, which also shows the usefulness of the interactive matrix $[g]$: the matrix $[g]$ components make it possible to determine the value of the angles $\{\langle\langle \gamma^b \rangle\rangle\}$ while representing at the same time angles $\{\langle\langle \gamma^t \rangle\rangle\}$ and vice versa.

In order to show the obtained data more clearly it is useful to introduce matrices $[\tilde{u}]$, $[\tilde{p}]$, and $[\tilde{g}]$ because their components can show how the nonuniform distribution of interply shear stresses affect the elastic characteristics of the interply shear in an RPL. The matrix components of $[\tilde{u}]$, $[\tilde{p}]$, and $[\tilde{g}]$ are the ratios of the matrix components of $[u]$, $[p]$, and $[g]$ to the corresponding conditional components obtained from the law of mixture supposing that interply shear stresses are uniformly distributed across the RPL thickness

$$\tilde{u}_{ij} = \frac{u_{ij}}{a_{ij}}; \qquad \tilde{p}_{ij} = \frac{p_{ij}}{a_{ij}}; \qquad \tilde{g}_{ij} = \frac{g_{ij}}{a_{ij}} \tag{9.19}$$

where

$$a_{ij} = \frac{1}{h^2} \sum_{r=1}^{n} \bar{S}_{ij}^r t_r$$

$$i, j = 4, 5$$

Here $r$ and $t_r$ denote the number and the thickness of the component plies respectively. It should be remembered that for homogeneous materials where shear stresses are distributed according to the law of the quadratic parabola [69], $\tilde{u} = \tilde{p} = \tilde{g} = 1, 2$. Fig. 9.5 illustrates theoretical relationships between matrix components $[\tilde{u}]$, $[\tilde{p}]$, and $[\tilde{g}]$ and the structural parameters of RPL.

As has been stated earlier, the strains of interply shear in RPL may result from self-balancing interply shear stresses determined by the third right-hand summand of Equation (9.11) though they do not create shear forces. In this case, determination of the strains of interply shear will be possible by using the equation of virtual work created by interply shear stresses corresponding to all the four shear forces $S^b_{yz}$, $S^b_{xz}$, $S^t_{yz}$ and $S^t_{xz}$ on one hand, and by the shear forces themselves on the virtual displacements under the action of external factors expressed in the form of derivatives $\partial N_x/\partial x = -\partial N_{xy}/\partial y$, $\partial N_y/\partial y = -\partial N_{xy}/\partial x$, $\partial N_x/\partial y$, $\partial N_y/\partial x$, $\partial M_x/\partial y$, $\partial M_y/\partial x$. The approach used already for obtaining the law of interply shear in Equation (9.13) will help us to arrive at the relation between angles $\langle\langle \gamma^b_{yz} \rangle\rangle$, $\langle\langle \gamma^b_{xz} \rangle\rangle$, $\langle\langle \gamma^t_{yz} \rangle\rangle$ and $\langle\langle \gamma^t_{xz} \rangle\rangle$ and the above derivatives. As a matter of fact, angles of interply shear should be observed in the RPL with these derivatives present

$$\langle\langle \gamma_{yz} \rangle\rangle = \langle\langle \gamma^b_{yz} \rangle\rangle + \langle\langle \gamma^t_{yz} \rangle\rangle$$

$$\langle\langle \gamma_{xz} \rangle\rangle = \langle\langle \gamma^b_{xz} \rangle\rangle + \langle\langle \gamma^t_{xz} \rangle\rangle$$

or

$$\{\langle\langle \gamma \rangle\rangle\} = \{\langle\langle \gamma^b \rangle\rangle\} + \{\langle\langle \gamma^t \rangle\rangle\}$$

Summation of angles is possible here (unlike angles determined by the law of interply shear in Equation (9.13)) because the angles due to

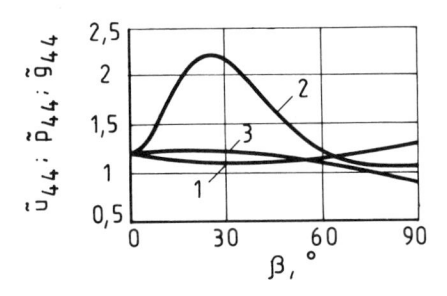

FIGURE 9.5. Dependence of $\tilde{u}_{44}$ (1), $\tilde{p}_{44}$ (2) and $\tilde{g}_{44}$ (3) of interply shear for an RPL of ($B/0°$) structure upon angle $B$. The initial data are those of Fig. 9.2.

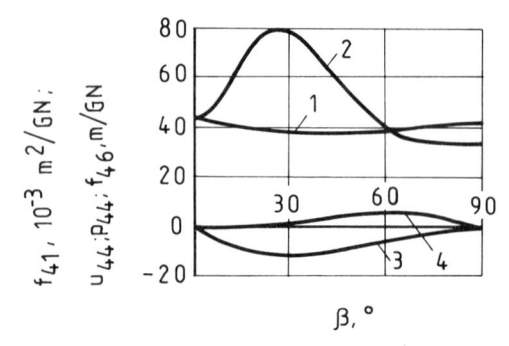

FIGURE 9.6. Dependence of $u_{44}$ (1), $p_{44}$ (2), $f_{46}$ (3) and $f_{41}$ (4) for an RPL of ($B/0°$) structure upon angle $B$. The initial data are those of Fig. 9.2.

bending and torsion under the action of the above mentioned factors are only determinable when taking the same type of shear forces into consideration. With this statement and Equation (9.18) in mind we arrive at the expression for the summed angles of interply shear in an RPL using all the force factors

$$\{\langle\langle\gamma\rangle\rangle\} = [u]\{S^b\} + [p]\{S^t\} + [f]\{\Phi\}$$

where

$$[f] = \begin{bmatrix} f_{41} & f_{42} & f_{43} & f_{44} & f_{45} & f_{46} \\ f_{51} & f_{52} & f_{53} & f_{54} & f_{55} & f_{56} \end{bmatrix}$$

$$= \int_{-h/2}^{h/2} [[V^b(z)]^T + [V^t(z)]^T][\bar{S}][V^\Phi(z)]dz$$

Fig. 9.6 presents theoretical relationships between some components of matrices $[u]$, $[p]$, and $[f]$ and the structural parameters of the RPL, which give an idea about their relative contributions to the interply shear in the RPL, shear forces, and other factors.

## 9.2 TORSION OF A FLAT LAMINATED BEAM

The problem of torsion for laminated rectangular cross-section beams has already been discussed in various papers [3, 41, 42]. In Mushelishvili

[42] we find a solution for a two-layer beam of isotropic bands having various elastic properties. In Arutyunyan & Abramou [3] and Lekhnitskii [41] we find a discussion of beams made of an arbitrary number of isotropic and orthotropic layers respectively and of arbitrary thicknesses. The results in these papers have been obtained by using stress functions and have the form of infinite series. Arutyunyan & Abramou [3] and Lekhnitskii [41] present simpler solutions for the specific case of three-ply materials having a symmetrical structure with respect to the midplane. The case of torsion of a non-homogeneous orthotropic beam of a rectangular cross-section, and with shear moduli varying along one edge according to the exponential law, is mainly of theoretical importance.

This paragraph includes the engineering approach to the solution of the beam torsion problem (Fig. 9.7) where the beam has a rectangular cross-section with the width $b$ and consists of monoclinic anisotropic layers oriented asymmetrically with respect to the midplane surface of the beam. The solution is sought on the basis of the assumption that the strains in the plane of the plies $\langle \epsilon_x \rangle$, $\langle \epsilon_y \rangle$ and $\langle \gamma_{xy} \rangle$ are linearly distributed across the beam thickness. This assumption makes it possible

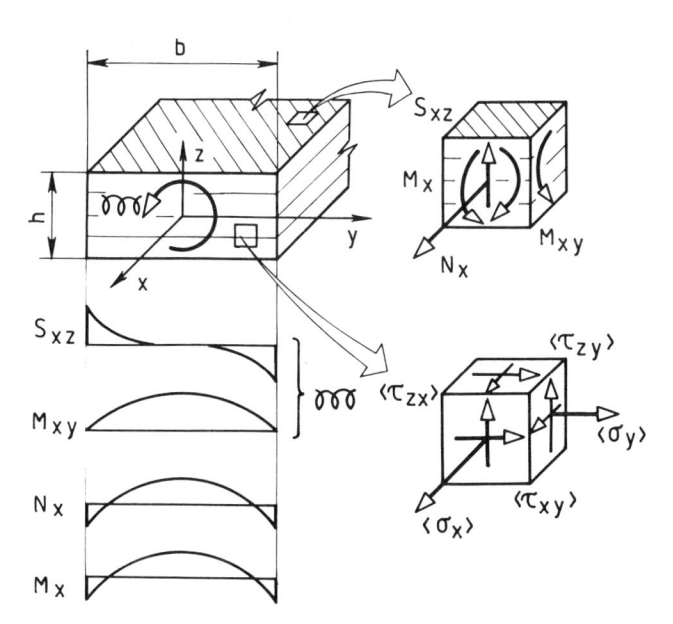

FIGURE 9.7. Calculation diagram for a beam.

to use the laminate theory [34, 57] in order to obtain theoretical relationships. Instead of examining the whole beam it is then sufficient to analyze its midplane surface with the shear forces $S_{xz}(y)$ (N/m) and torsional moments $M_{xy}(y)$ (N·m/m) distributed across the width, which are statically equivalent to the torsional moment $\mathcal{M}$ (N·m) acting in the cross-section of the beam (see Fig. 9.7)

$$-\int_{-b/2}^{b/2} M_{xy} dy + \int_{-b/2}^{b/2} S_{xz} y dy = \mathcal{M} \qquad (9.20)$$

The form of functions $M_{xy}(y)$ and $S_{xz}(y)$ are found on the basis of the assumption that the summed curvature $\bar{k}_{xy}$ is constant throughout the beam width, and every point of its middle surface is formed by curvature $k_{xy}$ which is proportional to moment $M_{xy}$, and by curvature $k_{xy}^s = -2\partial\langle\langle\gamma_{xz}\rangle\rangle/\partial y$ which is proportional to the derivative $\partial S_{xz}/\partial y$.

Taking into consideration that the constant curvature $\bar{k}_{xy}$ across the beam width corresponds to the double value of the relative torsion angle $2\theta'$ of the beam proper, we obtain

$$\bar{k}_{xy} = k_{xy}(y) + k_{xy}^s(y) = 2\theta' \qquad (9.21)$$

When determining functions $M_{xy}(y)$ and $S_{xz}(y)$ it is necessary to include the influence of additional forces arising in the beam due to its asymmetric structure (with respect to surfaces of both midplane and of the longitudinal section parallel to plane $xz$) and also due to the nonuniform distribution of moments $M_{xy}$. The role of additional forces depends upon the type of torsion, which depends in its turn upon a set of conditions imposed on the additional strains in the beam. The ultimate types of bar torsion may be called "free torsion" and "pure torsion." The "free torsion" case is supposed to include only torque, whereas all the additional strains of the longitudinal axis of the beam are freely implemented. In the free torsion case the beam axis may only undergo torsion but its possible additional longitudinal strains and curvatures are subjected to constraints.

The case of free torsion is the first to be discussed. It is assumed here that at some distance from the place where the torque is externally applied (in any possible way) all the cross-sections of a beam, which is sufficiently narrow and long, are in a uniform state of stress where

$$N_y = N_{xy} = M_y = S_{yz} = 0$$

$$\int_{-b/2}^{b/2} N_x dy = 0$$

$$\int_{-b/2}^{b/2} M_x dy = 0 \qquad (9.22)$$

$$\langle\langle \epsilon_x(y) \rangle\rangle = \text{const}$$

$$\bar{k}_x(y) = \text{const}$$

It follows from the above that forces $N_x$, $M_x$, $M_{xy}$ and $S_{xz}$ (see Fig. 9.7) act in the beam, vary in the beam width, and are constant throughout its length. Then there is only one relationship left from the equilibrium conditions of the beam element which is limited to two longitudinal and two infinitely close transversal sections:

$$\frac{\partial M_{xy}}{\partial y} = S_{xz}^t = S_{xz} \qquad (9.23)$$

where the superscript "$t$" indicates that the shear force corresponds to the derivative of the torque (see Section 9.1). The equilibrium condition of Equation (9.23) forms the basis of the following significant relationship obtainable by inserting Equation (9.23) into Equation (9.20) and by integrating Equation (9.20) in parts, keeping in mind that $M_{xy} = 0$ with $y = \pm b/2$

$$-\int_{-b/2}^{b/2} M_{xy} dy = \int_{-b/2}^{b/2} S_{xz}^t y\, dy = \frac{1}{2} \mathcal{M} \qquad (9.24)$$

As is evident from Equation (9.24), the torque sum across the beam width is equal to one half of the total torque $\mathcal{M}$ like the summed moment of all the shear forces.

In order to determine curvature $k_{xy}(y)$ given in Equation (9.21) the law of laminate deformation will be applied with $N_y = N_{xy} = M_y = 0$

$$\begin{Bmatrix} \langle\langle \epsilon_x^0 \rangle\rangle \\ k_x \\ k_{xy} \end{Bmatrix} = \begin{bmatrix} A'_{11} & B'_{11} & B'_{16} \\ B'_{11} & D'_{11} & D'_{16} \\ B'_{16} & D'_{16} & D'_{66} \end{bmatrix} \begin{Bmatrix} N_x \\ M_x \\ M_{xy} \end{Bmatrix} \qquad (9.25)$$

Integration of Equation (9.25) with respect to the beam width and some algebraic transformations considering Equations (9.22) and (9.24) yield the longitudinal strain and the curvature of the beam

$$\langle\langle \epsilon_x^0 \rangle\rangle = -\frac{B'_{16}}{2b} \mathcal{M}$$

$$k_x = -\frac{D'_{16}}{2b} \mathcal{M} \quad (9.26)$$

and also the curvature distribution

$$k_{xy}(y) = \frac{M_{xy}(y)}{\bar{\bar{D}}_{66}} + \frac{1}{2b} \frac{\bar{\bar{B}}_{16} B'_{16} + \bar{\bar{D}}_{16} D'_{16}}{\bar{\bar{D}}_{66}} \mathcal{M} \quad (9.27)$$

where

$$\begin{bmatrix} \bar{\bar{A}}_{11} & \bar{\bar{B}}_{11} & \bar{\bar{B}}_{16} \\ \bar{\bar{B}}_{11} & \bar{\bar{D}}_{11} & \bar{\bar{D}}_{16} \\ \bar{\bar{B}}_{16} & \bar{\bar{D}}_{16} & \bar{\bar{D}}_{66} \end{bmatrix} = \begin{bmatrix} A'_{11} & B'_{11} & B'_{16} \\ B'_{11} & D'_{11} & D'_{16} \\ B'_{16} & D'_{16} & D'_{66} \end{bmatrix}^{-1}$$

For determining the curvature $k_{xy}^s = -2\partial\langle\langle\gamma_{xz}\rangle\rangle/\partial y$ given in Equation (9.21) as a function of $y$, it is first necessary to find the relation between $\langle\langle\gamma_{xz}\rangle\rangle$ and $S_{xz}^t$. This is achieved by using the law of interply shear in Equation (9.18) which states that

$$\langle\langle\gamma_{xz}\rangle\rangle = p_{55} S_{xz}^t \quad (9.28)$$

It should be noted that the application of Equation (9.28) does not include the effect of interply shear stresses on the shear strain $\langle\langle\gamma_{xz}\rangle\rangle$. These stresses may appear in a RPL due to the presence of derivatives $\partial M_{xy}/\partial y$ and $\partial N_x/\partial y$ as follows from Equation (9.11). By inserting Equation (9.23) into Equation (9.28) and by integrating Equation (9.28) with respect to $y$ we obtain curvature $k_{xy}^s(y)$ expressed as the second derivative of $M_{xy}(y)$

$$k_{xy}^s(y) = -2\frac{\partial\langle\langle\gamma_{xz}\rangle\rangle}{\partial y} = -2p_{55}\frac{\partial^2 M_{xy}(y)}{\partial y^2} \quad (9.29)$$

By summarizing now the curvatures $k_{xy}(y)$ and $k_{xy}^s(y)$ expressed by

Equations (9.27) and (9.29), according to Equation (9.21), we arrive at the differential equation for determining the distribution of $M_{xy}(y)$

$$\frac{\partial^2 M_{xy}}{\partial y^2} - p^2 M_{xy} = \left(2\theta' + \frac{1}{2b}\frac{\bar{\bar{B}}_{16}B'_{16} + \bar{\bar{D}}_{16}D'_{16}}{\bar{\bar{D}}_{66}}\mathcal{M}\right)\frac{1}{2p_{55}} \quad (9.30)$$

where

$$p = \frac{1}{\sqrt{2p_{55}\bar{\bar{D}}_{66}}}$$

The solution of Equation (9.30) has the form

$$-M_{xy}(y) = C_1 \operatorname{ch} yp + C_2 \operatorname{sh} yp + 2D_{66}\theta'$$
$$+ \frac{1}{2b}(\bar{\bar{B}}_{16}B'_{16} + \bar{\bar{D}}_{16}D'_{16})\mathcal{M} \quad (9.31)$$

The determination of constants $C_1$ and $C_2$ is based on boundary conditions which state that $M_{xy} = 0$ with $y = \pm b/2$ and $\partial M_{xy}/\partial y = 0$ with $y = 0$. Then the solution of Equation (9.30) has the final form of

$$M_{xy}(y) = -\left[2\bar{\bar{D}}_{66}\theta' + \frac{1}{2b}(\bar{\bar{B}}_{16}B'_{16} + \bar{\bar{D}}_{16}D'_{16})\mathcal{M}\right]\left(1 - \frac{\operatorname{ch} yp}{\operatorname{ch} \frac{bp}{2}}\right) \quad (9.32)$$

The law of free torsion of the beam is obtained by integrating Equation (9.32) with respect to the beam width and considering Equation (9.24)

$$\theta' = d_t \mathcal{M} \quad (9.33)$$

where

$$d_t = \frac{1}{4b\bar{\bar{D}}_{66}}\left(\frac{2}{2 - \frac{\operatorname{th}\frac{bp}{2}}{bp}} - \bar{\bar{B}}_{16}B'_{16} - \bar{\bar{D}}_{16}D'_{16}\right)$$

In the above expression $d_t$ is torsional compliance of the beam cross-section.

It is useful to represent Equation (9.33) in combination with the laws of longitudinal deformation and bending in Equation (9.26) in the form

$$[\langle\langle\epsilon_x^0\rangle\rangle; k_x; \theta'] = [d_{t2}; d_{t1}; d_t]\mathcal{M} \tag{9.34}$$

where $d_{t2} = -B'_{16}/2b$ and $d_{t1} = -D'_{16}/2b$ are longitudinal torsion and bending torsion compliances of the beam cross-section respectively. The value

$$\bar{\bar{D}}_t = \frac{1}{d_t} \tag{9.35}$$

is then the rigidity of free torsion of the beam cross-section.

The law of $M_{xy}(y)$ distribution is obtained by inserting strain $\theta'$ from the torsional law of Equation (9.33) into Equation (9.32)

$$M_{xy}(y) = -\frac{f_1(y)}{2b}\mathcal{M} \tag{9.36}$$

Differentiation of Equation (9.36) then yields the law of distribution of shear forces

$$S_{xz}^t(y) = \frac{pf_2(y)}{2b}\mathcal{M} \tag{9.37}$$

Here

$$f_1(y) = \frac{1 - \dfrac{\operatorname{ch} yp}{\operatorname{ch} \dfrac{bp}{2}}}{1 - \dfrac{\operatorname{th} \dfrac{pb}{2}}{\dfrac{bp}{2}}}$$

$$f_2(y) = \frac{\operatorname{sh} yp}{\operatorname{ch} \frac{bp}{2}} \cdot \frac{1}{1 - \frac{\operatorname{th} \frac{pb}{2}}{\frac{bp}{2}}}$$

By expressing $N_x$ and $M_x$ by means of $\mathcal{M}$ and using Equations (9.25) and (9.26) it is easy to obtain the law of distribution for these forces

$$N_x(y) = -\frac{1}{2b\bar{\bar{D}}_{66}} [F_1 + \bar{\bar{B}}_{16} f_1(y)] \mathcal{M} \qquad (9.38)$$

$$M_x(y) = -\frac{1}{2b\bar{\bar{D}}_{66}} [F_2 + \bar{\bar{D}}_{16} f_2(y)] \mathcal{M} \qquad (9.39)$$

where

$$F_1 = \bar{\bar{A}}_{11} B'_{16} \bar{\bar{D}}_{66} + \bar{\bar{B}}_{11} D'_{16} \bar{\bar{D}}_{66} - \bar{\bar{B}}_{16} B'_{16} \bar{\bar{B}}_{16} - \bar{\bar{B}}_{16} D'_{16} \bar{\bar{D}}_{16}$$

$$F_2 = \bar{\bar{B}}_{11} B'_{16} \bar{\bar{D}}_{66} + \bar{\bar{D}}_{11} D'_{16} \bar{\bar{D}}_{66} - \bar{\bar{D}}_{16} B'_{16} \bar{\bar{B}}_{16} - \bar{\bar{D}}_{16} D'_{16} \bar{\bar{D}}_{16}$$

Fig. 9.8 contains curves plotted from Equations (9.36) through (9.39) representing the distribution of all the forces that appear in the cross-section of the carbon epoxy laminated beam with the structure $(0°/45°)_8$ in free torsion.

The pure torsion case involves complete restriction of longitudinal and bending strains in the beam axis, which means that the conditions of Equation (9.22) are replaced by

$$\int_{-b/2}^{b/2} N_x dy = N \neq 0$$

$$\int_{-b/2}^{b/2} M_x dy = M \neq 0 \qquad (9.40)$$

$$\langle\langle \epsilon_x^0(y) \rangle\rangle = k_x(y) = 0$$

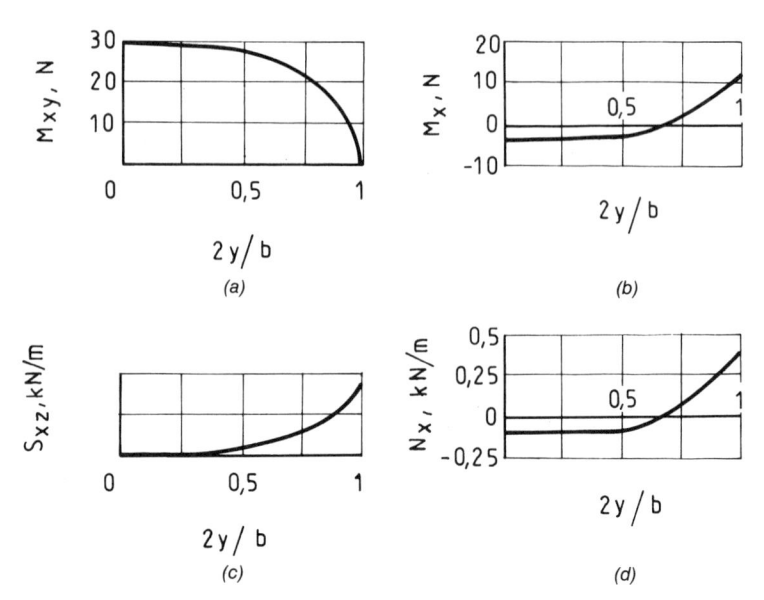

**FIGURE 9.8.** Curves showing forces $M_{xy}$ (a), $M_x$ (b), $S_{xz}$ (c), and $N_x$ (d) distributed across the RPL beam cross-section having the structure $(0°/45°)_8$ under unit torque. The curves have been plotted from formulas (9.36)–(9.39)

The conditions $N_y = N_{xy} = M_y = S_{yz} = 0$ remain in force, as the beam under consideration is sufficiently narrow and long and the distance from the concentrated force sections is also sufficiently large. The distribution of forces $N_x$, $N_x$, and $M_{xy}$ now depends only on the distribution of curvature $k_{xy}$

$$[N_x(y); M_x(y); M_{xy}(y)] = [\bar{\bar{B}}_{16}; \bar{\bar{D}}_{16}; \bar{\bar{D}}_{66}]k_{xy}(y) \quad (9.41)$$

and curvature $k_{xy}(y)$ is governed by the following relationship instead of Equation (9.27)

$$k_{xy}(y) = \frac{M_{xy}(y)}{\bar{\bar{D}}_{66}} \quad (9.42)$$

The curvature resulting from the changes in the beam force $k_{xy}^s(y)$ is obtained from the known Equation (9.29). Summarization of $k_{xy}(y)$ and $k_{xy}^s(y)$ yields a differential equation instead of Equation (9.30)

$$\frac{\partial^2 M_{xy}}{\partial y^2} - p^2 M_{xy} = \frac{\theta'}{p_{55}} \quad (9.43)$$

which has the following final solution

$$M_{xy}(y) = -2\bar{\bar{D}}_{66}\left(1 - \frac{\operatorname{ch} yp}{\operatorname{ch} \frac{bp}{2}}\right)\theta' \qquad (9.44)$$

The relationships showing the distribution of shear forces in the case of pure torsion are the same as in the case of free torsion. Integrating Equation (9.44) with respect to the beam width and taking into account Equation (9.24), we obtain the law of pure torsion, which differs from the law of free torsion given by Equation (9.33):

$$\mathcal{M} = D_t \theta' \qquad (9.45)$$

where

$$D_t = 4\bar{\bar{D}}_{66} b \left(1 - \frac{\operatorname{th}\frac{bp}{2}}{\frac{bp}{2}}\right)$$

$D_t$ here denotes the torsional rigidity in the beam cross-section.

By inserting the distribution $M_{xy}(y)$ into Equation (9.44) we can obtain the distribution of additional forces

$$N_x(y) = -2\bar{\bar{B}}_{16}\left(1 - \frac{\operatorname{ch} yp}{\operatorname{ch} \frac{bp}{2}}\right)\theta'$$

$$M_x(y) = -2\bar{\bar{D}}_{16}\left(1 - \frac{\operatorname{ch} yp}{\operatorname{ch} \frac{bp}{2}}\right)\theta' \qquad (9.46)$$

Integration of Equation (9.46) yields the total longitudinal force $N$ and the bending moment $M$ acting in the beam cross-section. It is possible to write these forces in the form of torsional strain $\theta'$ in combination with torque $\mathcal{M}$ and to obtain an expression different from Equation (9.34)

$$[N; M; \mathcal{M}] = [D_{t2}; D_{t1}; D_t]\theta' \qquad (9.47)$$

where

$$D_{t2} = -2\bar{\bar{B}}_{16}b\left(1 - \frac{\operatorname{th}\frac{bp}{2}}{\frac{bp}{2}}\right)$$

$$D_{t1} = -2\bar{\bar{D}}_{16}b\left(1 - \frac{\operatorname{th}\frac{bp}{2}}{\frac{bp}{2}}\right)$$

are longitudinal and bending torsion rigidities of the beam cross-section. Knowing the distribution of forces $N_x(y)$, $M_x(y)$ and $M_{xy}(y)$ it is easy to arrive at the stress distribution in the ply plane both for free and pure torsion

$$\left\{\begin{array}{c}\langle\sigma_x(z,y)\rangle\\\langle\sigma_y(z,y)\rangle\\\langle\tau_{xy}(z,y)\rangle\end{array}\right\} = \left\{\begin{array}{c}\bar{Q}_{1j}(z)\\\bar{Q}_{2j}(z)\\\bar{Q}_{6j}(z)\end{array}\right\}[(A'_{j1} + zB'_{1j})N_x(y)$$

$$+ (B'_{j1} + zD'_{j1})M_x(y) \qquad (9.48)$$

$$+ (B'_{j6} + zD'_{j6})M_{xy}(y)]$$

$$i = 1, 2, 6$$

In order to write the distribution of interply and shear stresses we shall use Equation (9.11). By inserting derivatives $\partial N_x/\partial y$, $\partial M_x/\partial y$ and $\partial M_{xy}/\partial y$ from Section 9.1 into Equation (9.11) we obtain the expression for both free and pure torsion

$$\left\{\begin{array}{c}\langle\tau_{yz}(y,z)\rangle\\\langle\tau_{xz}(y,z)\rangle\end{array}\right\} = \left\{\begin{array}{c}V_1(z)\\V_2(z)\end{array}\right\}\frac{pf_2(y)}{2b}\mathcal{M} \qquad (9.49)$$

where

$$\left\{\begin{array}{c}V_1(z)\\V_2(z)\end{array}\right\} = \left[\begin{array}{cc}A_{2i}(z) & B_{2i}(z)\\A_{6i}(z) & B_{6i}(z)\end{array}\right]\left\{\begin{array}{c}\bar{\bar{B}}_{16}A_{1i} + \bar{\bar{D}}_{16}B_{1i} + \bar{\bar{D}}_{66}B_{6i}\\\bar{\bar{B}}_{16}B_{1i} + \bar{\bar{D}}_{16}D_{1i} + \bar{\bar{D}}_{66}D_{6i}\end{array}\right\}$$

$A_{ij}(z)$ and $B_{ij}(z)$ are determined from the relationships presented in Chapter 8.

Frequently the RPL structure is symmetrical about the three planes in the coordinate system $x, y, z$. In such cases neither additional strains nor additional forces appear in a beam under torsion, whereas the free and pure torsion rigidities are equal, namely, $\bar{\bar{D}}_t = D_t$. The maximum stresses in the ply plane will be observed in the longitudinal midsection of the beam where $y = 0$. Depending upon the coordinate $z$ three stresses are determined according to a simplified form of Equation (9.48):

$$\left\{\begin{array}{c}\langle\sigma_x(z)\rangle\\ \langle\sigma_y(z)\rangle\\ \langle\tau_{xy}(z)\rangle\end{array}\right\} = z\left\{\begin{array}{c}\bar{Q}_{16}(z)\\ \bar{Q}_{26}(z)\\ \bar{Q}_{66}(z)\end{array}\right\}\frac{D'_{66}}{2b}\frac{1 - \dfrac{1}{\operatorname{ch}\dfrac{bp}{2}}}{1 - \dfrac{\operatorname{th}\dfrac{bp}{2}}{\dfrac{bp}{2}}}\mathcal{M} \qquad (9.50)$$

The following is the formula for determining the interply shear stresses at the free edges ($y = \pm b/2$) of the beam made of a symmetrically structured RPL

$$\left\{\begin{array}{c}\langle\tau_{yz}(z)\rangle\\ \langle\tau_{xz}(z)\rangle\end{array}\right\} = \left\{\begin{array}{c}B_{26}(z)\\ B_{66}(z)\end{array}\right\}\frac{pD'_{66}\bar{\bar{D}}_{66}}{2b}\frac{\operatorname{th}\left(\pm\dfrac{bp}{2}\right)}{1 - \dfrac{\operatorname{th}\dfrac{bp}{2}}{\dfrac{bp}{2}}}\mathcal{M} \qquad (9.51)$$

Fig. 9.9 is an evaluation of the theoretical relationships obtained through comparison between maximum shear stresses $\langle\tau_{xy}\rangle_{\max}$ and $\langle\tau_{xz}\rangle_{\max}$ according to Equations (9.50) and (9.51), as well as torsional rigidity $D_t$ obtained from Equation (9.45) on one hand, and the corresponding values presented by Lekhnitskii [41] and obtained by accurate estimation of an orthotropic material beam, on the other hand. It is evident from Fig. 9.9 that stresses $\langle\tau_{xy}\rangle_{\max}$ and rigidity $D_t$ agree satisfactorily with the exact solution and this agreement improves with the larger width-thickness ratio of the beam.

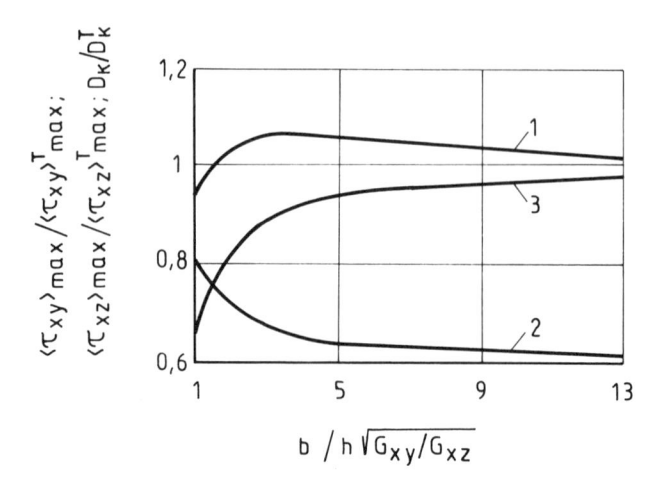

FIGURE 9.9. Comparison between maximum shear stresses and torsional rigidity calculated according to formulas (9.50), (9.51) and (9.45) and those presented in paper[41]:

$$1 - \frac{\langle \tau_{xy} \rangle_{max}}{\langle \tau_{xy} \rangle_{max}^T}; \quad 2 - \frac{\langle \tau_{xz} \rangle_{max}}{\langle \tau_{xz} \rangle_{max}^T}; \quad 3 - \frac{D_x}{D_x^T}$$

It should be remarked that the edge effect is not discussed here. Therefore the boundary conditions for stresses $\langle \sigma_y \rangle$, $\langle \tau_{xy} \rangle$ and $\langle \tau_{yz} \rangle$ are not satisfied, as these stresses should be equal to zero on the free edges because in Equations (9.48) through (9.51) they are represented by nonzero values. Boundary conditions are satisfied for stresses $\langle \sigma_x \rangle$ and $\langle \tau_{xz} \rangle$. Nevertheless the problem of the real distribution of stresses $\langle \sigma_x \rangle$ and $\langle \tau_{xz} \rangle$ near the free edge remains unsolved. The reason for this is the fact that the edge effects were neglected and also the fact the use of the hypothesis with respect to the sections of the beam near the free edge may not be effective. It is clear that the smaller the relative thickness of the component plies, the more limited is the distribution of the edge effect. Naturally, the normal and shear interply stresses concentrate near the free edges. In addition, the free edges appear to be the most vulnerable places in the RPL beam because the RPL has a low interply shear strength, and the edges therefore should be reinforced. This, for example, may be achieved by windings around the whole beam. Such winding, however, may serve as a structural part and at the same time increase considerably the torsional rigidity of the beam, especially when the angle between the winding and the longitudinal axis of the beam is ±45°.

We propose to analyze the performance of the winding layer by making two basic assumptions: the winding layer has a thickness that is much smaller than the beam thickness, and the winding layer is glued only to those surfaces of the beam that are parallel to the $xy$ plane. This allows us to suppose that the glued sections of the winding are to be regarded as the external plies of the RPL and at the same time as components of a thin-walled beam having a closed box profile in the cross section (Fig. 9.10). In addition, equal relative torsional angles $\theta'$ arise in a laminated part of the beam (Fig. 9.10) and in its external winding ply. Therefore the rigidities of the beam in the cases of free and pure torsion $\bar{D}_t^o$ and $D_t^o$ include the corresponding rigidities of the internal laminated part $\bar{D}_t$ and $D_t$ and of the external winding ply $D_t^k$

$$\bar{D}_t^o = \bar{D}_t + D_t^k; \qquad D_t^o = D_t + D_t^k \qquad (9.52)$$

where

$$D_t^k = 2Q_{66}^o \delta b^3 \frac{\left(\frac{h}{b}\right)^2}{1 + \frac{h}{b}} \qquad (9.53)$$

$Q_{66}^o$ is the shear component of the elasticity matrix for the winding layer; $\delta$ is the winding thickness, and $h$ is the beam thickness (see Fig. 9.10). Equation (9.53) has been obtained from the torsion law given by Ruditsin et al. [53] for closed thin-wall profiles made of an isotropic material.

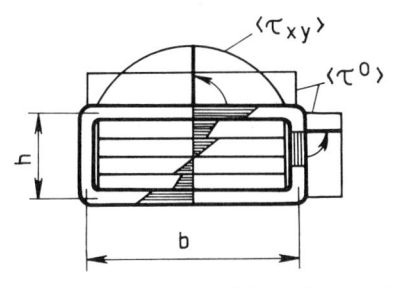

FIGURE 9.10. Calculation diagram for a winding-reinforced beam and the shear stresses acting in it.

As the external layers of the RPL are parts of the winding, there will be also shear stresses $\langle \tau^o \rangle$ in them in addition of those determined according to Equations (9.48) or (9.50). The distribution of the stresses $\langle \tau^o \rangle$ in the first approximation may be regarded as uniform across the thickness and throughout the whole winding loop (see Fig. 9.10). The stresses $\langle \tau^o \rangle$ are calculated according to formula

$$\langle \tau^o \rangle = \frac{2Q_{66}^o h}{1 + \frac{h}{b}} \theta' \qquad (9.54)$$

which is derived from the torsional law for a box-section beam

$$\mathcal{M}^k = \langle \tau^o \rangle \delta b h = D_t^k \theta' \qquad (9.55)$$

## 9.3 TORSION OF A THIN-WALL TABULAR LAMINATED BEAM

Let us consider a case where an axial force $N$ is applied to the beam in addition to the torque $\mathcal{M}$. The arbitrary ply $k$ is reinforced at angle $\beta_k$ to axis 1 which is parallel to the beam axis. The tangential direction will be marked as axis 2.

The beam strains will be determined from the following relationships:

$$\langle\langle \epsilon_1 \rangle\rangle = \frac{a_1}{2\pi R} N + \frac{a_3}{2\pi R^2} \mathcal{M}$$

$$\langle\langle \epsilon_2 \rangle\rangle = \frac{a_2}{2\pi R} N + \frac{a_4}{2\pi R^2} \mathcal{M} \qquad (9.56)$$

$$\langle\langle \gamma_{12} \rangle\rangle = \frac{a_5}{2\pi R} N + \frac{a_6}{2\pi R^2} \mathcal{M}$$

where

$$a_1 = A'_{11} - \frac{(B'_{11})^2}{D'_{11}}$$

$$a_2 = A'_{21} - \frac{B'_{21} B'_{11}}{D'_{11}}$$

$$a_3 = A'_{16} - \frac{B'_{11}B'_{61}}{D'_{11}}$$

$$a_4 = A'_{26} - \frac{B'_{21}B'_{61}}{D'_{11}}$$

$$a_5 = A'_{61} - \frac{B'_{61}B'_{11}}{D'_{11}}$$

$$a_6 = A'_{66} - \frac{(B'_{61})^2}{D'_{11}}$$

The average stresses in the arbitrary ply $k$ along axes 1 and 2 are found from the relationships

$$\langle \sigma_1 \rangle_k = \frac{r_1}{2\pi R} N + \frac{s_1}{2\pi R^3} \mathcal{M} \qquad (9.57)$$

where

$$r_1 = (\bar{Q}_{11})_k a_1 + (\bar{Q}_{12})_k a_2 + (\bar{Q}_{16})_k a_5$$

$$s_1 = (\bar{Q}_{11})_k a_3 + (\bar{Q}_{12})_k a_4 + (\bar{Q}_{16})_k a_6 \qquad (9.58)$$

$$\langle \sigma_2 \rangle_k = \frac{r_2}{2\pi R} N + \frac{s_2}{2\pi R^2} \mathcal{M}$$

where

$$r_2 = (\bar{Q}_{12})_k a_1 + (\bar{Q}_{22})_k a_2 + (\bar{Q}_{26})_k a_5$$

$$s_2 = (\bar{Q}_{12})_k a_3 + (\bar{Q}_{22})_k a_4 + (\bar{Q}_{26})_k a_6 \qquad (9.59)$$

$$\langle \tau_{12} \rangle_k = \frac{r_3}{2\pi R} N + \frac{s_3}{2\pi R} \mathcal{M}$$

where

$$r_3 = (\bar{Q}_{16})_k a_1 + (\bar{Q}_{26})_k a_2 + (\bar{Q}_{66})_k a_5$$

$$s_3 = (\bar{Q}_{16})_k a_3 + (\bar{Q}_{26})_k a_4 + (\bar{Q}_{66})_k a_6$$

Along the elastic symmetry axes of ply $k$ the average stresses are given by

$$\langle\sigma_\parallel\rangle_k = \langle\sigma_1\rangle_k \cos^2\beta_k + \langle\sigma_2\rangle_k + \langle\sigma_2\rangle_k \sin^2\beta_k$$

$$+ 2\langle\tau_{12}\rangle_k \sin\beta_k \cos\beta_k = \frac{F_7}{2\pi R} N + \frac{F_8}{2\pi R^2} \mathcal{M}$$

(9.60)

where

$$F_7 = r_1 \cos^2\beta_k + r_2 \sin^2\beta_k + 2r_3 \sin\beta_k \cos\beta_k$$

$$F_8 = s_1 \cos^2\beta_k + s_2 \sin^2\beta_k + 2s_3 \sin\beta_k \cos\beta_k$$

$$\langle\sigma_\perp\rangle_k = \langle\sigma_1\rangle_k \sin^2\beta_k + \langle\sigma_2\rangle_k \cos^2\beta_k$$

(9.61)

$$- 2\langle\tau_{12}\rangle_k \sin\beta_k \cos\beta_k = \frac{F_9}{2\pi R} N + \frac{F_{10}}{2\pi R^2} \mathcal{M}$$

where

$$F_9 = r_1 \sin^2\beta_k + r_2 \cos^2\beta_k - 2r_3 \sin\beta_k \cos\beta_k$$

$$F_{10} = s_1 \sin^2\beta_k + s_2 \cos^2\beta_k - 2s_3 \sin\beta_k \cos\beta_k$$

$$\langle\tau_{\parallel\perp}\rangle_k = (\langle\sigma_2\rangle_k - \langle\sigma_1\rangle_k) \sin\beta_k \cos\beta_k$$

(9.62)

$$+ \langle\tau_{12}\rangle_k (\cos^2\beta_k - \sin^2\beta_k) = \frac{F_{11}}{2\pi R} N + \frac{F_{12}}{2\pi R^2} \mathcal{M}$$

where

$$F_{11} = (r_2 - r_1) \sin\beta_k \cos\beta_k + r_3(\cos^2\beta_k - \sin^2\beta_k)$$

$$F_{12} = (s_2 - s_1) \sin\beta_k \cos\beta_k + s_3(\cos^2\beta_k - \sin^2\beta_k)$$

By inserting the expressions for determination of stresses $\langle\sigma_\perp\rangle_k$ and $\langle\tau_{\parallel\perp}\rangle_k$ into the general strength criteria of ply $k$ we obtain specific criteria under torsion and axial loading combined:

*Torsion of a Laminated Beam*    235

- When the matrix fails under torsion in ply $k$

$$(1 - \nu_A^2)\left(\frac{\bar{\sigma}_r}{R_A^+}\right)^2 \left(\frac{\mathcal{N}F_9}{2\pi R} + \frac{\mathcal{M}F_{10}}{2\pi R^2}\right)^2$$

$$+ \frac{1}{T_A^2}\left(\frac{\mathcal{N}F_{11}}{2\pi R} + \frac{\mathcal{M}F_{12}}{2\pi R^2}\right)^2 = 1 \qquad (9.63)$$

- When the matrix fails under shear

$$(1 + \nu_A^2)\left(\frac{\mathcal{N}F_9}{2\pi R} + \frac{\mathcal{M}F_{10}}{2\pi R^2}\right)^2 + 2(1 + \nu_A)$$

$$\times \left(\frac{\mathcal{N}F_{11}}{2\pi R} + \frac{\mathcal{M}F_{12}}{2\pi R^2}\right)^2 + \left(\frac{\mathcal{N}F_9}{2\pi R} + \frac{\mathcal{M}F_{10}}{2\pi R^2}\right)^2 (1 + \nu_A) \qquad (9.64)$$

$$\times \sqrt{\left(\frac{\mathcal{N}F_9}{2\pi R} + \frac{\mathcal{M}F_{10}}{2\pi R^2}\right)^2 (1 - \nu_A^2) + 4\left(\frac{\mathcal{N}F_{11}}{2\pi R} + \frac{\mathcal{M}F_{12}}{2\pi R^2}\right)^2}$$

$$= 2(R_A^+)^2$$

- when the bond fails between the fibers and the matrix

$$\left(\frac{\mathcal{N}F_9}{2\pi R} + \frac{\mathcal{M}F_{10}}{2\pi R^2}\right) T_b^2 \bar{\sigma}_r + \left(\frac{\mathcal{N}F_{11}}{2\pi R} + \frac{\mathcal{M}F_{12}}{2\pi R^2}\right)^2 R_b \bar{\tau}_{rz}^2 = T_b^2 R_b \qquad (9.65)$$

- when ply $k$ fails under transversal compression

$$\left(\frac{\mathcal{N}F_9}{2\pi R} + \frac{\mathcal{M}F_{10}}{2\pi R^2}\right)^2 + 2(1 + \nu_{\perp\!\perp})$$

$$\times \left(\frac{\mathcal{N}F_{11}}{2\pi R} + \frac{\mathcal{M}F_{12}}{2\pi R^2}\right)^2 + \left(\frac{\mathcal{N}F_9}{2\pi R} + \frac{\mathcal{M}F_{10}}{2\pi R^2}\right) \qquad (9.66)$$

$$\times \sqrt{\left(\frac{\mathcal{N}F_9}{2\pi R} + \frac{\mathcal{M}F_{10}}{2\pi R^2}\right)^2 + 4\left(\frac{\mathcal{N}F_{11}}{2\pi R} + \frac{\mathcal{M}F_{12}}{2\pi R^2}\right)^2}$$

$$= 2(R_\perp^-)^2$$

- when the fibers fail under axial tension or compression

$$\langle\langle\epsilon_1\rangle\rangle \cos^2 \beta_k + \langle\langle\epsilon_2\rangle\rangle \sin^2 \beta_k$$
$$+ \langle\langle\gamma_{12}\rangle\rangle \sin \beta_k \cos \beta_k = \frac{r_4}{2\pi R} N + \frac{s_4}{2\pi R^2} \mathcal{M} = \epsilon_{BR}^{\pm} \quad (9.67)$$

where

$$r_4 = a_1 \cos^2 \beta_k + a_2 \sin^2 \beta_k + a_5 \sin \beta_k \cos \beta_k$$

$$s_4 = a_3 \cos^2 \beta_k + a_4 \sin^2 \beta_k + a_6 \sin \beta_k \cos \beta_k$$

If the beam consists of unidirectionally reinforced plies oriented at various angles $\beta$, the plies do not fail simultaneously, and in determining $A_{ij}$ the stepwise character of the laminate should be taken into consideration.

CHAPTER 10

# Elastic Properties of Laminated Thin-Wall Beams with Open Profile

## 10.1 FORMULATION OF THE PROBLEM AND THE MAIN ASSUMPTIONS

Reinforced plastic laminates (RPL) are increasingly being used for the manufacture of thin-wall open-profile beams. The state of the art permits design of beams with varying structure, i.e. the beam structure is not the same in every profile element and in every beam section.

The choice of RPL structures and beam profiles is usually based on empirical experience. The RPL structure is generally made symmetrical with respect to the midsurface and to surfaces of the longitudinal and the transversal cross-sections of the beam. It follows that the elementary section of an RPL between two longitudinal and two cross-sections of the beam (Fig. 10.1) is usually orthotropic with respect to the forces and strains in the ply planes. The symmetry of the RPL about the midplane may be desirable to present the warping of the beam wall. In an asymmetrically loaded beam, however, when the load is not symmetrical to the longitudinal axis, a structure that is not symmetrical (not orthotropic) about the longitudinal and cross-sectional planes may not appear to be the optimum structure for an RPL beam. The utilization of unorthotropic-structure RPL beams is limited through the lack of an engineering design theory for determining the stress-strain states and also the elastic properties of such beams. It is only in the paper by Lekhnitskii [40] that we find the discussion of a homogeneous strip made of a monoclinic anisotropic material loaded in its own plane, which is a specific case. The calculations by this author are used to estimate the results of the present chapter.

Presently we are going to discuss a straight thin-wall beam of open profile and a cylindrical (in the mathematical sense) surface made of

238  MECHANICS OF ELASTIC COMPOSITE BEAM

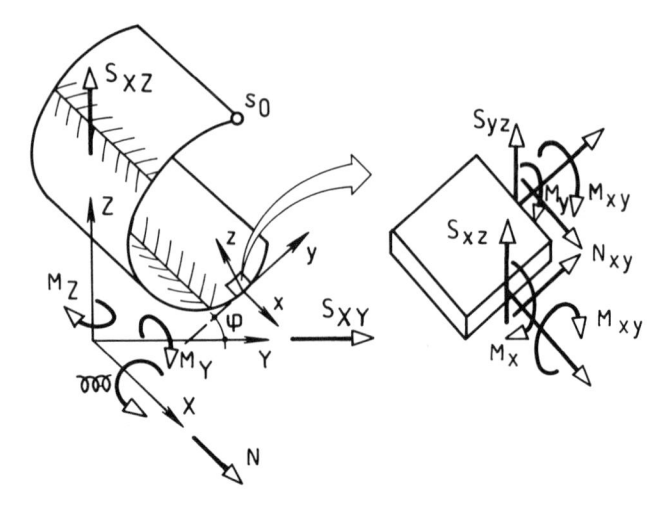

FIGURE 10.1. Calculation diagram for a beam.

an RPL having unorthotropic and symmetrical structure about the midsurface of the laminate. The structure of the beam profile may vary continuously according to a certain law, or it may be regular in repetitive parts, or may just be regular. It is assumed that the profile has a constant structure along the beam length. A beam having repetitively regular profile structure is then said to consist of varying-width strips that are characterized by varying material structure, and these parts are joined in such a manner that their midsurfaces form a continuous midsurface of the whole beam wall with a cross-sectional outline $s$ (see Fig. 10.1).

The structure and the stress-strain state of an RPL in every point of a beam wall are analyzed in the right-hand orthogonal system of coordinates $x$, $y$, $z$ where the initial point is on the RPL midsurface and the axes are oriented as shown in Fig. 10.1. Another right-hand orthogonal system of coordinates $X$, $Y$, $Z$ is applied for examining the whole beam; the initial point of this system is arbitrarily located on the cross-sectional plane and the axes $Y$, $Z$ have an arbitrary orientation in this plane (see Fig. 10.1). Thus the cross-sectional profile $s$ of the beam is described by axes $X$ and $Y$.

The principal aim of the present work is to determine the elastic characteristics of the beam cross-section, i.e. rigidities or compliances that form the deformation law for an elementary area of the beam depending upon the geometry and the lay-up of the profile and upon the elastic properties of the RPL. This deformation law has to show the relations between the longitudinal force $N$, bending moments $M_Y$,

$M_Z$, shear forces $S_{YX}$, $S_{ZX}$ and torque $\mathcal{M}$ that are active in the beam sections (see Fig. 10.1), and the corresponding deformations of the beam $\epsilon$, $k_Y$, $k_Z$, $\gamma_{YX}$, $\gamma_{ZX}$ and $\theta'$ defined as

$$\epsilon = \frac{\partial u_0}{\partial X}; \quad k_Y = -\frac{\partial^2 \omega_0}{\partial X^2}; \quad k_Z = -\frac{\partial^2 v_0}{\partial X^2};$$

$$\gamma_{YX} = \frac{\partial \omega_0}{\partial X}; \quad \gamma_{ZX} = \frac{\partial v_0}{\partial X}; \quad \theta' = \frac{\partial \theta_0}{\partial X} \tag{10.1}$$

where $u_0$, $v_0$, $\omega_0$ are the displacements of the initial point of the beam coordinate system in the given sectional plane along the $X$, $Y$, $Z$ axes respectively; $\theta$ is the rotation angle of the section about axis $X$ in the counterclockwise direction.

RPL are discussed by using the classical laminate theory [34, 56] where instead of examining the realistic beam, only its midsurface is considered, which may generally be acted upon by membrane forces $\{N\}^T = [N_x, N_y, N_{xy}]$, moments $\{M\}^T = [M_x, M_y, M_{xy}]$ and shear forces $S_{xz}$ and $S_{yz}$ (see Fig. 10.1). The relations among membrane forces and relations between the moments and midsurface deformations of the RPL $\{\langle\langle\epsilon\rangle\rangle\}^T = [\langle\langle\epsilon_x\rangle\rangle, \langle\langle\epsilon_y\rangle\rangle, \langle\langle\gamma_{xy}\rangle\rangle]$ and $\{k\}^T = [k_x, k_y, k_{xy}]$ are stated by the well-known deformation law in Equation (4.11); for the symmetrical structure of the RPL about its midsurface, with $B_{ij} = 0$, $i, j = 1, 2, 6$, this equation has the form

$$\{N\} = [A]\{\langle\langle\epsilon\rangle\rangle\}; \quad \{M\} = [D]\{k\} \tag{10.2}$$

The interply shear of the RPL is analyzed in detail in Section 9.1.

In the general case, the relations between the forces $N$, $M_y$, $M_z$, $S_{yx}$, $S_{Zx}$ and $\mathcal{M}$ resulting across the beam section, and forces $N_x$, $M_x$, $M_{xy}$ and $S_{xz}$ distributed over the profile $s$, are given by

$$N = \int_s N_x ds \tag{10.3}$$

$$M_Y = \int_s N_x Z ds + \int_s M_x \cos\varphi ds \tag{10.4}$$

$$M_Z = \int_s N_x Y ds - \int_s M_x \sin\varphi ds \tag{10.5}$$

240   MECHANICS OF ELASTIC COMPOSITE BEAM

$$\mathcal{M} = -\int_s N_{xy} Z \cos \varphi ds + \int_s N_{xy} Y \sin \varphi ds$$

$$-\int_s M_{xy} ds + \int_s S_{xz} Z \sin \varphi ds + \int_s S_{xz} Y \cos \varphi ds \quad (10.6)$$

$$S_{YX} = \int_s N_{xy} \cos \varphi ds - \int_s S_{xz} \sin \varphi ds \quad (10.7)$$

$$S_{ZX} = \int_s N_{xy} \sin \varphi ds + \int_s S_{xz} \cos \varphi ds \quad (10.8)$$

where $\varphi$ is the angle between the tangent, the profile outline $s$ and axis $Y$ (see Fig. 10.1). Equations (10.3) through (10.8) have been obtained assuming that the forces causing positive strains in the direction of their action are considered to be positive (see Fig. 10.1 where all the forces are positive). A number of simplifications will be necessary, however, on account of the thin-walled structure.

It is supposed that the flexural rigidity of the RPL is so insignificant in comparison to the flexural rigidity of the whole section that moments $M_x$ may be neglected. Similarly, the contribution of shear forces $S_{xz}$ distributed across the section profile outline $s$ to the formation of the resulting shear forces $S_{YZ}$ and $S_{ZX}$ may also be neglected. It is important to remember that the present chapter will be limited to the analysis of only those external conditions that guarantee free development of deplanation of beam cross-sections under torsion and other kinds of loading due to shear strains in the beam wall. Taking into consideration the above statement, Equations (10.4) through (10.8) are made considerably simpler:

$$M_Y = \int_s N_x Z ds \quad (10.9)$$

$$M_X = \int_s N_x Y ds \quad (10.10)$$

$$\mathcal{M} = -\int_s M_{xy} ds + \int_s S_{xz} Z \sin \varphi ds + \int_s S_{xz} Y \cos \varphi ds \quad (10.11)$$

$$S_{YX} = \int_s N_{xy} \cos\varphi \, ds \qquad (10.12)$$

$$S_{ZX} = \int_s N_{xy} \sin\varphi \, ds \qquad (10.13)$$

In addition, it is supposed that the section under consideration is so far removed from the beam ends and the areas of application of the concentrated loads that according to the principle of Saint-Venant any local effects may be neglected. For this purpose the following conditions are adopted with respect to the midsurface curvature of the RPL

$$k_x(s) = 0 \qquad (10.14)$$

$$k_{xy}(s) = \text{const} = -2\theta' \qquad (10.15)$$

Equation (10.14) has been adopted considering the thin-wall structure of the beam, whereas Equation (10.15) corresponds to the free torsion of the open-profile beam [39]. In addition to the conditions established in Equations (10.14) and (10.15), conditions should be adopted also concerning the state of the RPL in the longitudinal sections of the beam. According to the classical approach [12, 15] it is assumed that the cross-sectional profile outline of the beam does not deform, namely,

$$\langle\langle \epsilon_y \rangle\rangle = k_y = 0 \qquad (10.16)$$

Practically, this refers to the case when stiff diaphragms are laid along the beam length or the beam works in combination with sheathing or light filling. Otherwise, Equation (10.16) should be replaced by the assumption

$$N_y(s) = M_y(s) = 0 \qquad (10.17)$$

which is considered correct if no distributed load is applied to the beam section under discussion or the load is sufficiently small compared with the forces distributed across the profile outline. The effect of such a load upon the deformation of a homogeneous monoclinic anisotropic strip has been analyzed by Lekhnitskii [40].

## 10.2 FREE TORSION

If the RPL structure is symmetrical about the midsurface, the beam torsion (such that it produces free deplanation of its cross-sections) under moment $\mathcal{M}$ may be discussed independently from other types of loading. This follows from the deformation law in Equation (10.2) and from Equation (10.11), and also from the fact that with membrane-bending rigidities $B_{i6} = 0$, $i = 1, 2, 6$ the torque $M_{xy}$ which is distributed across the profile outline does not cause normal and shear strains of the midsurface of the RPL and consequently of the whole beam.

As follows from Equation (10.11) the moment $\mathcal{M}$ is formed by torques $M_{xy}$ as well as by shear forces $S_{xz}$ distributed across profile outline $s$. For thin-walled beams the law of distribution of forces $M_{xy}$ and $S_{xz}$ has to be determined, because the calculated value of the beam torsional rigidity is governed by it [14, 39]. In determining the elastic properties of thin-walled beams the distribution of $M_{xy}$ within the outline areas $s$ with a constant RPL structure may be considered to be uniform, i.e. it is possible to neglect the actual curved shape of the $M_{xy}$ diagram, which is more pronounced near the free edges of the beam or in the joints between the different strips (Fig. 10.2), should they exist. This assumption prevents the application of the following relationship for the specific case

$$S_{xz} = \frac{\partial M_{xy}}{\partial y} \qquad (10.18)$$

In order to take into account the effect of the shear force $S_{xz}$, Kruklinsh & Kalvish [39] propose an indirect approach stating that the contribution of shear forces to the formation of the external torque $\mathcal{M}$ is equal to the contribution of torques $M_{xy}$ distributed across the cross-sectional outline. By inserting Equation (10.18) into Equation (10.11) and integrating Equation (10.11) by parts we obtain

$$\mathcal{M} = -2 \int_s M_{xy} ds \qquad (10.19)$$

It follows from Equations (10.14) through (10.16) that moments $M_{xy}$ are distributed across the outline $s$ in direct proportion to the torsional rigidity $D_{66}$ of the RPL. By inserting $M_{xy}$ from the deformation law of Equation (10.2) into Equation (10.19) and taking into Equation

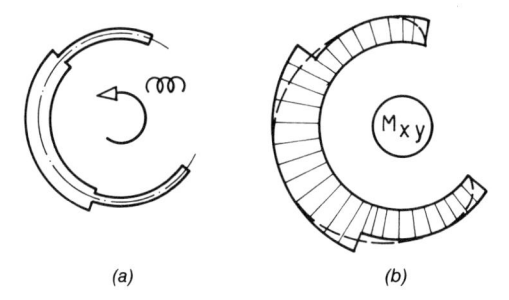

FIGURE 10.2. The outline of the beam cross-section acted upon by the applied torque $\mathcal{M}$ (a) and the diagram showing the conditional replacement of virtual distribution (----) of forces by repetitively regular segments (b).

(10.15) we obtain the law of free torsion for a thin-walled beam having a non-uniform profile structure

$$\mathcal{M} = D_t \theta' \qquad (10.20)$$

where $D_t = 4 \int_s D_{66} ds$ is the free torsion rigidity.

## 10.3 AXIAL LOADING AND PURE BENDING

When axial loading combines with pure bending, shear forces $S_{YX}$ and $S_{ZX}$ are absent in the beam section and only forces $N$, $M_Y$ and $M_Z$ may act; the latter will be collectively denoted by $\{N\}$. These forces are made up by only force $N_x$ distributed across the outline $s$ as stated in Equations (10.3), (10.9) and (10.10) which, in their turn, will be collectively denoted as

$$\{N\} = \int_s N_x \{J\}^T ds \qquad (10.21)$$

where

$$\{J\}^T = [1, Z, Y]$$

With a free deplanation of the beam section due to shear strains in its wall (when $N_{xy} = 0$) and with delayed transverse strains ($\langle\langle \epsilon_y \rangle\rangle = 0$) the force $N_x$, as follows from the deformation law in Equation

(10.2), causes longitudinal strains $\langle\langle\epsilon_x\rangle\rangle$ determinable according to the law

$$\langle\langle\epsilon_x\rangle\rangle = \bar{\bar{a}}_{11} N_x \qquad (10.22)$$

where $\bar{\bar{a}}_{11}$ is determined from matrices

$$\begin{bmatrix} \bar{\bar{a}}_{11} & \bar{\bar{a}}_{16} \\ \bar{\bar{a}}_{16} & \bar{\bar{a}}_{66} \end{bmatrix} = \begin{bmatrix} A_{11} & A_{16} \\ A_{16} & A_{66} \end{bmatrix}$$

In conditions of pure bending, the strain $\langle\langle\epsilon_x\rangle\rangle$ is linearly distributed across the outline $s$ and is expressed as longitudinal strain $\epsilon$ and curvatures of the beam $k_Y$, $k_Z$ (or $\{\epsilon\}$, for short) in the following way

$$\langle\langle\epsilon_x\rangle\rangle = \epsilon + k_y Z + k_z Y \qquad (10.23)$$

where

$$\langle\langle\epsilon_x\rangle\rangle = \{J\}^T \{\epsilon\}$$

By subsequent replacement of Equations (10.23), (10.22) and (10.21) one into another we obtain the following deformation law for the elementary area of the beam

$$\begin{Bmatrix} N \\ M_Y \\ M_Z \end{Bmatrix} = \begin{bmatrix} K_{11} & K_{12} & K_{13} \\ K_{21} & K_{22} & K_{23} \\ K_{31} & K_{32} & K_{33} \end{bmatrix} \begin{Bmatrix} \epsilon \\ k_Y \\ k_Z \end{Bmatrix} \qquad (10.24)$$

or

$$\{N\} = [K]\{\epsilon\}$$

where

$$[K] = \int_s \frac{1}{\bar{\bar{a}}_{11}} \{I\}\{J\}^T ds \qquad (10.25)$$

is the stiffness matrix of the beam. As seen from Equation (10.25), $[K]^T = [K]$, i.e. matrix $[K]$ is symmetric, or $K_{ij} = K_{ji}$, $i,j = 1, 2, 3$.

As stated in Section 10.1, the coordinate axes $X$, $Y$ and $Z$ are arbitrary. In order to determine the positions of principal central axes $X'$, $Y'$ and

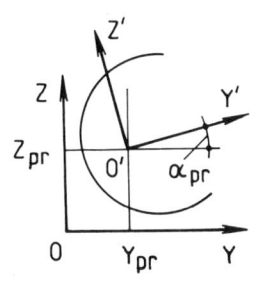

*Elastic Properties of Laminated Thin-Wall Beams with Open Profile* 245

FIGURE 10.3. Diagram for determining the position of the principal central axes.

$Z'$ where the acting forces do not cause additional strains in the beam axis, coordinates of the initial point of the principal axes $Y_{pr}$, $Z_{pr}$ and their rotation angle $\alpha_{pr}$ about the initial axes will be introduced (see Fig. 10.3). Determination of $Y_{pr}$, $Z_{pr}$ and $\alpha_{pr}$ is possible by using the principle stating that all the non-diagonal components of the beam stiffness matrix in the principal central axes are equal to zero. Then we proceed to express the primary coordinates $Y$, $Z$ (included in matrix $\{J\}$) in Equation (10.25) by means of principal central coordinates $Y'$ and $Z'$ through the relationships

$$Y = Y_{pr} + Y' \cos \alpha_{pr} - Z' \sin \alpha_{pr}$$

$$Z = Z_{pr} + Y' \sin \alpha_{pr} + Z' \cos \alpha_{pr}$$

(10.26)

arriving at formulas for calculating the parameters determining the position of the principal central axes where the beam stiffnesses are included in the initial axes

$$Y_{pr} = \frac{K_{13}}{K_{11}}$$

$$Z_{pr} = \frac{K_{12}}{K_{11}}$$

(10.27)

$$\alpha_{pr} = \frac{1}{2} \text{arctg } 2 \frac{K_{12}K_{13} - K_{11}K_{23}}{K_{11}(K_{22} - K_{33}) - K_{12}^2 - K_{13}^2}$$

## 10.4 BEAM DEFORMATION LAW SUBJECTED TO SHEAR FORCES

According to Equations (10.12) and (10.14) the resultant shear forces $S_{YX}$ and $S_{ZX}$ make up forces $N_{xy}$ which are distributed across the profile outline $s$. In order to determine this distribution an equilibrium condition for the beam elementary area will be used

$$\frac{\partial N_x}{\partial x} + \frac{\partial N_{xy}}{\partial y} = 0$$

where coordinates $x, y$ will be substituted for $X$ and $s$ owing to the thin-walled structure of the beam, and this yields

$$N_{xy}(s) = -\int_{s_0}^{s} \frac{\partial N_x}{\partial X} ds \qquad (10.28)$$

In the above expression, integration constants are absent because of the absence of forces $N_{xy}$ on the free edge of the beam where $s = s_0$. The distribution of $N_{xy}(s)$ can be expressed also through the resultant shear forces $S_{YZ}$ and $S_{ZX}$ by inserting the corresponding values of Equations (10.22) through (10.24) into Equation (10.28) and taking into account that

$$\frac{\partial N}{\partial X} = 0; \qquad \frac{\partial M_y}{\partial X} = S_{ZX}; \qquad \frac{\partial M_Z}{\partial X} = S_{YZ} \qquad (10.29)$$

or, in a more concise way,

$$\left\{\frac{\partial N}{\partial X}\right\}^T = \{S\}^T = [0; S_{ZX}; S_{YX}]$$

The expression becomes more convenient by writing the shear force vector $\{S\}$ in the above form. The final expression for the determination

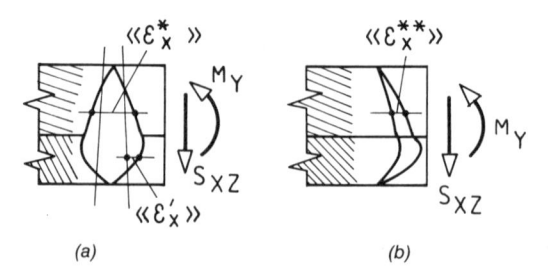

FIGURE 10.4. Conditional diagram of ruptures of the first (a) and the second (b) type.

of $N_{xy}(s)$ will be given further where the determination of other forces is discussed.

The presence of force $N_{xy}$ primarily causes additional shear strains in the beam wall resulting in additional deplanation of its cross-sections and in additional transverse shear strains of the beam under the action of shear forces. Besides this direct effect, the action of force $N_{xy}$ and the force derivative $\partial N_x/\partial X$ associated with it (see Equation (10.28)) for an unorthotropic RPL causes additional effects of the first and of the second types respectively.

The additional effect of the first type is due to the fact that force $N_{xy}$ tends to cause normal strains (to be denoted by $\langle\langle \epsilon_x^* \rangle\rangle$ here) in addition to shear strains, and if these normal strains could develop, they would be determined by the following relationship:

$$\langle\langle \epsilon_x^* \rangle\rangle = \bar{\bar{a}}_{16} N_{xy} \tag{10.30}$$

The conditional development of $\langle\langle \epsilon_x^* \rangle\rangle$, however, would lead to the formation of "ruptures" (Fig. 10.4(a)) or "beddings" depending upon the sign of $\bar{\bar{a}}_{16}$ and $N_{xy}$. Then the continuity condition leads to the existence of additional forces of the first type $N_x'$ that balance themselves across the outline $s$ and delay the conditional ruptures leading to additional strain $\epsilon'$ and producing first-type curvatures $k_Y'$, $k_Z'$ of the whole beam.

Due to the additional effect of the second-type, additional forces $N_x''$ also appear, which balance themselves across the outline $s$ and delay the conditional ruptures of the second type, leading to the additional strain $\epsilon''$ and beam curvatures $k_Y''$, $k_Z''$. The cause of the additional effect is the tendency of forces $N_x$ to produce shear strains in addition to normal strains as stated by the law

$$\langle\langle \gamma_{xy} \rangle\rangle = \bar{\bar{a}}_{16} N_x \tag{10.31}$$

This leads to the deplanation of the beam cross-section which is characterized by longitudinal displacements $u(s)$ given by

$$u(s) = \int_{s_0}^{s} \langle\langle \gamma_{xy} \rangle\rangle ds \tag{10.32}$$

If $\partial N_x/\partial X \neq 0$, the section will tend to a different degree of deplanation. If an arbitrary condition is made that this degree of deplanation would be a reality, this would result in the second type of ruptures (or beddings) as seen in Fig. 10.4(b). The relative width of the conditional ruptures are characterized by longitudinal strain

$$\langle\langle \epsilon_x^{**} \rangle\rangle = -\frac{\partial u}{\partial X} \qquad (10.33)$$

where the minus sign strictly assigns negative values to strains $\langle\langle \epsilon_x^* \rangle\rangle$ and $\langle\langle \epsilon_x^{**} \rangle\rangle$ in rupture formation (see Fig. 10.3) and positive strain values in cases of bedding. By inserting Equations (10.32) and (10.31) into Equation (10.33), a conditional rupture strain of the second type is obtained

$$\langle\langle \epsilon_x^{**} \rangle\rangle = -\int_{s_0}^{s} \bar{\bar{a}}_{16} \frac{\partial N_x}{\partial X} ds \qquad (10.34)$$

Additional longitudinal forces $N'_x$ and $N''_y$ correspond to the additional strains $\langle\langle \epsilon'_x \rangle\rangle$ and $\langle\langle \epsilon''_x \rangle\rangle$ caused by the difference between the conditional position of the beam section that would set in if $\langle\langle \epsilon_x^* \rangle\rangle$ and $\langle\langle \epsilon_x^{**} \rangle\rangle$ were developed, and its virtual position determined by the strains and curvatures of the beam $\epsilon'$, $k'_Y$, $k'_Z$ (see Fig. 10.4(a)) and $\epsilon''$, $k''_Y$, $k''_Z$. In the mathematical form it is written as

$$\epsilon' + k'_Y Z + k'_Z Y - \langle\langle \epsilon_x^* \rangle\rangle = \bar{\bar{a}}_{11} N'_x = \langle\langle \epsilon'_x \rangle\rangle \qquad (10.35)$$

$$\epsilon'' + k''_Y Z + k''_Z Y - \langle\langle \epsilon_x^{**} \rangle\rangle = \bar{\bar{a}}_{11} N''_x = \langle\langle \epsilon''_x \rangle\rangle \qquad (10.36)$$

In order to determine total additional forces $N'_x + N''_x$ distributed across the profile outline $s$ and caused by the presence of shear forces $S_{YX}$ and $S_{ZX}$ in the beam it is still necessary to adopt the condition of self-equilibrium of $N'_x$ and $N''_x$ (see Equation (10.21))

$$\int_s (N'_x; N''_x)\{J\} = 0 \qquad (10.37)$$

The above statement enables us to obtain the expression for the distribution of normal and shear forces across the profile outline $s$ under the action of forces $\{N\}^T = [N, M_Y, M_Z]$ and $\{S\}^T = [0, S_{ZX}, S_{XY}]$ in the beam

$$N_x(s) = \{J_1(s)\}^T\{N\} + \{J_2(s)\}^T\{S\}$$

$$N_{xy}(s) = \{J_3(s)\}^T\{S\} \qquad (10.38)$$

Inserting Equation (10.38) into the deformation law of Equation (10.2) and considering that transversal strains are absent ($\langle\langle \epsilon_y \rangle\rangle = 0$) we obtain expressions for the distribution of strains $\langle\langle \epsilon_x \rangle\rangle$ and $\langle\langle \gamma_{xy} \rangle\rangle$ across the profile outline $s$

$$\langle\langle \epsilon_x(s) \rangle\rangle = \{J_4(s)\}^T \{N\} + \{J_5(s)\}^T \{S\}$$

$$\langle\langle \gamma_{xy}(s) \rangle\rangle = \{J_6(s)\}^T \{N\} + \{J_7(s)\}^T \{S\} \quad (10.39)$$

In Equations (10.38) and (10.39) the symbols $\{J_i(s)\}$, $i = 1$–$7$ have the following meaning

$$\{J_1(s)\} = \frac{1}{\bar{\bar{a}}_{11}} [K]^{-1T} \{J\}$$

$$\{J_2(s)\} = \frac{1}{\bar{\bar{a}}_{11}} [K]^{-1T} \{[U]^T [K]^{-1T} \{J\} - \bar{\bar{a}}_{16} \{U_1\} - \{U_2\}\}$$

$$\{J_3(s)\} = [K]^{1T} \{U_1\}$$

$$\{J_4(s)\} = \bar{\bar{a}}_{11} \{J_1(s)\}$$

$$\{J_5(s)\} = \bar{\bar{a}}_{11} \{J_2(s)\} + \bar{\bar{a}}_{16} \{J_3(s)\} \qquad (10.40)$$

$$\{J_6(s)\} = \bar{\bar{a}}_{16} \{J_1(s)\}$$

$$\{J_7(s)\} = \bar{\bar{a}}_{16} \{J_2(s)\} + \bar{\bar{a}}_{66} \{J_3(s)\}$$

$$[U] = \int_{s_0} \left( \frac{\bar{\bar{a}}_{16}}{\bar{\bar{a}}_{11}} \{J\} \{U_1\}^T + \frac{1}{\bar{\bar{a}}_{11}} \{J\} \{U_2\}^T \right) ds$$

$$\{U_1\} = \int_{s_0}^{s} \frac{1}{\bar{\bar{a}}_{11}} \{J\} ds$$

$$\{U_2\} = \int_{s_0}^{s} \frac{\bar{\bar{a}}_{16}}{\bar{\bar{a}}_{11}} \{J\} ds$$

Having determined the distribution of forces and deformations across the profile outline $s$, it is easy to determine elastic characteristics of the whole beam section by equalizing the parts of work done by internal and external forces in the elementary beam area having the length $dX$

$$\int_s (N_X \langle\langle \epsilon_x \rangle\rangle + N_{xy} \langle\langle \gamma_{xy} \rangle\rangle) ds dX = \begin{Bmatrix} \{N\} \\ \{S\} \end{Bmatrix}^T \begin{Bmatrix} \{\epsilon\} \\ \{\gamma\} \end{Bmatrix} dX \quad (10.41)$$

where

$$\{\gamma\}^T = [0; \gamma_{ZX}; \gamma_{YX}]$$

By inserting Equation System (10.39) into Equation (10.41) and expressing the strain and the curvature of the beam through the corresponding forces and by means of the deformation law containing a supposedly symmetrical compliance matrix

$$\begin{Bmatrix} \{\epsilon\} \\ \{\gamma\} \end{Bmatrix} = \begin{bmatrix} [F'] & [G'] \\ [G']^T & [H'] \end{bmatrix} \begin{Bmatrix} \{N\} \\ \{S\} \end{Bmatrix} \quad (10.42)$$

the sought compliance submatrices of the beam expressed by means of its geometrical and structural parameters and the RPL characteristics are

$$[F'] = \int_s \{J_1(s)\} \{J_4(s)\}^T ds$$

$$[H'] = \int_s (\{J_2(s)\} \{J_5(s)\}^T + \{J_3(s)\} \{J_7(s)\}^T) ds$$

$$(10.43)$$

$$[G'] = \int_s \{J_1(s)\} \{J_5(s)\}^T ds$$

$$= \int_s (\{J_4(s)\} \{J_2(s)\}^T + \{J_6(s)\} \{J_3(s)\}^T) ds$$

This confirms the symmetric character of the complete compliance matrix of the beam elementary area. It should also be noted that the submatrix $[F'] = [K]^{-1T}$.

The obtained deformation law of Equation (10.42) should be transformed, because the fourth term in the vector is equal to zero ($\partial N / \partial X = 0$). It becomes possible to eliminate the fourth line and the fourth column of the complete compliance matrix, and torsional compliance $D'_t = D_t^{-1}$ may be filled in the place of their intersection; the latter expression is the link between the relative torsion $\theta$ and the moment $\mathcal{M}$. The deformation law for the elementary beam area is proposed in the following form

$$\begin{Bmatrix} \varepsilon \\ k_Y \\ k_Z \\ \hdashline \theta' \\ \hdashline \gamma_{ZX} \\ \gamma_{YX} \end{Bmatrix} \begin{bmatrix} F'_{11} & F'_{12} & F'_{13} & \vdots & & & \vdots & G'_{12} & G'_{13} \\ F'_{21} & F'_{22} & F'_{23} & \vdots & & & \vdots & G'_{22} & G'_{23} \\ F'_{31} & F'_{32} & F'_{33} & \vdots & & & \vdots & G'_{32} & G'_{33} \\ \hdashline & & & \vdots & D'_t & \vdots & & & \\ \hdashline E'_{12} & E'_{22} & E'_{32} & \vdots & & & \vdots & H'_{22} & H'_{23} \\ E'_{13} & E'_{23} & E'_{33} & \vdots & & & \vdots & H'_{32} & H'_{33} \end{bmatrix} \begin{Bmatrix} N \\ M_Y \\ M_Z \\ \mathcal{M} \\ S_{ZX} \\ S_{YX} \end{Bmatrix} \quad (10.44)$$

where the compliance matrix is generally symmetrical, i.e. $F'_{ij} = F'_{ji}$; $H'_{ij} = H'_{ji}$; $G'_{23} \neq G'_{32}$. By reversing Equation (10.44), the deformation law is obtained containing the beam stiffness characteristics $F$, $G$, $D_t$ and $H$.

In determining the elastic properties of the beam according to Equations (10.43) and (10.40) it is easy to include the eventually different conditions concerning the realization of transversal strains $\langle\langle \epsilon_y \rangle\rangle$ in various areas of the outline $s$. If, for example, in a certain profile outline area the condition $\langle\langle \epsilon_y(s) \rangle\rangle = 0$ (see Equation (10.16)) may be replaced by the assumption that $N_y(s) = 0$, then the RPL area is characterized by compliances $\bar{a}_{ij}$ instead of compliances $a_{ij}$, the former obtained as a result of reversing the complete matrix of membrane stiffness [$A$].

To test the validity of the theoretical results obtained, a specific case of a thin-walled beam was examined—a flat homogeneous cantilevered strip loaded at the end by a concentrated force. For this particular case the formulas expressing the distribution of forces $N_x$ and $N_{xy}$ in the strip (see Equation (10.38)) turned out to be identical to the distribution formulas for normal and shear stresses in an anisotropic cantilever obtained by an exact solution in paper [40]. Fig. 10.5(a) illustrates the distribution of forces $N_x$ and $N_{xy}$ in a homogeneous anisotropic strip plotted according to Equation System (10.38) under the action of a unit moment and a unit shear force.

The new results obtained in the present paper are illustrated in Fig. 10.5(b), which shows the distribution of forces $N_x$ and $N_{xy}$ in a beam having a non-homogeneous profile structure. The diagrams are based on the same formulas and the same loads. The distributions of forces

252    MECHANICS OF ELASTIC COMPOSITE BEAM

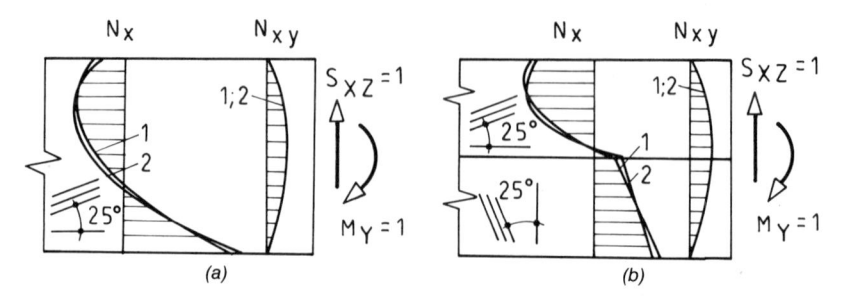

**FIGURE 10.5.** Distribution of forces in the strip having a uniform (a) and a repeatedly uniform (b) structure of the reinforced plastic under a unit moment and the unit shear force. Theoretical curves have been built according to formulas (10.38) with $N_y = 0$ (1) and $\langle\langle \epsilon_y \rangle\rangle = 0$ (2). The initial data are those of Fig. 9.2.

in Fig. 10.5 refer to two opposite cases of external conditions, i.e. $\langle\langle \epsilon_y \rangle\rangle = 0$ and $N_y = 0$.

In order to determine the position of the bending center, crossed by the shear forces without causing torsion of the beam, the torque $\mathcal{M}$ should be equalized to zero due to the action of forces $N_{xy}$ with respect to a point having unknown coordinates $Y_b$ and $Z_b$

$$\mathcal{M} = -\int_s (Z - Z_b) N_{xy} \cos \varphi ds + \int_s (Y - Y_b) N_{xy} \sin \varphi ds = 0 \quad (10.45)$$

By expressing $N_{xy}$ through Equation (10.38) in the form of shear forces, a condition is obtained for the determination of the bending center coordinates $Y_b$ and $Z_b$

$$\begin{Bmatrix} Z_b \\ Y_b \end{Bmatrix} = \left[ \int_s \{\bar{\bar{J}}_3(s)\} \begin{Bmatrix} -\cos \varphi \\ \sin \varphi \end{Bmatrix}^T ds \right]^{-1}$$

$$\times \left\{ \int_s \{\bar{\bar{J}}_3(s)\} \begin{Bmatrix} -\cos \varphi \\ \sin \varphi \end{Bmatrix}^T \begin{Bmatrix} Z \\ Y \end{Bmatrix} ds \right\} \quad (10.46)$$

where $\{\bar{\bar{J}}_3(s)\}$ is the vector consisting of the second and the third components of the vector $\{J_3(s)\}$.

The results obtained lead to the conclusion that the basic feature characterizing the elastic properties of open-profile thin-walled beams made of unorthotropic material is the presence of compliances and stiffnesses $G'_{ij}$ and $G_{ij}$, $i, j = 1$ which characterize the interaction between the longitudinal deformation and the bending of the beam, on one part, and its transversal shear, on the other part.

# Part III
# MECHANICS OF BEAM SYSTEMS

# CHAPTER 11
# Elastic Displacement of Laminated Beam Systems

## 11.1 POTENTIAL ENERGY OF A LAMINATED BEAM

The potential energy of an elastic laminated beam is expressed as

$$U = \frac{1}{2}\int_0^l N_x d\delta + \frac{1}{2}\int_0^l M_x d\varphi + \frac{1}{2}\int_0^l S_x d\gamma \qquad (11.1)$$

with the displacements $d\delta$, $d\varphi$ and $d\gamma$ for a beam element of length $dx$ given by

$$d\delta = d\langle\langle\epsilon_x^0\rangle\rangle = A'_{11}N_x dx + B'_{11}M_x dx$$

$$d\varphi = dk_x = B'_{11}N_x dx + D'_{11}M_x dx \qquad (11.2)$$

$$d\gamma = \frac{\kappa}{A_{55}}S_x dx$$

where $A_{55} = G_{xz}h$.

The following symbols have been used in Equation System (11.2):

$N_x$, $M_x$ and $S_x$—internal force and moment resultants acting on a beam of unit width

$\kappa$—non-dimensional coefficient of shearing strain depending on the cross-sectional shape and the lay-up configuration of the beam

$G_{xz}$—laminate shear modulus

$A_{55}$—component of laminate stiffness matrix

Inserting the expressions for the displacements $d\delta$, $d\varphi$ and $d\gamma$ into Equation (11.1) we obtain

$$U = \frac{1}{2}\int_0^l N_x^2 A'_{11}\,dx + \int_0^l N_x M_x B'_{11}\,dx$$
$$+ \frac{1}{2}\int_0^l M_x^2 D'_{11}\,dx + \frac{1}{2}\int_0^l \frac{\kappa}{A_{55}} S_x^2\,dx \qquad (11.3)$$

Equation (11.3) is valid for the case when the width of the rectangular cross-section of a beam is unity. If the width is $b$, then

$$U = \frac{1}{2b}\int_0^l N_p^2 A'_{11}\,dx + \frac{1}{b}\int_0^l N_p M_p B'_{11}\,dx$$
$$+ \frac{1}{2b}\int_0^l M_p^2 D'_{11}\,dx + \frac{1}{2b}\int_0^l \frac{\kappa}{A_{55}} S_p^2\,dx \qquad (11.4)$$

If the beam is symmetrically laminated with respect to its midplane, Equation (11.4) is written as

$$U = \frac{1}{2b}\int_0^l N_p^2 a_{11}\,dx + \frac{1}{2b}\int_0^l M_p^2 d_{11}\,dx + \frac{1}{2b}\int_0^l \frac{\kappa}{A_{55}} S_p^2\,dx \qquad (11.5)$$

where

$$a_{11} = \frac{1}{\Delta_A}(A_{22}A_{66} - A_{26}^2) \qquad d_{11} = \frac{1}{\Delta_D}(D_{22}D_{66} - D_{26}^2)$$

$$\Delta_A = \begin{vmatrix} A_{11} & A_{12} & A_{16} \\ A_{21} & A_{22} & A_{26} \\ A_{61} & A_{62} & A_{66} \end{vmatrix} \qquad \Delta_D = \begin{vmatrix} D_{11} & D_{12} & D_{16} \\ D_{21} & D_{22} & D_{26} \\ D_{61} & D_{62} & D_{66} \end{vmatrix}$$

The methods for determination of the shearing strain coefficient $\kappa$ have been discussed, for example, by Dharmarjan & McSutchnen [79] and Teh & Huang [97].

## 11.2 CASTIGLIANO'S METHOD

In accordance with the theorem of Castigliano, the generalized displacements $\Delta_i$ are equal to the partial derivatives of the potential energy $U$ taken with respect to the corresponding generalized forces $P_i$

$$\Delta_i = \frac{\partial U}{\partial P_i} \qquad (11.6)$$

When calculating displacements according to the theorem of Castigliano, the potential energy should be expressed as a function of generalized forces only.

It will be observed that the theorem of Castigliano allows for the determination of displacements only at those points of the system at which generalized forces are applied (e.g. a concentrated force or a concentrated moment). Determination of the displacement at an arbitrary point in an elastic system is made in the following artificial way. An imaginary additional concentrated force or moment is applied to the point where the displacement is to be determined and the potential energy is expressed as a function of the given and additionally superimposed load $P_g$ (or $M_g$). The unknown displacement is obtained from the equations

$$\Delta_k = \left(\frac{\partial U}{\partial P_g}\right)_{P_g=0} \qquad (11.7)$$

or

$$\theta_k = \left(\frac{\partial U}{\partial M_g}\right)_{M_g=0}$$

Using the potential energy expression for a laminated plastic in the form of Equation (11.4), Equation (11.6) may be developed in the following manner

$$\Delta_i = \frac{1}{b}\int_0^l N_P A'_{11} \frac{\partial N_P}{\partial P_i} dx + \frac{1}{b}\int_0^l N_P B'_{11} \frac{\partial M_P}{\partial P_i} dx$$

$$+ \frac{1}{b}\int_0^l M_P B'_{11} \frac{\partial N_P}{\partial P_i} dx + \frac{1}{b}\int_0^l M_P D'_{11} \frac{\partial M_P}{\partial P_i} dx \qquad (11.8)$$

$$+ \frac{1}{b}\int_0^l S_P \frac{\kappa}{A_{55}} \frac{\partial S_P}{\partial P_i} dx$$

For a symmetric configuration, Equation (11.8) is simplified as follows

$$\Delta_i = \frac{1}{b} \int_0^l N_P A'_{11} \frac{\partial N_P}{\partial P_i} dx$$
$$+ \frac{1}{b} \int_0^l M_P D'_{11} \frac{\partial M_P}{\partial P_i} dx + \frac{1}{b} \int_0^l S_P \frac{\kappa}{A_{55}} \frac{\partial S_P}{\partial P_i} dx \quad (11.9)$$

## 11.3 MOHR'S METHOD

Mohr's method is based on the principle of virtual work. Let us assume that a unit load is applied to a laminated beam with the load acting at the location and in the direction corresponding to the displacement, which is to be determined. Subsequently, a virtual displacement is applied to the beam, resulting in the displacement of the unit load by an increment $\Delta$ and in the virtual work $1\Delta$. The virtual work of the unit load will be equal to the total work of internal forces (resulting from the action of unit load) due to infinitesimally small deformations of the beam when the virtual displacement is applied

$$1\Delta = \int_0^l \bar{N}_x d\delta + \int_0^l \bar{M}_x d\varphi + \int_0^l \bar{S}_x d\gamma \quad (11.10)$$

where $\bar{N}$, $\bar{M}$, and $\bar{S}$ are internal force and moment resultants caused by the action of the unit load; $d\delta$, $d\varphi$ and $d\gamma$ are infinitesimal displacements of the beam elements. Equation System (11.2) will be used to determine the displacements of a laminated beam of asymmetric construction.

Inserting expressions for $d\delta$, $d\varphi$ and $d\gamma$ into Equation (11.10) and taking into account that the width of the beam's cross-section is $b$, we obtain the following formula for the determination of displacements in a laminated beam

$$\Delta_{iP} = \sum \frac{A'_{11}}{b} \int_{l_i} \bar{N}_i N_P dx + \sum \frac{B'_{11}}{b} \int_{l_i} \bar{N}_i M_P dx$$
$$+ \sum \frac{B'_{11}}{b} \int_{l_i} \bar{M}_i N_P dx + \sum \frac{D'_{11}}{b} \int_{l_i} \bar{M}_i M_P dx \quad (11.11)$$
$$+ \sum \frac{\kappa}{A_{55} b} \int_{l_i} \bar{S}_i S_P dx$$

For transversal bending, Equation (11.11) is given by

$$\Delta_{iP} = \sum \frac{B'_{11}}{b} \int_{l_i} \bar{N}_i N_P dx + \sum \frac{D'_{11}}{b} \int_{l_i} \bar{M}_i M_P dx$$
$$+ \sum \frac{\kappa}{A_{55}b} \int_{l_i} \bar{S}_i S_P dx \quad (11.12)$$

If the beam is symmetrically laminated about its midplane, Equations (11.11) and (11.12) simplify to

$$\Delta_{iP} = \sum \frac{a_{11}}{b} \int_{l_i} \bar{N}_I N_P dx + \sum \frac{d_{11}}{b} \int_{l_i} \bar{M}_i M_P dx$$
$$+ \sum \frac{\kappa}{A_{55}b} \int_{l_i} \bar{S}_i S_P dx \quad (11.13)$$

$$\Delta_{iP} = \sum \frac{d_{11}}{b} \int_{l_i} \bar{M}_i M_P dx + \sum \frac{\kappa}{A_{55}b} \int_{l_i} \bar{S}_i S_P dx \quad (11.14)$$

where

$$a_{11} = \frac{1}{\Delta_A}(A_{22}A_{66} - A_{26}^2)$$

$$\Delta_A = \begin{bmatrix} A_{11} & A_{12} & A_{16} \\ A_{21} & A_{22} & A_{26} \\ A_{61} & A_{62} & A_{66} \end{bmatrix}$$

$$d_{11} = \frac{1}{\Delta_D}(D_{22}D_{66} - D_{26}^2)$$

$$\Delta_D = \begin{bmatrix} D_{11} & D_{12} & D_{16} \\ D_{21} & D_{22} & D_{26} \\ D_{61} & D_{62} & D_{66} \end{bmatrix}$$

As an example, consider bending under the action of uniformly distributed load of a hinge-supported beam of rectangular cross-section

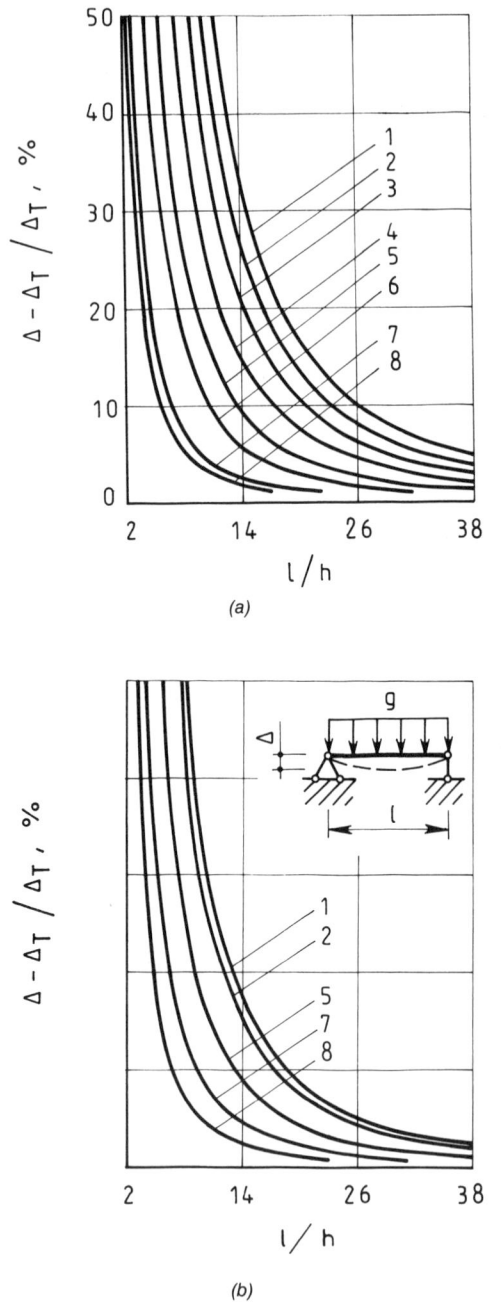

**FIGURE 11.1.** Dependence of displacement relative differences in the center of a hinge-supported beam of rectangular cross-section made of symmetric boron-epoxy (a) and carbon-epoxy (b) angle-ply upon the structure geometry: $B = 0°$ (1), $15°$ (2), $20°$ (3), $25°$ (4), $30°$ (5), $35°$ (6), $45°$ (7), $90°$ (8).

of height $h$, made of angle-ply boron-epoxy ($E_B/E_A = 100$; $\nu_A = 0.35$; $\nu_B = 0.23$; $\psi = 0.6$) and carbon-epoxy ($E_{Br}/E_A = 2$; $E_{Bz}/E_{Br} = 20$; $\nu_A = 0.35$; $\psi = 0.6$) laminates. Fig. 11.1 shows the calculated results of the relative difference between the deflections in the beam center determined with shear forces included and those determined with shear forces disregarded ($\Delta_T$ is the corresponding displacement determined, disregarding the shear forces). As follows from the results presented in Fig. 11.1, it is important to take into account shear forces in analyzing the strained state of beam systems if the beams have relatively small lengths.

The idea about relative shortness of beams (the ratio between the beam length $l$ and its height $h$) is basically dependent upon the type and properties of the material. Specifically, for reinforced plastic laminates the shortness of the beam is associated with the strain properties of its monolayers, lay-up orientation and reinforcement design (the symmetry of the material). The relative contribution of shear forces to the displacements of beam systems made of reinforced plastic laminates depends upon the symmetry properties of the ply lay-up design. It was shown by an example of [$\pm\beta$] and [0°/90°] type structures, where typical glass-, carbon- and boron fiber reinforced plastics were analyzed on the basis of numerical calculation, that this contribution is more remarkable for symmetrical structures than it is for asymmetric ones. Therefore, the effect of shear forces on the internal forces and the strained state of a beam system will be estimated for symmetric structures.

The greatest differences in the beam deflections calculated with shear forces in regards, compared with the corresponding deflections calculated by classical methods (where only bending moments are taken into account), are seen (see Fig. 11.1) for the most anisotropic reinforced plastics ($\beta = 0°$). With a layer lay-up angle $\beta$ this difference becomes smaller (and consequently, the effect of shear forces also decreases) because the bending stiffness of the reinforced plastic undergoes considerable changes ($D'_{11}$ is growing). We should note that the effect of shear forces is more pronounced for the boron fiber reinforced plastic than it is for the carbon fiber reinforced plastic, the former being a material with a higher strain anisotropy of plies, whereas for the glass fiber reinforced plastic this effect is still less pronounced.

To summarize, it may be said that the displacements in beam systems made of reinforced plastic laminates should be determined with shear forces taken into account, which is not the rule for displacement determination in homogeneous isotropic materials.

# CHAPTER 12
# Viscoelastic Displacements of Laminated Beam Systems

## 12.1 STRUCTURAL METHOD FOR THE DETERMINATION OF RHEOLOGICAL RESPONSE CHARACTERISTICS

In investigating the stress-strain state of laminated structures subjected to continuous loading it is necessary to know the rheological properties of the materials used, which may be determined either experimentally or analytically. The response functions obtained experimentally are valid only for one particular laminate configuration and are thus not applicable to laminated plastics composed of similar unidirectionally reinforced plies if the lay-up scheme is different. This explains why it is necessary to determine the visco-elastic properties of laminated reinforced plastics by structural methods, e.g., by taking into account the deformation characteristics of the individual layers and their stacking geometry. We should note that the structural method can be applied at the microstructural level of the individual layers. In fact, the problem of determining the visco-elastic properties of unidirectionally reinforced plies on the basis of the microstructural level of the individual components has been successfully solved, for instance, by Bulavs & Radinsh [9]. With this in mind, it will be further assumed for simplicity that the elastic and rheological deformation characteristics of the individual plies are known.

According to Equation (8.5) the deformation law for a linear visco-elastic laminate can be written in a generalized way as

$$\begin{bmatrix} \langle\langle \epsilon(t) \rangle\rangle \\ k(t) \end{bmatrix} = \begin{bmatrix} A'_* & | & B'_* \\ \hline B'_{*t} & | & D'_* \end{bmatrix} \begin{bmatrix} N(t) \\ M(t) \end{bmatrix} \qquad (12.1)$$

The components of the matrices $A'_*$, $B'_*$ and $D'_*$ are linear operators determined from the following relations

$$A'_{*ij}[F(t)] = A'_{*ij}\left\{F(t) + \int_0^t F(\xi)\tilde{A}'_{ij}(t-\xi)d\xi\right\}$$

$$B'_{*ij}[F(t)] = B'_{*ij}\left\{F(t) + \int_0^t F(\xi)\tilde{B}'_{ij}(t-\xi)d\xi\right\} \quad (12.2)$$

$$D'_{*ij}[F(t)] = D'_{*ij}\left\{F(t) + \int_0^t F(\xi)\tilde{D}'_{ij}(t-\xi)d\xi\right\}$$

The values of $A'_{ij}$, $B'_{ij}$, $D'_{ij}$, $i,j = 1, 2, 6$ are obtained from the elastic problem and represent the elastic matrix elements of membrane, membrane-bending, and bending compliances of laminated reinforced plastics.

For uniaxial loads we have

$$\langle\langle \epsilon_i(t) \rangle\rangle_{,N_j} = A'_{ij*}[N_j(t)]$$

$$\langle\langle \epsilon_i(t) \rangle\rangle_{,M_j} = B'_{ij*}[M_j(t)]$$

$$k_i(t)_{,N_j} = B'_{ji*}[N_j(t)] \quad (12.3)$$

$$k_i(t)_{,M_j} = D'_{ij*}[M_j(t)]$$

Here $\langle\langle \epsilon_i(t) \rangle\rangle_{,N_j}$; $\langle\langle \epsilon_i(t) \rangle\rangle_{,M_j}$ ... corresponds to the $i$-th component of the laminate deformation caused by the action of the $j$-th loading component ($N_j$, $M_j$). Hence, it is easy to determine the rheological properties of laminated plastics subjected to sustained loading ($N_j(t)$ = constant; $M_j(t)$ = constant) with the aid of Equation System (12.2)

$$\int_0^t \tilde{A}'_{ij}(t-\xi)d\xi = \left[\frac{\langle\langle \epsilon_i(t) \rangle\rangle_{,N_j}}{A'_{ij}N_j} - 1\right] \quad (12.4)$$

$$\int_0^t \tilde{B}'_{ij}(t-\xi)d\xi = \left[\frac{\langle\langle \epsilon_i(t) \rangle\rangle_{,M_j}}{B'_{ij}M_j} - 1\right]$$

$$= \left[\frac{k_j(t)_{,N_i}}{B'_{ij}N_i} - 1\right] \quad (12.5)$$

$$\int_0^t \tilde{D}'_{ij}(t-\xi)d\xi = \left[\frac{K_i(t)_{,M_j}}{D'_{ij}M_j} - 1\right] \quad (12.6)$$

By differentiating Equations (12.4) through (12.6) with respect to the variable $t$, and taking into account the following

$$\int_0^b F(b-x)dx = \int_0^b F(x)dx \quad (12.7)$$

we arrive at

$$\tilde{A}'_{ij}(t) = \frac{d\langle\langle \epsilon_i(t)\rangle\rangle_{,N_j}}{dt} \frac{1}{A'_{ij}N_j} \quad (12.8)$$

$$\tilde{B}'_{ij}(t) = \frac{d\langle\langle \epsilon_i(t)\rangle\rangle_{,M_j}}{dt} \frac{1}{B'_{ij}M_j}$$

$$= \frac{dk_j(t)_{,N_i}}{dt} \frac{1}{B'_{ij}N_i} \quad (12.9)$$

$$\tilde{D}'_{ij}(t) = \frac{dk_i(t)_{,M_j}}{dt} \frac{1}{D'_{ij}M_j} \quad (12.10)$$

Note that Equation (12.9) possesses the same type of kernels $\tilde{B}_{ij}(t - \xi)$ and is chosen as the defining expression depending on the availability of experimental (computational) data, convenience of computation, and other specific conditions.

The determination of creep kernels representing the time-dependent deformation characteristics of laminated reinforced plastics on the basis of Equations (12.8) through (12.10) requires knowledge of the corresponding creep curves $\langle\langle \epsilon_i(t)\rangle\rangle_{,N_j}$; $\langle\langle \epsilon_i(t)\rangle\rangle_{,M_j}$; and so on. If the time-dependent visco-elastic deformation relations for plastics are analytically prescribed, the corresponding creep kernels are obtained as a result of analytical differentiation of Equations (12.8) through (12.10). An illustrative example will be a balanced angle-ply reinforced plastic; its time-dependent stress-strain response under constant uniaxial tension

**266** MECHANICS OF BEAM SYSTEMS

has been analyzed in Section 5.3. Hence, Equation (12.8) for, say, $\tilde{A}'_{11}(t - \xi)$ yields

$$\tilde{A}'_{11}(t - \zeta) = C'_{22} \exp\left[-\alpha_{22}(t - \xi)\right]$$

$$+ C'_{66} \exp\left[-\alpha_{66}(t - \xi)\right] \quad (12.11)$$

$$+ \sum_{i=1}^{2} C'_i \exp\left[P_i(t - \xi)\right]$$

Here

$$C'_{22} = \frac{\sin^2 \beta}{hA'_{11}} \left\{ S_{22}C_{22} \sin^2 \beta - \frac{A_3}{B_1 P_1 P_2} S_{22}C_{22} \sin 2\beta \right.$$

$$\left. - \frac{S_{22}C_{22} \sin 2\beta \alpha_{22}}{B_1} \sum_{i=1}^{2} \frac{[A_1 P_i^2 + A_2 P_i + A_3]}{P_i(P_i - P_j)(\alpha_{22} + P_i)} \right\}$$

$i \neq j$

$$C'_{66} = \frac{1}{hA'_{11}} \left\{ S_{66}C_{66} \frac{\sin^2 2\beta}{4} + \frac{A_3}{B_1 P_1 P_2} S_{66}C_{66} \frac{\sin 4\beta}{4} \right.$$

$$\left. - \frac{S_{66}C_{66}\alpha_{66} \sin 4\beta}{4B_1} \sum_{i=1}^{2} \frac{[A_1 P_i^2 + A_2 P_i + A_3]}{P_i(P_i - P_j)(\alpha_{66} + P_i)} \right\} \quad (12.12)$$

$i \neq j$

$$C'_i = \frac{[A_1 P_i^2 + A_2 P_i + A_3]}{B_1 h A'_{11}} \left\{ S^*_{16} - \frac{S_{66}C_{66} \sin 4\beta}{4(\alpha_{66} + P_i)} \right.$$

$$\left. - \frac{2 S_{22}C_{22} \sin^3 \beta \cos \beta}{(\alpha_{22} + P_i)} \right\}$$

The values of the terms $P_1$, $P_2$, $A_1$, $D_1$, ..., $A_3$, $D_3$ have been determined in Section 5.3.

The determination of the creep kernels $\tilde{A}'_{22}(t - \xi)$, $\tilde{A}'_{12}(t - \xi)$, $\tilde{A}'_{66}(t - \xi)$ is analogous to the determination of $\tilde{A}'_{11}(t - \xi)$.

The rheological properties computed on the basis of Equations (12.11) and (12.12) hold true for any balanced cross-ply reinforced plastic regardless of the properties of the individual plies. Knowledge of these properties permits us to describe the deformation of a material in any type of plane stress state—both time-dependent and time-independent—by applying the deformation law given by Equation (12.1).

Note that analogous analytical solutions have been obtained also for balanced orthogonally reinforced plastics on the basis of results from Gurvich [21]. Similar to the case of orthogonally reinforced plastics, time-dependent variations in the stress state of the monolayers were taken into account here.

Analytical solutions to the problem of visco-elastic deformation of cross-ply and angle-ply reinforced plastics given in the previous subsections were obtained by Laplace transformations. In the general case of arbitrarily unbalanced configurations subjected to different combined loads, the application of the Laplace transformations to the various expressions is a laborious task. The solution becomes considerably more complicated when the creep kernels of the individual layers have been represented by a more complicated form than the exponential form. Therefore, the analysis of three-directional schemes of reinforcement, or schemes involving an even larger number of directions, will be more conveniently carried out by numerical computation creep curves for laminated reinforced plastics with a prescribed lay-up geometry. These methods will also permit numerical evaluation of the corresponding rheological properties of the material.

By using the numerical approach to determine the stress-strain response of laminated reinforced plastics under combined continuously applied loads where the distribution of time-dependent stresses within the individual plies is taken into account, we obtain a set of discrete values that correspond to deformations at prescribed periods of time: $\langle\langle \epsilon'_i(t_1) \rangle\rangle$, $\langle\langle \epsilon'_i(t_2) \rangle\rangle$, ..., $\langle\langle \epsilon'(t_k) \rangle\rangle$, where $k$ is the number of calculated points. The above values may be regarded as being experimental, (that is, may be viewed as the results of a computer experiment, which differs from a real experiment in that the actual properties of the prescribed configuration are predicted by means of numerical simulation instead of using real material). Therefore, the mechanism of computing the rheological properties of reinforced plastics does not differ, in principle, from that used for experimentally obtained creep kernels.

Let the unknown kernels be some functions of the material constants prescribed earlier:

$$\tilde{A}'_{ij}(t - \xi) = \tilde{A}'_{ij}(a_l, t - \xi)$$

$$\tilde{B}'_{ij}(t - \xi) = \tilde{B}'_{ij}(b_l, t - \xi) \qquad (12.13)$$

$$\tilde{D}'_{ij}(t - \xi) = \tilde{D}'_{ij}(d_l, t - \xi)$$

where $a_l, b_l, d_l, l = 1, \ldots, M$ are the unknown parameters of the rheological properties of the laminate. Subsequently, using Equation System (12.3), it is possible to obtain the corresponding deformation values for any prescribed time period in terms of the parameters $a_l, b_l, d_l$:

$$\langle\langle \epsilon_i(a_l, b_l, d_l, t_1) \rangle\rangle, \ldots, \langle\langle \epsilon_i(a_l, b_l, d_l, t_k) \rangle\rangle$$

By comparing the above deformations with those computed by the numerical approach, we find the unknown parameters of the rheological response functions of laminated reinforced plastics. The requirement for minimizing the error vector $S$ is used as a condition for determining $a_l, b_l, d_l$:

$$S = \sqrt{\sum_{j=1}^{k} \left[ \frac{\langle\langle \epsilon'_i(t_j) \rangle\rangle - \langle\langle \epsilon_i(a_l, b_l, d_l, t_j) \rangle\rangle}{\langle\langle \epsilon'_i(t_j) \rangle\rangle} \right]^2} \to \min \qquad (12.14)$$

Equation (12.14), which is analogous to the least-squares method, permits the determination of the unknown parameters $a_l, b_l, d_l$ for any type of creep kernels. An appropriate standard program in FORTRAN-IV and compiled on an ES-series computer has been developed using a series representation for the creep kernels in terms of exponential functions. As an algorithm for finding the minimum value of the error vector $S$ and its representative values $a_l, b_l$ and $d_l$, a grid search method has been employed which proved more effective in saving computer time than the minimax method, which used derivative computation.

## 12.2 THE EFFECT OF LOADING DURATION ON THE STRESS-STRAIN RESPONSE OF LAMINATED BEAM SYSTEMS

The determination of displacements in laminated beam systems, consisting of elements with expressed rheological properties, subjected

Viscoelastic Displacements of Laminated Beam Systems 269

to sustained loads, is conveniently carried out by employing the Volterra principle, which states that in the short-term solution, elastic constants can be replaced by the corresponding operators. Let the load applied to the beam system be a function of time, which results in the following force and moment resultants:

$$M_p(t, x) = \sum_{i=1}^{m} f_i(t) M_p^{(i)}(x)$$

$$N_p(t, x) = \sum_{i=1}^{m} f_i(t) N_p^{(i)}(x) \qquad (12.15)$$

$$S_p(t, x) = \sum_{i=1}^{m} f_i(t) N_p^{(i)}(x)$$

where $m$ is the number of various time-varying generalized loads acting on the system.

Consequently, taking Equation (11.12) into account, the function of the time variation of the displacements will be determined as [25]

$$\Delta_{ij}(t) = \sum \int_{(l)} \frac{1}{b} \sum_{i=1}^{m} \{ \bar{N} N_p^{(i)} A'_{11*}[f_i(t)]$$

$$+ \bar{N} M_p^{(i)} B'_{11*}[f_i(t)] + \bar{M} N_p^{(i)} B'_{11*}[f_i(t)] \qquad (12.16)$$

$$+ \bar{M} M_p^{(i)} D'_{11*}[f_i(t)] + \bar{S} S_p^{(i)} F_*[f_i(t)] \} dx$$

Here $A'_{11*}$, $B'_{11*}$, $D'_{11*}$ are operators defined by Equation (12.2) whereas the operator $F_*$ is given by

$$F_*[f_i(t)] = F \left\{ f_i(t) + \int_0^t f_i(\xi) \tilde{F}(t - \xi) d\xi \right\} \qquad (12.17)$$

and its rheonomic characteristics are determined by analogous methods described in Section 12.1. The value $F$ denotes the compliance of the beam to shear [23] and according to Equation (11.12) is given by

$$F = \frac{\kappa}{A_{55}} \qquad (12.18)$$

Taking into account Equation (12.16) when constant load is unchangeable through time ($f_i(t)$ = const = 1, $m$ = 1) we obtain

$$\Delta_{ij}(t) = \sum \int_l \frac{1}{b} [\bar{N}N_p A'_{11*}(1) + \bar{N}M_p B'_{11*}(1)$$

$$+ \bar{M}N_p B'_{11*}(1) + \bar{M}M_p D'_{11*}(1) + \bar{S}S_p F_*(1)] dx \quad (12.19)$$

Equation (12.16) permits us to analyze the relative contribution of the individual long-term acting internal forces to the evolution of time-dependent displacements in beam systems constructed of reinforced plastics. Such an analysis leads to the following conclusion: the influence of the shear force increases as the ratio of the axial modulus of elasticity to the longitudinal shear modulus increases. This ratio becomes larger with time, and so does, consequently, the contribution of the shear force to the deformation of the system. By way of an example, consider a simply-supported beam acted upon by the centrally applied concentrated force $P$. Fig. 12.1 shows the results of calculation of the relative

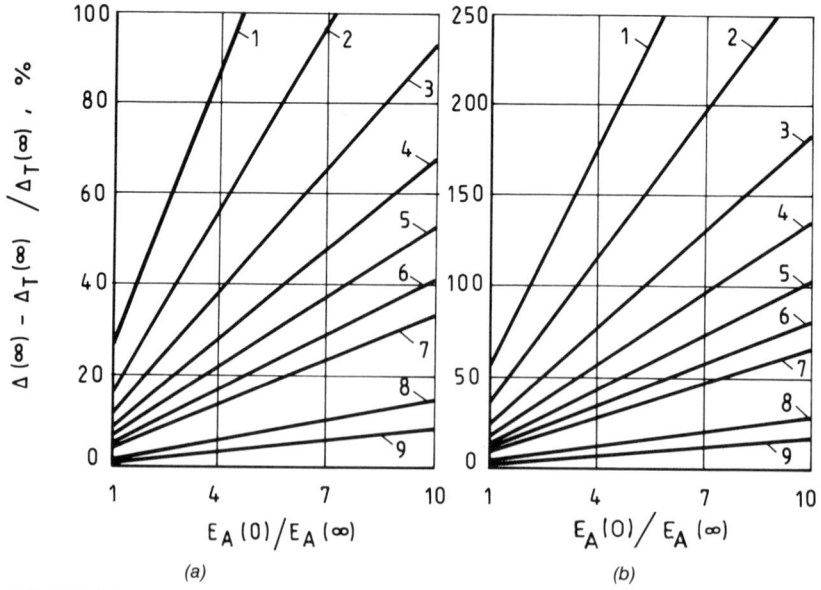

**FIGURE 12.1.** Dependence of the relative difference of vertical displacements of a hinge-supported unidirectionally reinforced beam's centre upon the degree of creep of the polymer matrix in the case of (a) glass-epoxy and (b) carbon-epoxy with $l/h$ = 8 (1), 10 (2), 12 (3), 14 (4), 16 (5), 18 (6), 20 (7), 30 (8), 40 (9).

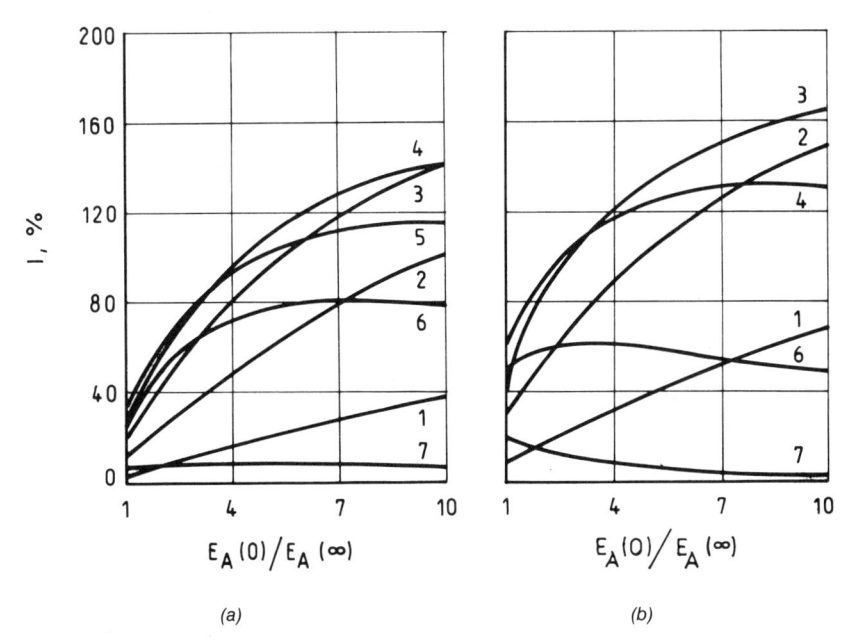

**FIGURE 12.2.** Dependence of parameter $J$ at $t \to \infty$ for a long beam made of angle-ply glass-epoxy (a) and carbon-epoxy (b) upon the degree of creep of the polymer matrix: $B = 5°$ (1), 10 (2), 15 (3), 20 (4), 25 (5), 30 (6), 45° (7).

difference in the vertical displacement of the beam's center when the action of the shear forces is neglected. The results are presented for two types of unidirectionally reinforced glass- and carbon-epoxy with $\psi = 0.6$; $\Delta_T(\infty)$ are deflections at $t \to \infty$ calculated without considering shear forces, and $\Delta(\infty)$ are deflections at $t \to \infty$ calculated by including the action of shear forces. It follows from Fig. 12.1 that the relative difference in the vertical displacement introduced by neglecting the weak shear resistance of laminated reinforced plastics actually increases in a linear way that depends on the degree of creep in the polymer matrix. This difference can reach considerable magnitudes. The definition of a 'short beam' changes with increased loading time, depending on the rheological properties of the plastic components. Even for relatively long beams ($l/h = 20$ or more) it is necessary to consider the action of shear forces when calculating the deformed state at $t \to \infty$. Note that the response of plastics with a higher degree of anisotropy is more sensitive to low shear resistance (in the given example, carbon-epoxy as compared to glass-epoxy).

Another factor for structuring reinforced plastic laminates that changes radically with the loading duration is the symmetry (or asymmetry) of the reinforcement design of the materials. Let us introduce parameter $J$ designating the dependence of strains in beam systems made of reinforced plastic laminates, upon their structural symmetry (in percentage) according to the rule

$$J = \frac{\Delta_{\text{asym}} - \Delta_{\text{sym}}}{\Delta_{\text{sym}}} \qquad (12.20)$$

where $\Delta_{\text{sym}}$ is the strain of the beam system under fixed load and with symmetrical structural design (for angle-ply laminates—an infinite number of ply pairs laid in directions $\pm\beta$); $\Delta_{\text{asym}}$ is the corresponding strain of the beam system under the same fixed load and with asymmetrical structural design (for angle-ply laminates—a single pair of plies $\pm\beta$).

Fig. 12.2 shows the calculated results of parameter $J$ for beam deflections with $t \to \infty$, where the beam is made of angle-ply glass-epoxy and carbon-epoxy laminates with $\psi = 0.6$ depending upon the degree of creep of the polymer matrix and upon the lay-up direction of the fibers. The beam is regarded as sufficiently long, which makes it possible to neglect the action of shear forces. The effect of the structural asymmetry of the material structure on the displacements varies with the changes in the lay-up angle of the monolayers. Thus, for angle-ply laminates with $\beta \geq 45°$ the difference in the maximum displacements of beams of symmetric and asymmetric structures respectively does not exceed 20 percent, but with the ply lay-up angles of $10° \leq \beta \leq 25°$ the above difference may be as large as 150 percent.

CHAPTER 13

# Numerical Analysis of Statically Indeterminate Beam Systems Under Long-Term Loading

## 13.1 ANALYSIS BY THE METHOD OF FORCES OF BEAM SYSTEMS CONSISTING OF REINFORCED PLASTICS LAMINATES

The specific nature of the numerical analysis of statically indeterminate beam systems made of reinforced plastics, if compared to the conventional methods of structural mechanics, is based on the specificity of the beam material where an asymmetric structure is possible and the longitudinal shear stiffness is low. Another essential characteristic, as indicated in Chapter 12, is the significant dependence of the strains in beam systems made of reinforced plastics upon the loading time, and as a result there is the time-dependence of the stress state of statically indeterminate systems.

Let us consider an arbitrary statically indeterminate beam system where each member is made of a reinforced plastic laminate, under the action of short-term load. With the formation of a statically determinate primary structure by substituting the unknown responses $X$ ($X = X_1$, ..., $X_k$) where $X_i$ denotes generalized forces; $k$ is the degree of static indeterminacy of the system according to the method of forces) for "redundant constraints", the forces are written as

$$M = M_p + \sum_{i=1}^{k} X_i \bar{M}_i$$

$$N = N_p + \sum_{i=1}^{k} X_i \bar{N}_i$$

273

274   MECHANICS OF BEAM SYSTEMS

$$S = S_p + \sum_{i=1}^{k} X_i \bar{S}_i \qquad (13.1)$$

where $\bar{M}_i$, $\bar{N}_i$ and $\bar{S}_i$ are the respective forces on the primary structure under the action of the force $X_i = 1$.

The system of superposition equations derived from the condition of zero displacement on the primary structure in the directions of the action of the unknown redundants has the form $X_i$

$$\Delta_{ii} = 0, \qquad i = 1, \ldots, k \qquad (13.2)$$

where $\Delta_{ii}$ is the displacement of the primary structure in the direction of the action of the $i$-th redundant in the $i$-th point. By inserting Equation (11.12) into Equation (13.2) we obtain the matrix

$$[\delta][X] + [\Delta_p] = [0] \qquad (13.3)$$

where $[X]$ is the column matrix of the redundants sought. The elements of the square matrix $[\delta]$ are given by

$$\delta_{ij} = \sum \int_{(l)} \frac{1}{b} \{ \bar{N}_i \bar{N}_j A'_{11} + (\bar{N}_i \bar{M}_j + \bar{N}_j \bar{M}_i) B'_{11} \\ + \bar{M}_i \bar{M}_j D'_{11} + \bar{S}_i \bar{S}_j F \} dx \qquad (13.4)$$

The elements $\Delta_{ip}$ of the column matrix $[\Delta_p]$ are given by Equation (11.12).

After solving Equation (13.3), the unknown redundants are determined as

$$[X] = -[\delta]^{-1}[\Delta_p] \qquad (13.5)$$

The values of the redundants in the removed restraints calculated by using Equation System (13.1) make it possible to determine the forces in the beams of statically indeterminate systems where the structure and the strain properties of the reinforced plastic laminates are taken into account.

As an example, consider a supported cantilever beam (structure is indeterminate to the first degree) bent under a uniformly distributed load. Applying the classical solution through the method of forces for a homogeneous isotropic beam, the values of internal forces are determined by its dimensions and the magnitude of the external load. Thus, in the case under consideration, the redundant of the right-hand support is given by the expression

$$X_T = \tfrac{3}{8} gl \qquad (13.6)$$

and the redundant of the rigid fixing is given by

$$M_T = \tfrac{1}{8} gl^2 \qquad (13.7)$$

where the subscript "$T$" refers to the redundants determined regardless of the action of shear forces. Applying the method of numerical determination, however, which enables us to include the effect of the shear stiffness of the laminates, their anisotropy and structural inhomogeneity, i.e., determining the respective characteristics of $X$ or $M$ and according to Equation (13.6) while also taking into account Equations (13.4) and (13.5), we have

$$X = \left[ \frac{1 + 4\eta \left(\dfrac{h}{l}\right)^2}{1 + 3\eta \left(\dfrac{h}{l}\right)^2} \right] X_T \qquad (13.8)$$

Here the dimensionless parameter $\eta$ is determined by the structure of the reinforced plastic laminate as given by

$$\eta = \frac{F}{D'_{11}} \frac{1}{h^2} = \frac{\kappa}{A_{55} D'_{11}} \frac{1}{h^2} \qquad (13.9)$$

where $h$ is the height of the beam. It may be shown that in the rigid fixing of the simply-supported cantilever the moment is determined by using Equation (13.1):

$$M = \left[ \frac{1}{1 + 3\eta \left(\dfrac{h}{l}\right)^2} \right] M_T \qquad (13.10)$$

Figs. 13.1 and 13.2 show the dependence of unknown reactions upon the structure of the material and the geometry of a rectangular cross-section beam (with the relative height $l/h$). The beams under discussion are made of symmetrical angle-ply boron ($E_B/E_A = 100$, $\nu_A = 0.35$, $\nu_B = 0.23$) and carbon-epoxy ($E_{Br}/E_A = 2$, $E_{Bz}/E_{Br} = 20$, $\nu_A = 0.35$) with $\psi = 0.6$. The action of shear forces is only effective in relatively short beams (with practically $l/h \leq 7$), this effect being better pronounced in more anisotropic materials ($\beta = 0°$). With a large packaging angle $\pm\beta$ the value of parameter $\eta$ decreases and reduces the difference between $X_T$ and $X$ ($M_T$ and $M$) to the minimum. For relatively short beams ($l/h \leq 7$), the difference between the corresponding values of the diagram ordinates of the bending moments $M$ and $M_T$ does not exceed 20 percent. Therefore the action of shear forces should be taken into account in designing statically indeterminate systems by means of the method of forces as required by the accuracy of the design.

It should be noted that the effect of shear forces is most profound for symmetrical structures, because of an asymmetric design of a laminate structure leads to higher values of $D'_{11}$ and does not significantly influence the value $F$ that is responsible for the decreasing of parameter $\eta$.

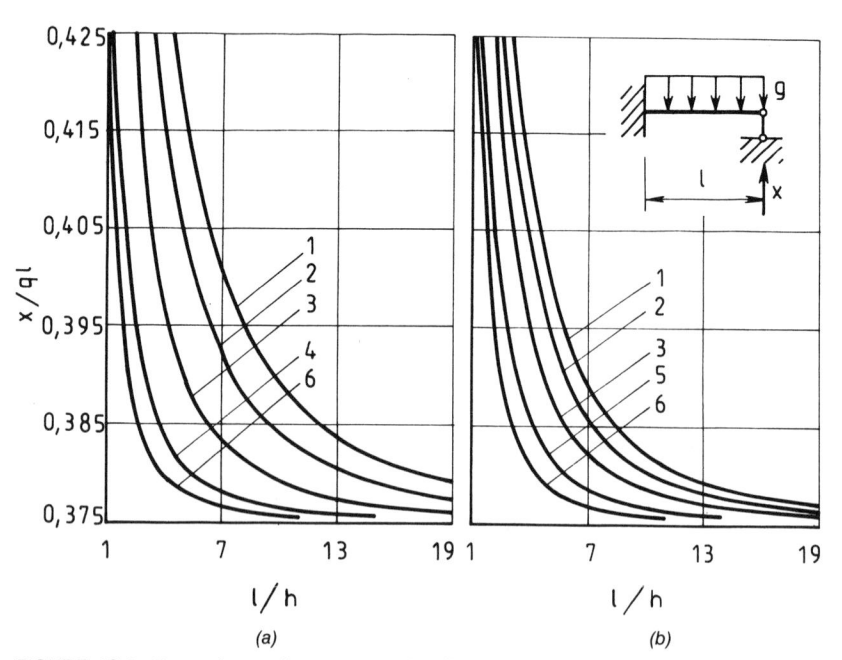

**FIGURE 13.1.** Dependence of support reaction $X$ upon the structure geometry of the beam made of angle-ply symmetrical boron- (a) and carbon-epoxy (b): $B = 0°$ (1), 20 (2), 30 (3), 40 (4), 45 (5), 90° (6).

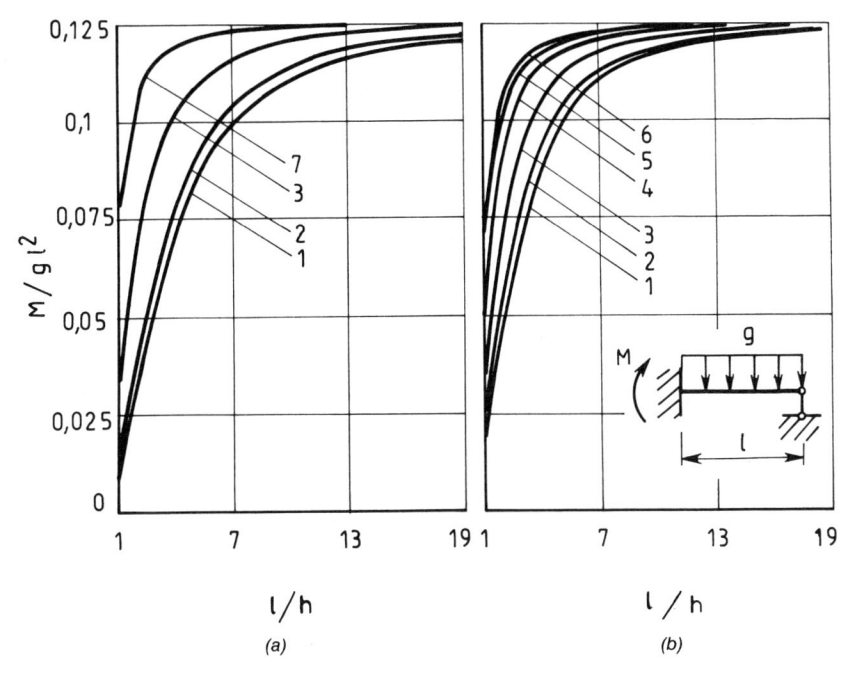

**FIGURE 13.2.** Dependence of moment reaction $M$ upon the structure geometry of the beam made of angle-ply symmetrical boron- (a) and carbon-epoxy (b): $B = 0°$ (1), 15 (2), 30 (3), 45 (4), 60 (5), 75 (7), 90° (7).

The simplest type of a statically indeterminate beam having one unknown reaction was chosen as the most effective example for analyzing the influence of the material structure upon the stress state of the system. Investigations of statically indeterminate systems having many unknown reactions (frame calculation with $k = 2,3$) have given analogous results.

## 13.2 SPECIFIC APPLICATION OF THE METHOD OF DISPLACEMENTS

The defining expressions of the method of forces developed in Section 13.1 make it possible to include the specific behaviour of reinforced plastics when the method of displacements is used. We shall proceed to show the results of designing elementary statically indeterminate beams under unit displacements or transverse loading. The results given here are needed for calculating coefficients of superposition equations

**278** MECHANICS OF BEAM SYSTEMS

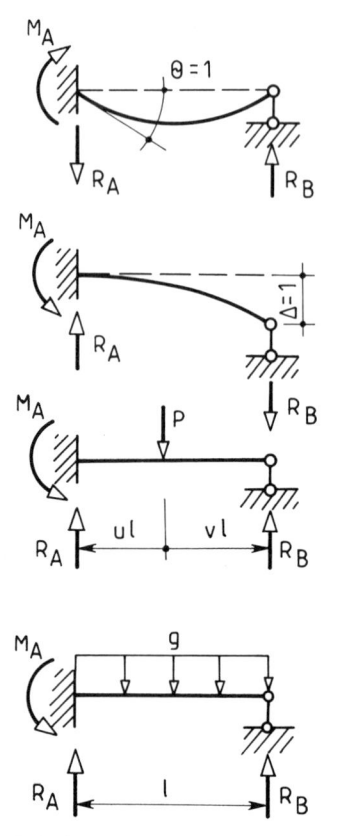

**FIGURE 13.3.** Diagram of unit displacements and simple loads of a simply-supported cantilever.

of the displacement method, its algorithm complying fully with the conventional approach.

Consider a simply-supported cantilever

- under a unit angular displacement (Fig. 13.3)

$$M_A = \frac{3}{D'_{11}l\left[1 + 3\eta\left(\frac{h}{l}\right)^2\right]}$$

$$R_A = R_B = \frac{3}{D'_{11}l^2\left[1 + 3\eta\left(\frac{h}{l}\right)^2\right]}$$

(13.11)

- under a unit linear displacement (see Fig. 13.3)

$$M_A = \frac{3}{D'_{11}l^2\left[1 + 3\eta\left(\frac{h}{l}\right)^2\right]}$$

$$R_A = R_B = \frac{3}{D'_{11}l^3\left[1 + 3\eta\left(\frac{h}{l}\right)^2\right]}$$

(13.12)

- under concentrated force $P$ (see Fig. 13.3)

$$M_A = Pl\left\{\frac{uv(u + 2v)}{2\left[1 + 3\eta\left(\frac{h}{l}\right)^2\right]}\right\}$$

$$R_A = P\frac{\left[v(3 - v^2) + 6\eta\left(\frac{h}{l}\right)^2(1 + u)\right]}{2\left[1 + 3\eta\left(\frac{h}{l}\right)^2\right]}$$

(13.13)

$$R_B = P\frac{u\left\{u(2 + v) + 6\eta\left(\frac{h}{l}\right)^2\right\}}{2\left[1 + 3\eta\left(\frac{h}{l}\right)^2\right]}$$

- under uniformly distributed load $g$ (see Fig. 13.3)

$$M_A = \frac{gl^2}{8}\left\{\frac{1}{1 + 3\eta\left(\frac{h}{l}\right)^2}\right\}$$

$$R_A = \frac{gl^2}{8}\left\{\frac{5 - 12\eta\left(\frac{h}{l}\right)^2}{1 + 3\eta\left(\frac{h}{l}\right)^2}\right\}$$

(13.14)

$$R_B = \frac{gl}{8}\left\{\frac{3 + 12\eta\left(\frac{h}{l}\right)^2}{1 + 3\eta\left(\frac{h}{l}\right)^2}\right\}$$

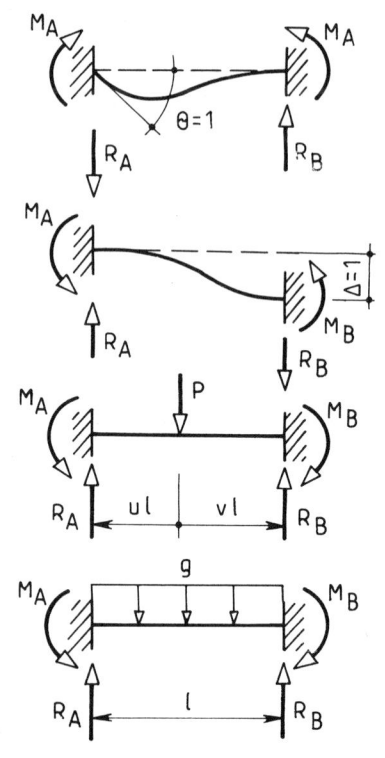

**FIGURE 13.4.** Diagrams of unit displacements and simple loads of a restrained beam.

Consider a beam with fixed supports
- under a unit angular displacement (Fig. 13.4)

$$M_A = \frac{1}{D'_{11}l} \left\{ \frac{4 + 12\eta\left(\frac{h}{l}\right)^2}{1 + 12\eta\left(\frac{h}{l}\right)^2} \right\}$$

$$M_B = \frac{1}{D'_{11}l} \left\{ \frac{2 - 12\eta\left(\frac{h}{l}\right)^2}{1 + 12\eta\left(\frac{h}{l}\right)^2} \right\} \quad (13.15)$$

$$R_A = R_B = \frac{6}{D'_{11}l^2 \left[1 + 12\eta\left(\frac{h}{l}\right)^2\right]}$$

# Numerical Analysis of Statically Indeterminate Beam Systems 281

- under a unit linear displacement (see Fig. 13.4)

$$M_A = M_B = \frac{6}{D'_{11}l^2\left[1 + 12\eta\left(\frac{h}{l}\right)^2\right]}$$

$$R_A = R_B = \frac{12}{D'_{11}l^3\left[1 + 12\eta\left(\frac{h}{l}\right)^2\right]}$$

(13.16)

- under concentrated force $P$ (see Fig. 13.4)

$$M_A = Pl\left\{\frac{uv\left[v + 6\eta\left(\frac{h}{l}\right)^2\right]}{1 + 12\eta\left(\frac{h}{l}\right)^2}\right\}$$

$$M_B = Pl\left\{\frac{uv\left[u + 6\eta\left(\frac{h}{l}\right)^2\right]}{1 + 12\eta\left(\frac{h}{l}\right)^2}\right\}$$

$$R_A = P\left\{\frac{v\left[v + 2uv + 12\eta\left(\frac{h}{l}\right)^2\right]}{1 + 12\eta\left(\frac{h}{l}\right)^2}\right\}$$

$$R_B = P\left\{\frac{u\left[u + 2uv + 12\eta\left(\frac{h}{l}\right)^2\right]}{1 + 12\eta\left(\frac{h}{l}\right)^2}\right\}$$

(13.17)

- under uniformly distributed load $g$ (see Fig. 13.4)

$$M_A = M_B = \frac{gl^2}{12}$$

$$R_A = R_B = \frac{gl}{2}$$

(13.18)

## 13.3 THE EFFECT OF LOADING DURATION ON INTERNAL STRESSES

As has already been shown, the creep in reinforced plastic laminates is the cause of increasing displacement in beam systems made of these materials. It is thought, nevertheless, that the displacement does not become significantly large. Taking into account the time-dependent linear relationship existing between strains and stresses, it may be concluded that the principle of virtual displacements also remains and, consequently, the theorems of reciprocity between displacements and work are valid. We consider that the principle of independence of acting forces, as well as the principle of superposition, exist at any time. Thus it follows from the hypotheses proposed that the design methods for statically indeterminate beam systems, according to the method of forces where the loading time is included, are based on the same principle as the design of systems characterized by elastic performance. It is to be noted that superposition equations also express the equality to zero of the complete displacements along one of the unknowns, the displacements being caused by the action of all the unknowns and of the external load at any time. On account of the creep in reinforced plastic laminates, however, there are time-dependent changes in displacements and subsequently a continuous variation in unknown redundants. Therefore, the superposition equations for the method of forces are written regardless of the creep character of the material as

$$X_1 \delta_{i1}(t) + \int_0^t \frac{dX_i}{d\tau} \delta_{i1}(t, \tau) d\tau + \ldots + X_k \delta_{ik}(t)$$

$$+ \int_0^t \frac{dX_k}{d\tau} \delta_{ik}(t, \tau) d\tau + P\Delta_{ip}(t) + \int_0^t \frac{dP}{d\tau} \Delta_{ip}(t, \tau) d\tau = 0 \qquad (13.19)$$

where $X_1 \ldots X_k$ are initial values of the unknowns determined at the time instant $t = 0$, $\delta_{i1}(t), \ldots, \delta_{ik}(t)$ are displacements along the $i$-th unknown at the time instant $t$ caused by unit values of the unknowns $\bar{X}_1, \ldots, \bar{X}_k$ applied at the initial time instant; $\delta_{i1}(t, \tau), \ldots, \delta_{ik}(t, \tau)$ are displacements in the $i$-th unknown direction at the time instant $t$ caused by unit values of the unknowns applied at any time instant $\tau$; $P$ is the initial value of the parameter defining the external load; $\Delta_{ip}(t)$ is the displacement in the direction of the $i$-th unknown at the instant $t$ caused by the load applied at $t = 0$; $\Delta_{ip}(t, \tau)$ are analogous displacements caused by a constant load applied at any time instant. The products $X_m\delta_{im}(t)$ and $P\Delta_{ip}(t)$ are displacements of the primary structure at the time instant $t$ caused by initial values of the unknowns $X$ or the initial value of the load; the integrals of the Equation (13.19) are the displacements at the time instant $t$ caused by variations in the unknowns and the loads during the time interval between 0 and $t$.

Investigation of the rheonomic properties of reinforced plastics has proved that this type of material is characterized by pronounced viscoelasticity, and its deformative properties under sustained loading are described by the Boltzmann-Volterra integral equations with marked creep kernels, shown in Equation (12.2). In this case, Equation (13.19) may be written in the matrix form instead of a superposition equation of the method of forces

$$[\delta][X(t)] + \int_0^t [\delta'(t - \tau)][X(\tau)]d\tau = -[\Delta_p(t)] \quad (13.20)$$

where elements of matrices $[\delta']$ and $[\Delta p]$ for beam systems will be given by the expressions [25]:

$$\delta'_{ij}(t - \tau) = \sum \int_{(l)} \frac{1}{b} \{A'_{11}\tilde{A}'_{11}(t - \tau)\bar{N}_i\bar{N}_j$$
$$+ B'_{11}\tilde{B}'_{11}(t - \tau)(\bar{N}_i M_j + \bar{N}_j\bar{M}_i) \quad (13.21)$$
$$+ D'_{11}\tilde{D}'_{11}(t - \tau)\bar{M}_i\bar{M}_j + F\tilde{F}(t - \tau)\bar{S}_i\bar{S}_j\}dx$$

$$\Delta_{ip}(t) = \sum \sum_{l=1}^{m} \int_{(l)} \frac{1}{b} \{\bar{N}_i N_{pl} A'_{11*}[f_l(t)]$$
$$+ (\bar{N}_i M_{pl} + \bar{M}_i N_{pl})B'_{11*}[f_l(t)] \quad (13.22)$$
$$+ \bar{M}_i M_{pl} D'_{11*}[f_l(t)] + \bar{S}_i S_{pl} F_*[f_l(t)]\}dx$$

where the internal forces in the primary structure caused by the applied load are determined according to Equation (12.15) and the functionals $A'_{11*}$, $B'_{11*}$, $D'_{11*}$, $F_*$ are determined according to Equations (12.2) and (12.17).

For a constant load, Equation (12.19) yields

$$\Delta_{ip}(t) = \sum \int_{(l)} \frac{1}{b} \{A'_{11*}[1]\bar{N}_i N_p + B'_{11*}[1](\bar{N}_i M_p$$
$$+ \bar{M}_i N_p) + D'_{11*}[1]\bar{M}_i M_p + F_*[1]\bar{S}_i S_p\}dx \quad (13.23)$$

## 284 MECHANICS OF BEAM SYSTEMS

The system of integral superposition equations derived from Equation (13.20) can be solved either numerically by a computer or analytically. The latter approach is more conveniently applied by using the Laplace transformation method. The application of the Laplace transformation in Equation (13.20) yields

$$[\delta][X(P)] + [\delta'(P)][X(P)] = -[\Delta_p(P)] \quad (13.24)$$

Hence

$$[X(P)] = -[[\delta] + [\delta'(P)]]^{-1}[\Delta_p(P)] \quad (13.25)$$

Performing a reverse transformation of Equation (13.25), we arrive at the solution sought. Note that with a higher degree of static indeterminacy of the investigated beam system, the reverse transformations become many times more laborious. That is why numerical methods are recommended in a number of cases.

The solution to Equation (13.20) yields the law of changes in the unknowns $X_1, \ldots, X_k$ during the loading duration. In this case, as stated by the relationships

$$M(t) = M_p + \sum_{i=1}^{k} X_i(t)\bar{M}_i$$

$$N(t) = N_p + \sum_{i=1}^{k} X_i(t)\bar{N}_i \quad (13.26)$$

$$S(t) = S_p + \sum_{i=1}^{k} X_i(t)\bar{S}_i$$

the internal forces in the beam change with time even under a constant externally applied load. The rate and the limits of these changes depend upon both the rheonomic properties of the reinforced plastics components and the design structure of the laminate.

As an example, consider a structure indeterminate to the first degree where all the beams consist of the same material and are rectilinear in shape. In the case of a constant load, ignoring the action of the axial force, Equation (13.20) yields the following equation with respect to $X(t)$:

$$\sum D'_{11}\left[\frac{1}{b}\int_0^l \bar{M}_1^2(x)dx\right]\left\{X(t) + \int_0^t X(\tau)\tilde{D}'_{11}(t-\tau)d\tau\right\}$$

$$+ \sum F\left[\frac{1}{b}\int_0^l \bar{S}_1^2(x)dx\right]\left\{X(t) + \int_0^t X(\tau)\tilde{F}(t-\tau)d\tau\right\}$$

$$= \sum D'_{11}\left[\frac{1}{b}\int_0^l \bar{M}_1(x)M_p(x)dx\right]\left\{1 + \int_0^t \tilde{D}'_{11}(t-\tau)d\tau\right\}$$

(13.27)

$$+ \sum F\left[\frac{1}{b}\int_0^l \bar{S}_1(x)S_p(x)dx\right]\left\{1 + \int_0^t \tilde{F}(t-\tau)d\tau\right\}$$

After applying the Laplace transformation the expression for the unknown function $X$ in the transformation parameter resign takes the form

$$X(P) = \frac{(R_1 + R_2) + [R_1\tilde{D}'_{11}(P) + R_2\tilde{F}(P)]}{(R_3 + R_4) + [R_3\tilde{D}'_{11}(P) + R_4\tilde{F}(P)]}\frac{1}{P} \quad (13.28)$$

where

$$R_1 = \sum D'_{11}\left[\frac{1}{b}\int_0^l \bar{M}_1(x)M_p(x)dx\right]$$

$$R_2 = \sum F\left[\frac{1}{b}\int_0^l \bar{S}_1(x)S_p(x)dx\right]$$

(13.29)

$$R_3 = \sum D'_{11}\left[\frac{1}{b}\int_0^l \bar{M}_1^2(x)dx\right]$$

$$R_4 = \sum F\left[\frac{1}{b}\int_0^l \bar{S}_1^2(x)dx\right]$$

Performing the reverse transformation of the Equation (13.28) requires the choice of specific creep kernels $\tilde{D}'_{11}(t - \tau)$ and $\tilde{F}(t - \tau)$. When these kernels are defined by functions

$$\tilde{D}'_{11}(t - \tau) = C_D \exp[-\alpha_D(t - \tau)]$$
$$\tilde{F}(t - \tau) = C_F \exp[-\alpha_F(t - \tau)]$$
(13.30)

their respective representations are

$$\tilde{D}'_{11}(P) = \frac{C_D}{P + \alpha_D}; \quad \tilde{F}(P) = \frac{C_F}{P + \alpha_F} \quad (13.31)$$

whereas the Equation (13.28) has the form

$$X(P) = \frac{a_2 P^2 + a_1 P + a_0}{b_2 P^2 + b_1 P + b_0} \frac{1}{P} \quad (13.32)$$

where

$$a_0 = (R_1 + R_2)\alpha_D \alpha_F + R_1 C_D \alpha_F + R_2 C_F \alpha_D$$

$$a_1 = R_1 C_D + R_2 C_F + (R_1 + R_2)(\alpha_D + \alpha_F)$$

$$a_2 = (R_1 + R_2)$$

$$b_0 = (R_3 + R_4)\alpha_D \alpha_F + R_3 C_D \alpha_F + R_4 C_F \alpha_D$$

$$b_1 = R_3 C_D + R_4 C_F + (R_3 + R_4)(\alpha_D + \alpha_F)$$

$$b_2 = (R_3 + R_4)$$

A reverse manipulation of Equation (13.32) yields

$$X(t) = A_1 \exp(l_1 t) + A_2 \exp(l_2 t) + A_3 \quad (13.33)$$

where

$$A_i = \frac{l_i^2 a_2 + l_i a_1 + a_0}{b_2 l_i (l_i - l_j)}$$

$$i, j = 1, 2, \quad i \neq j$$

$$A_3 = \frac{a_0}{b_2 l_1 l_2}$$

and $l_1$, $l_2$ are the roots of the quadratic equation

$$b_2 l^2 + b_1 l + b_0 = 0$$

It becomes evident from the analysis of Equation (13.33) that even permanent loading of statically indeterminate systems of reinforced plastic laminates result in time-variant reactions of these systems and consequently of internal forces within them.

To estimate numerically the time-variant internal forces of statically indeterminate beam systems made of reinforced plastic laminates, we shall consider a simply supported cantilever, i.e., a system that is statically indeterminate to the first degree, acted upon by a uniformly distributed load. Selecting the right-hand support redundant as the unknown redundant, we have

$$R_1 = D'_{11} \frac{gl^4}{8}$$

$$R_2 = F \frac{gl^2}{2}$$

$$R_3 = D'_{11} \frac{l^3}{3}$$

$$R_4 = Fl$$

(13.34)

and for the rigid fixing it will be

$$R_1 = D'_{11} \frac{gl^3}{24}$$

$$R_2 = 0$$

$$R_3 = D'_{11} \frac{l}{3}$$

$$R_4 = \frac{F}{l}$$

(13.35)

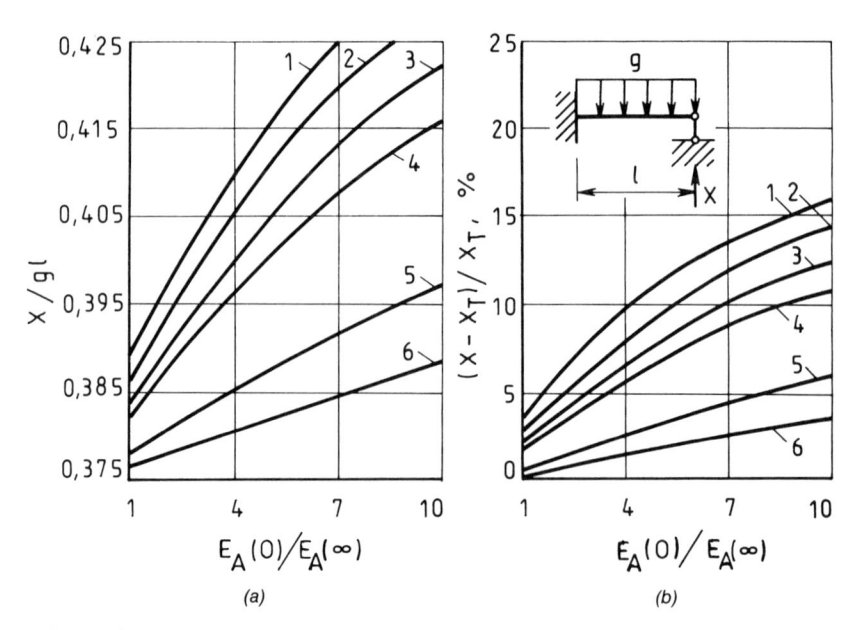

**FIGURE 13.5.** Dependence of absolute (a) and relative (b) values of the reactive force $X$ upon the degree of creep in the polymer matrix in the case of a unidirectionally carbon fiber reinforced material beam: $l/h = 7$ (1), 8 (2), 9 (3), 10 (4), 15 (5), 20 (6).

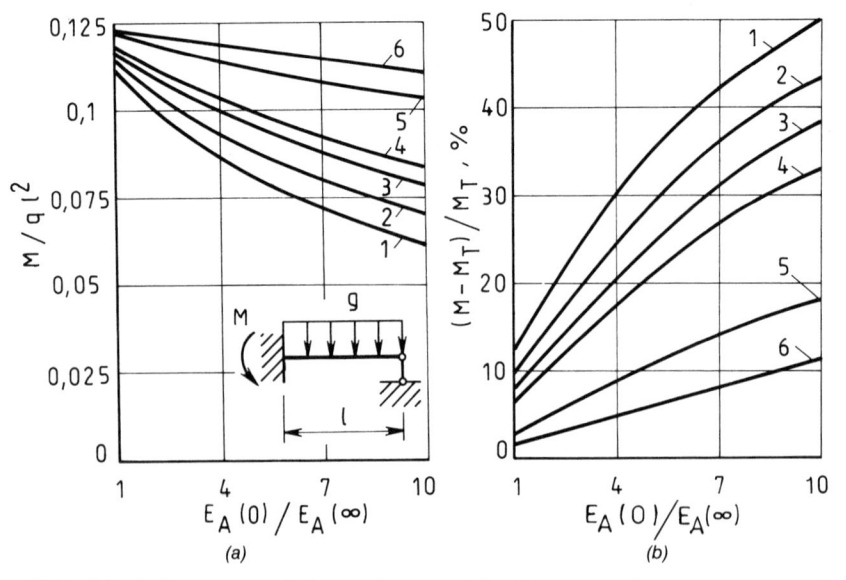

**FIGURE 13.6.** Dependence of absolute (a) and relative (b) values of the support moment $M$ upon the degree of polymer matrix creep in a beam made of a unidirectionally carbon fiber reinforced material: $l/h = 7$ (1), 8 (2), 9 (3), 10 (4), 15 (5), 20 (6).

The ultimate values of support redundants $X(t)$ subject to Equation (13.33) are the following

when $t = 0$

$$X(0) = A_1 + A_2 + A_3$$

when $t \to \infty$

$$X(\infty) = A_3$$

Figs. 13.5 and 13.6 include the dependence of the calculation data of the support redundants of the given statically indeterminate system with $t \to \infty$ upon the degree of creep in the polymer matrix and the structural geometry of the beam. The beams under discussion have been made of an unidirectionally carbon fiber reinforced material ($E_{Br}/E_A = 2$, $E_{Bz}/E_{Br} = 20$, $\nu_A = 0.35$, $\psi = 0.6$). For $E_A(0)/E_A(\infty) = 1$ we have short-term values $X(\infty) = X(0)$ ($M(\infty) = M(0)$). Variations in the values of $X(\infty)$ as compared with $X(0)$ depend substantially not only upon the rheonomical characteristics of the polymer matrix but upon the structure geometry of the beams as well. According to the principles of structural analysis, which do not take into account the effect of shear forces, the internal forces of statically indeterminate systems consisting of constant-section beams do not depend upon the properties of the material and the duration of loading.

For reinforced plastics, however, the energy contribution by the shear forces is comparable to the deformation energy governed by the bending moment and, as has already been shown, it increases with the loading duration. This explains the major relationship between internal forces and the rheonomical properties of the material in relatively short beams, because the effect of shear forces becomes smaller with a larger length ($l/h$ ratio) of the material. In the discussed examples, the deviation from the theoretical values (without including the action of shear forces) of the internal forces is as much as 40 percent for the reactive moments ($M_T = \frac{1}{8}gl^2$) and 20 percent for reactive forces ($X_T = \frac{3}{8}gl$). This proves the necessity of taking the shear forces into account when the method of forces is applied in designing beam systems made of reinforced plastic laminates and acted upon by sustained loading.

The reason why the bearing capacity of beam systems like frames or frameworks may be exhausted is either the stability loss of the whole system or its component parts. The methodology for the design of beam systems for stability are discussed, for example, by Volmir [16] and

Smirnov et al. [66]. Dealing with systems made of reinforced plastic laminates, such peculiarities as low longitudinal shear strength, possible disbalance of the plies within the laminate, and the significant effect of the loading duration upon the strain properties of the material components are to be taken into account. These peculiarities are discussed in Chapters 11–13 with respect to the estimation of the stress-strain state of beam systems made of reinforced plastic laminates. The results obtained may serve as a basis for formulation and solution of design problems for the stability of beam systems made of reinforced plastic laminates [22].

APPENDIX

# List of Basic Symbols

1, 2, 3—elastic symmetry axes of a laminate
$o, y, z$—elastic symmetry axes of a fabric reinforced plastic
$\|, \perp, \underline{\perp}$—elastic symmetry axes of a unidirectionally reinforced plastic
$r, z$—elastic symmetry axes of transverse isotropic fibres
$o, y$—warp and weft directions in a fabric
$r, z, \theta$—axes of a cylindrical coordinate system
$\langle\langle\sigma\rangle\rangle, \langle\langle\tau\rangle\rangle, \langle\langle\epsilon\rangle\rangle, \langle\langle\gamma\rangle\rangle$—mean stresses and strains of a laminate
$\langle\sigma\rangle, \langle\tau\rangle, \langle\epsilon\rangle, \langle\gamma\rangle$—mean stresses and strains of a unidirectionally reinforced plastic
$\langle\sigma_\|\rangle, \langle\sigma_\perp\rangle, \langle\sigma_{\underline{\perp}}\rangle, \langle\epsilon_\|\rangle, \langle\epsilon_\perp\rangle, \langle\epsilon_{\underline{\perp}}\rangle$—mean normal stresses and strains in a unidirectionally reinforced plastic parallel and perpendicular to the reinforcement direction
$\langle\tau_{\|\perp}\rangle, \langle\tau_{\|\underline{\perp}}\rangle, \langle\tau_{\perp\underline{\perp}}\rangle, \langle\gamma_{\|\perp}\rangle, \langle\gamma_{\|\underline{\perp}}\rangle, \langle\gamma_{\perp\underline{\perp}}\rangle$—mean stresses and strains under longitudinal shear in a unidirectionally reinforced plastic
$\sigma_r, \sigma_z, \sigma_\theta$—normal stresses in the matrix
$\tau_{rz}, \tau_{r\theta}$—shear stresses in the matrix
$\bar{\sigma}_r, \bar{\sigma}_z, \bar{\tau}_{rz}$—dimensionless structural parameters of a composite showing stress concentration in the structural components
$\nu_{ij}$—Poisson's ratio determining transverse strain in $j$-direction under $i$-directional loading
$E_{Bz}, E_{Br}$—elasticity moduli of anisotropic fibers in axial and radial directions
$G_{Brz}$—shear modulus of anisotropic fibers
$\nu_{Brz}, \nu_{Br\theta}$—Poisson's ratio of anisotropic fibers
$E_B, G_B, \nu_B$—elastic characteristics of isotropic fibers
$r_B$—fiber radius
$l, p$—fiber packaging parameters
$\psi$—relative fiber volume content

$\Psi$—relative energy dissipation

$E_A, G_A, \nu_A$—elastic characteristics of the matrix

$E_{\|}, E_\perp, G_{\|\perp}, \nu_{\|\perp}, \nu_{\perp\|}, \nu_{\perp\perp}$—technical elastic characteristics of a unidirectionally reinforced plastic

$E_1, E_2, G_{12}, \nu_{12}, \nu_{21}$—technical elastic characteristics of a laminate

$E_o, E_y, G_{oy}, \nu_{oy}, \nu_{yo}$—technical elastic characteristics of a fabric-reinforced plastic before continuity loss

$E_o^+, E_y^+, \nu_{oy}^+, \nu_{yo}^+$—elasticity moduli and Poisson's ratios for a fabric-reinforced plastic under tension after cracking in the weft direction

$E_o^{(+)}, E_y^{(+)}, \nu_{oy}^{(+)}, \nu_{yo}^{(+)}$—elasticity moduli and Poisson's ratios of a fabric-reinforced plastic under tension after cracking in the warp direction

$G_{oy}^+, G_{oy}^{(+)}, G_{oy}^{+(+)}$—shear modulus of a fabric-reinforced plastic after cracking in the direction of weft, in the direction of warp and in the weft-warp directions

$E_o^{+(+)}, E_y^{+(+)}, \nu_{oy}^{+(+)}, \nu_{yo}^{+(+)}$—elasticity moduli and Poisson's ratios of a fabric-reinforced plastic after cracking in the warp and the weft directions

$R_A^+, T_A$—tensile and shear strengths of the bond

$R_b, T_b$—tensile and shear strengths of the matrix

$R_{Bz}^+, R_{Br}^+, T_{Brz}$—axial and transverse tensile strengths and longitudinal shear strength of the fibers

$\epsilon_{BR}^+, \epsilon_{BR}^-$—ultimate strains of fibers under tension and compression

$\langle\langle \sigma_o^+ \rangle\rangle, \langle\langle \sigma_o^{+(+)} \rangle\rangle, \langle\langle \sigma_o^{+++} \rangle\rangle$—mean tensile stresses of a fabric-reinforced plastic in the warp direction resulting in continuity loss of the first, the second and the third type

$T_o, T_y, C_o, C_y, \beta_o, \beta_y$—parameters of weave

$h$—thickness of the laminate

$\beta$—packaging angles of unidirectionally reinforced plies

$m$—specific thickness of a unidirectionally reinforced ply

$S$—shear force

$M$—bending moment

$\mathcal{M}$—torque

$N$—axial force

$t$—time

$T$—temperature

Subscripts "$A$", "$B$", and "$C$" denote characteristics of the polymer matrix, fibers and high-modulus fibers respectively in three-component materials. The signs "+" and "−" show behaviour under tension and compression. The symbol "$R$" is used to denote critical characteristics of the material at the instant of failure. Subscripts "$o$" and "$y$" characterize the properties of a fabric-reinforced plastic in the warp and weft directions.

# REFERENCES

1. Antans, V. P. and A. M. Skudra. 1967. "The Effect of Glass Fiber Inelasticity upon Long-Term Strength of Reinforced Plastics under Uniaxial Tension in Fiber Direction" (in Russian), *Mehanika Polimerov*, (4):719–725.
2. Arutyunyan, N. H. 1952. *Some Problems of Creep Theory* (in Russian). Moscow: Gostehizdat Publishers, 323 p.
3. Arutyunyan, N. H. and B. L. Abramov. 1963. *Torsion of Elastic Bodies* (in Russian). Moscow: Fizmatgiz Publishers, 688 p.
4. Auzukalns, Ya. V. 1974. "Long-Term Strength of Reinforced Plastics under Compression" (in Russian), Cand. of Eng. Sc. Diss., Riga.
5. Bulavs, F. Ya. and M. R. Gurvich. 1984. "Continuity Loss in Reinforced Plastic Laminates under Sustained Flat Stress State" (in Russian), in *Mehanika Kompozitnih Materialov*. Riga: Riga Polytechnical Institute, pp. 11–20.
6. Bulavs, F. Ya. and M. R. Gurvich. 1984. "Efficient Design of Reinforced Plastic Laminate Structures According to Strength Conditions" (in Russian), in *Proyektirovaniye i Optimizatsia Konstruktsiy Inzhenernih Sooruzheniy*. Riga: Riga Polytechnical Institute, pp. 111–118.
7. Bulavs, F. Ya. and M. R. Gurvich. 1982. "Long-Term Strength of Unidirectionally Reinforced Plastics under Combined Tensile and Shear Loading" (in Russian), in *Mehanika Kompozitnih Materialov*. Riga: Riga Polytechnical Institute, pp. 33–41.
8. Bulavs, F. Ya. and I. G. Radinsh. 1980. "The Effect of Non-Linearity of Strain Properties of the Polymer Matrix in the Strain Properties of Composites under Sustained Transversal Loading" (in Russian), in *Mehanika Kompozitnih Materialov*. Riga: Riga Polytechnical Institute, pp. 56–65.
9. Bulavs, F. Ya. and I. G. Radinsh. 1980. "Strain Properties of Unidirectionally Reinforced Plastics under Transversal Loading" (in Russian), in *Voprosi Dinamiki i Prochnosti*. Riga: Zinatne Publishers, pp. 73–81.
10. Bulavs, F. Ya. and I. G. Radinsh. 1980. "Time-Dependent Variations in Stress–Strain State of Unidirectionally Reinforced Plastics Components under Sustained Static Loading" (in Russian), in *Mehanika Kompozitnih Materialov*, Riga: Riga Polytechnical Institute, pp. 5–18.
11. Bulavs, F. Ya. and A. A. Skudra. 1983. "Strength of Plastic Laminates" (in Russian), in *Mehanika Armirovannih Plastikov*. Riga: Riga Polytechnical Institute, pp. 4–18.

12. Bichkov, D. V. 1962. *Structural Mechanics of Thin-Wall Bar Structures* (in Russian). Moscow: Gosstroiizdat Publishers, 476 p.
13. Van Fo Fi, G. A. 1981. *Reinforced Plastics Structures* (in Russian). Kiev: Tehnika Publishers, 220 p.
14. Vasilyeva, A. V. 1988. "Free Torsion of Composite Thin-Wall Bars of Circular Cross-Section" (in Russian), in *Mehanika Kompozitnih Materialov*. Riga: Riga Polytechnical Institute, pp. 53–59.
15. Vlasov, V. Z. 1940. *Elastic Thin-Wall Bars* (in Russian). Moscow: Leningrad: Stroiizdat Publishers, 276 p.
16. Volmir, A. S. 1969. *Stability of Elastic Systems* (in Russian). Moscow: Fizmatgiz Publishers, 879 p.
17. Vishvanyuk, V. I., V. T. Alimov and R. A. Turusov. 1985. "Thermal Expansion of Unidirectional Composites Having a Low Thermal Linear Expansion Coefficient" (in Russian), in *Mehanika Kompozitnih Materialov*, (2):357–360.
18. Gorbatkina, Yu. A. and V. G. Ivanova-Mumzhiyeva. 1979. "The Effect of Loading Rate upon the Strength of Glass-Plastics and Reinforced Structure Elements" (in Russian). *Fizika Prochnosti Kompozitnih Materialov; Proceedings of the 3rd All-Union Seminar*, Leningrad, pp. 44–49.
19. Gurvuch, M. R. 1982. "Long-Term Strength of the Polymer Matrix under Combined Varying State of Stress" (in Russian), in *Mehanika Kompozitnih Materialov*. Riga: Riga Polytechnical Institute, pp. 25–32.
20. Gurvich, M. R. 1983. "Long-Term Strength of Unidirectionally Reinforced Plastics with Fiber-Polymer Matrix Bond Failure" (in Russian), in *Mehanika Armirovannih Plastikov*. Riga: Riga Polytechnical Institute, pp. 32–43.
21. Gurvich, M. R. 1983. "Structural Long-Term Strength Criteria of Reinforced Plastics Laminates" (in Russian). Cand. of Eng. Sc. diss., Riga.
22. Gurvich, M. R. 1986. "The Effect of Anisotropic Materials Structure on the Stability of Flat Bar Systems" (in Russian), in *Proyektirovanye i Optimizatsia Konstruktsiy Inzhenernih Sooruzheniy*. Riga: Riga Polytechnical Institute, pp. 32–41.
23. Gurvich, M. R. 1986. "Estimation of the Effect of Transversal Forces on the Deformed State of Reinforced Plastics Systems" (in Russian), in *Mehanika Kompozitnih Materialov*. Riga: Riga Polytechnical Institute, pp. 84–97.
24. Gurvich, M. R. 1985. "A Numerical Method for Investigating Viscoelastic Properties of Reinforced Plastic Laminates under Combined Loading" (in Russian), in *Mehanika Armirovannih Plastikov*. Riga: Riga Polytechnical Institute, pp. 49–60.
25. Gurvich, M. R. and F. Ya. Bulavs. 1985. "The Effect of Loading Time upon the Stress–Strain State of Statically Indeterminate Bar Systems Made of Reinforced Plastics" (in Russian), in *Mehanika Armirovannih Plastikov*. Riga: Riga Polytechnical Institute, pp. 61–71.
26. Zinovyev, P. A. and Yu. N. Yermakov. 1985. "Anisotropy of Dissipation Properties in Fiber Composites" (in Russian), *Mehanika Kompozitnih Materialov*, (5):816–825.
27. Zinovyev, P. A. and Yu. N. Yermakov. 1985. "Dissipation Characteristics of a Unidirectional Fiber Material" (in Russian), *Izvestia Vuzov. Mashinostroyeniye*, (12):16–21.
28. Zinovyev, P. A. and Yu. N. Yermakov. 1986. "Energy Dissipation in Multiple Fiber Composites under Bending" (in Russian), *Izvestia Vuzov. Mashinostroyeniye*, (4):15–20.
29. Zinovyev, P. A. and Yu. N. Yermakov. 1986. "Matrices of Elastic Dissipation Charac-

teristics of a Unidirectional Composite. Coordinate Transformation" (in Russian), *Izvestia Vuzov. Mashinostroyeniye*, (2):46–51.

30. Ilyushin, A. A. 1986. "An Approximation Method for Structure Design According to Linear Theory of Thermal Viscoelasticity" (in Russian). *Mehanika Polimerov*, (2):210–218.

31. Kirulis, B. A. 1977. "Bond Strength in Reinforced Plastics" (in Russian), Cand. of Eng. Sc. diss., Riga.

32. Koltunov, M. A. 1966. "Some Aspects of Kernel Selection for Problem Solution Including Creep and Relaxation" (in Russian), *Mehanika Polimerov*, (4):483–497.

33. Korf, O. Ya. and A. M. Skudra. 1966. "Long-Term Strength of Isotropic Polymer Materials in a Flat Stress State" (in Russian), *Mehanika Polimerov*, (6):837–844.

34. Christensen, R. M. 1979. *Mechanics of Composite Materials*. John Wiley and Sons, 334 p.

35. Kruklinsh, A. A. 1984. "Rigidity Characteristics of Fabric-Reinforced Plastics" (in Russian), in *Mehanika Kompozitnih Materialov*. Riga: Riga Polytechnical Institute, pp. 75–88.

36. Kruklinsh, A. A. 1984. "Structural Strength Characteristics of Fabric-Reinforced Plastics" (in Russian), in *Mehanika Kompozitnih Materialov*. Riga: Riga Polytechnical Institute, pp. 57–74.

37. Kruklinsh, A. A. 1985. "Structural Theory of Fabric-Reinforced Plastics" (in Russian). Cand. of Eng. Sci. diss., Riga.

38. Kruklinsh, A. A. and K. K. Kalvish. 1988. "The Complex Nature of Interply Shear in Reinforced Plastic Laminates" (in Russian), in *Mehanika Kompozitnih Materialov*. Riga: Riga Polytechnical Institute, pp. 45–55.

39. Kruklinsh, A. A. and K. K. Kalvish. 1987. "Free Torsion of a Reinforced Plastic Laminate Bar" (in Russian), in *Mehanika Armirovannih Plastikov*. Riga: Riga Polytechnical Institute, pp. 42–53.

40. Lekhnitskii, S. G. 1968. *Anisotropic Plates*. Gordon and Breach Science Publishers, 464 p.

41. Lekhnitskii, S. G. 1971. *Torsion of Anisotropic and Non-Homogeneous Bars* (in Russian). Moscow: Nauka Publishers, 240 p.

42. Mushelishvili, N. I. 1966. *Some Basic Problems of Mathematical Theory of Elasticity* (in Russian). 5th edition. Moscow: Nauka Publishers, 321 p.

43. Panovko, Ya. G. 1960. *Internal Friction in the Vibrations of Elastic Systems* (in Russian). Moscow: Fizmatgiz Publishers, 193 p.

44. Peschanskaya, N. N. and V. A. Stepanov. 1974. "Durability of Polymers under Tension and Torsion" (in Russian), *Mehanika Polimerov*, (6):1003–1006.

45. Piskozub, L. G. 1984. "On the Influence of Anisotropy on Damping in Reinforced Laminated Composites" (in Russian). *Non-Classical Problems of Composite Mechanics and Structures Made of Composites*. Report at the 2nd All-Union Science and Technology Seminar. Lvov, Sept. 1984. Kiyev: p. 40.

46. Pobedrya, B. E. 1973. "Mathematical Theory of Non-Linear Viscoelasticity," *Elasticity and Non-Elasticity* (in Russian). Moscow: Moscow State University, (3):95–173.

47. Rabotnov, Yu. N. 1966. *Creep in Structural Elements* (in Russian). Moscow: Nauka Publishers, 752 p.

48. Rabotnov, Yu. N. 1979. "Strength of Laminated Composites" (in Russian), *Transactions of the Academy of Sciences of the USSR*, Mehanika Tvyordogo Tela, (1):113-119.
49. Rabotnov, Yu. N. 1948. "Equilibrium of Elastic Media Having an Aftereffect" (in Russian), *Prikladnaya Matematika i Mehanika*, 12(1):12-62.
50. Rabotnov, Yu. N. and L. H. Papernik. 1969. *Tables of Exponential Fractional Function of Negative Parameters and Its Integrals* (in Russian). Moscow: 132 p.
51. Rzhanitsin, A. R. 1968. *Theory of Creep* (in Russian). Moscow: Stroiizdat Publishers, 416 p.
52. Rosen, B. W., S. V. Kulkarni and P. V. McLaughlin. 1975. "Failure and Fatique Mechanisms in Composite Materials," in *Inelastic Behaviour of Composite Materials*, AMD—Vol. 13, The American Society of Mechanical Engineers, pp. 17-38.
53. Ruditsin, M. N., P. Ya. Artemov and M. I. Lyuboshin. 1985. *A Reference Book on Strength of Materials* (in Russian). Minsk: Visheish Shkola Publishers, 509 p.
54. Sih, G. C. 1979. "Fracture Mechanics of Composite Materials," in *Fracture of Composite Materials*. The Netherlands: Sijthoff and Noordhoff, pp. 111-130.
55. Skudra, A. M. and F. Ya. Bulavs. 1982. *Strength of Reinforced Plastics* (in Russian). Moscow: Himiya Publishers, 213 p.
56. Skudra, A. M., F. Ya. Bulavs and K. A. Rocens. 1971. *Creep and Statical Fatigue in Reinforced Plastics* (in Russian). Riga: Zinatne Publishers, 235 p.
57. Skudra, A. M. and F. Ya. Bulavs. 1978. *Structural Theory of Reinforced Plastics* (in Russian). Riga: Zinatne Publishers, 192 p.
58. Skudra, A. M. and K. K. Kalvish. 1986. "Strength of a Laminated Reinforced Plastic Bar" (in Russian), in *Mehanika Kompozitnih Materialov*. Riga: Riga Polytechnical Institute, pp. 98-110.
59. Skudra, A. M. and O. V. Sbitnevs. 1981. "Coefficient of Linear Expansion of Laminated Plastics" (in Russian), in *Mehaniak Armirovannih Plastikov*. Riga: Riga Polytechnical Institute, pp. 65-75.
60. Skudra, A. M. and O. V. Sbitnevs. 1982. "Temperature Dependence of Thermal Expansion Coefficient in Reinforced Plastics" (in Russian), in *Mehanika Kompozitnih Materialov*. Riga: Riga Polytechnical Institute, pp. 12-24.
61. Skudra, A. M. and O. V. Sbitnev. 1984. "Function of Thermal Expansion of Reinforced Plastics" (in Russian), in *Mehanika Kompozitnih Materialov*. Riga: Riga Polytechnical Institute, pp. 88-99.
62. Sokolov, E. A. 1980. "Possibilities of Predicting Creep in a Laminated Organic Plastic According to the Properties of a Unidirectionally Reinforced Material" (in Russian), in *Mehanika Kompozitnih Materialov*, (1):142-147.
63. Sokolov, E. A. 1980. "Strain and Strength Properties of High-Strength Organic Plastics" (in Russian), Eng. M. diss., Riga.
64. Sorokin, E. S. 1956. *Dynamic Calculation of Bearing Structures of Buildings* (in Russian). Moscow: Gosstroiizdat, 340 p.
65. Stepanov, V. A. 1975. "Strain and Failure of Polymers" (in Russian). *Mehanika Polimerov*, (1):95-106.
66. Smirnov, A. F., A. V. Aleksandrov, B. Ya. Lashchenikov and N. N. Shaposhnikov. 1984. *Structural Mechanics. Dynamics and Stability of Structures* (in Russian). Moscow: Stroiizdat, 416 p.

67. Tarnopol'skii, Yu. M. 1987. "Engineering Mechanics of Fiber Composites" (in Russian). *Academy of Sciences of the Latvian SSR*, (11):90–97.
68. Tarnopol'skii, Yu. M. and T. Ya. Kincis. 1985. *Static Test Methods for Composites*. Van Nostrand Reinhold Company Inc., 298 p.
69. Timoshenko, S. P. and J. M. Gere. 1979. *Mechanics of Materials*. Moscow: Nauka Publishers, 560 p.
70. Chamis, C. C. 1974. "Micromechanical Theories of Strength," in *Composite Materials*. New York, NY: Acad. Press, (5): Fracture and Fatigue.
71. Schapery, R. A. 1974. "Viscoelastic Behavior of Composite Materials," in *Composite Materials*. New York, NY: Acad. Press, (4): Mechanics of Composite.
72. Yakovlev, A. P. 1985. *Dissipative Properties of Non-Homogeneous Materials and Systems* (in Russian). Kiev: Nauk. Dumka, 120 p.
73. Adams, D. F. 1968. "Micro- and Macromechanics Analysis of Composite Materials," *Advanced Fibrous Reinforced Composites: SAMPE 10th Nat. Symp.*, 10:G1–G13.
74. Adams, R. D. and D. G. Bacon. 1973. "The Dynamic Properties of Unidirectional Fibre Reinforced Composites in Flexure and Torsion," *J. Composite Materials*, 7:53–67.
75. Adams, R. D. and D. G. Bacon. 1973. "Effect of Fibre Orientation and Laminate Geometry and the Dynamic Properties of CFRP," *J. Composite Materials*, 7:402–428.
76. Agarwal, B. D. and L. J. Broutman. 1980. *Analysis and Performance of Fiber Composites*, New York, NY: Wiley-Intersci. Publ., 297 p.
77. Bax, I. 1970. "Deformation Behaviour and Failure of Glass-Fiber-Reinforced Resin Material," *Plastic and Polymers*, 38(133):27–30.
78. Brinson, H. F., W. I. Griffith and D. H. Morris. 1981. "Creep Rupture of Polymer-Matrix Composites," *Experimental Mech.*, 21(9):329–335.
79. Dharmarjan, S. and H. McSutchnen. 1973. "Shear Coefficients for Orthotropic Beams," *J. Composite Materials*, 7:530–535.
80. *Handbook of Composites. Vol. 2.—Structures and Design*. C. T. Herakovich and Yu. Tarnopolskii, eds., Amsterdam, North-Holland, 672 p.
81. *Handbook of Composites. Vol. 3.—Failure Mechanics of Composites*. G. C. Sih and A. M. Skudra, eds., Amsterdam, North Holland, 444 p.
82. Hartwig, G. and A. Puck. 1974. "Termische Kontraction von Glassfaserverstarkten Epoxydharzen bis zu Tiefsten Temperaturen," *Kunststoffe*, 64(1):32–35.
83. Kalnin, I. L. 1972. "Evaluation of Unidirectional Glass-Graphite Fiber Epoxy Resin Composites," *Composite Materials: Testing and Design. 2nd Conf. ASTM STP 497 Amer. Soc. Testing and Materials*, pp. 551–563.
84. Knappe, W. and W. Schneider. 1972. "Bruchkriterien fur Unidirektionalen Glasfer: Kunststoff unter Ebener Kurzzeit- und Langzeitbeanspruchung," *Kunststoffe*, 62(12):864–868.
85. Kourtides, D. A. 1984. "Processing and Flammability Parameters of Bismatermide and Some Other Thermallen Stable Resin Matrices for Composites," *Polymer Composites*, 5(2):143–166.
86. Lauke, B. and R. Barthel. 1983. "Theoretischer Beitrag zum Deformationsmechanischen Verhalten Mehrschichtiger Verbunplatten unter Beruchsichtigung von Schereffekten," *Techn. Mech.*, 4(3):38–46.

87. Lifshitz, J. M. and A. Rotem. 1970. "An Observation on the Strength of Unidirectional Fibrous Composites," *J. Composite Materials*, 4:133-134.
88. Ni, R. G. and R. D. Adams. 1984. "A Rational Method for Obtaining the Dynamic Mechanical Properties of Laminae for Predicting the Stiffness and Damping of Laminated Plates and Beams," *Composites*, 15:193-199.
89. Ni, R. G. and R. D. Adams. 1984. "The Damping and Dynamic Moduli of Symmetric Laminated Composite Beams—Theoretical and Experimental Results," *J. Composite Materials*, 18:104-121.
90. Owen, M. I. and D. I. Rice. 1984. "Biaxial Strength Behaviour of Glass-Fabric-Reinforced Plyester Resin," *Composites*, 15:13-25.
91. Raghava, R. S. 1984. "Prediction of Thermal and Mechanical Properties of Glass-Epoxy Composite Laminates," *Polymer Composites*, 5(3):173-178.
92. Rogers, K. F., L. M. Philip and D. M. Kingston-Lee, et al. 1977. "The Thermal Expansion of Carbon Fibre-Reinforced Plastics," *J. Materials Sci.*, (12):718-734.
93. Schapery, R. A. 1962. "Approximate Methods of Transform Inversion for Viscoelastic Stress Analysis," *Proc. 4th U.S. Nat. Congr. Appl. Mech. New York*, 2:1075-1085.
94. Schneider, W. 1971. "Warmeansdehnungs Koeffizienten und Warmspannungen Glasfaser: Kunststoffverbunden aus Unidirektionalen Schichten," *Kunststoffe*, 61(4): 273-277.
95. Shapery, A. A. 1968. "Thermal Expansion Coefficient of Composite Materials Based on Energy Principles," *J. Composite Materials*, 2:358-365.
96. Skudra, A. M. and F. Ya. Bulavs. 1982. "Strength and Creep Micromechanics," in *Mechanics of Composites*, J. F. Obraztsov and V. V. Vasilev, eds., Moscow: Mir Publ., pp. 77-109.
97. Teh, K. K. and C. C. Huang. 1979. "Shear Deformation Coefficient for Generally Orthotropic Beams," *Fibre Sci. Technol.*, 12(1):73-80.
98. Weidmann, G. W. and R. M. Ogorkiewiez, R. M. 1974. "Tensile Creep of a Unidirectional Glass-Fibre-Epoxy Laminate," *Composites*, 5(5):117-121.
99. Wu, E. M. and D. C. Ruhman. 1975. "Stress Rupture of Glass-Epoxy Composites: Environmental and Stress Effects," *Composite Reliability. ASTM STP 580*, pp. 263-287.

# BIOGRAPHIES

## ALBERTS SKUDRA

Alberts Skudra, professor, doctor of technical sciences, is chief of the Structural Mechanics Department at Riga Technical University, a leader of the laboratory of composite micromechanics as well as the author of 5 monographs on composite materials (in Russian). He is an editor and co-author of the international edition of the *Handbook of Composites* (Vol. 3 and 4), North-Holland Publishers. From 1977 to 1981 he was a member of the editorial advising board of the *Journal of Composite Materials* and a member of the editorial board of *Mechanics of Composite Materials* (in Russian).

## FELIX BULAVS

Felix Bulavs, doctor of technical sciences, is a professor in the Structural Mechanics Department at Riga Technical University, as well as the author of 4 monographs on composite materials. Since 1978 he has been a member of the editorial board of *Mechanics of Composite Materials* (in Russian).

## MARK GURVICH

Mark Gurvich is an Associate Professor, Cand. of Technical Sciences in the Structural Mechanics Department of Riga Technical University, as well as the author of a monograph on composite materials (in Russian).

## AIVARS KRUKLINSH

Aivars Kruklinsh is an Associate Professor, Cand. of Technical Sciences in the Structural Mechanics Chair at Riga Technical University, as well as the author of a monograph on composite materials (in Russian).